The United Nations University Series on Regionalism, launched by UNU-CRIS and Springer, offers a platform for innovative work on (supra-national) regionalism from a global and inter-disciplinary perspective. It includes the World Reports on Regional Integration, published in collaboration with other UN agencies, but it is also open for theoretical, methodological and empirical contributions from academics and policy-makers worldwide.

Book proposals will be reviewed by an International Editorial Board.

The series editors are particularly interested in book proposals dealing with:

- comparative regionalism;
- comparative work on regional organizations;
- inter-regionalism;
- the role of regions in a multi-level governance context;
- the interactions between the UN and the regions;
- the regional dimensions of the reform processes of multilateral institutions;
- the dynamics of cross-border micro-regions and their interactions with supranational regions;
- methodological issues in regionalism studies.

Accepted book proposals can receive editorial support from UNU-CRIS for the preparation of manuscripts.

Please send book proposals to: pdelombaerde@cris.unu.edu and lvanlangenhove@cris.unu.edu.

United Nations University Series on Regionalism

Volume 24

Graham Hassall

The United Nations and the Pacific Islands

Springer

Graham Hassall
School of Government
Victoria University of Wellington
Wellington, New Zealand

ISSN 2214-9848 ISSN 2214-9856 (electronic)
United Nations University Series on Regionalism
ISBN 978-3-031-34154-0 ISBN 978-3-031-34155-7 (eBook)
https://doi.org/10.1007/978-3-031-34155-7

This Springer imprint is published by the registered company Springer Nature Switzerland AG
The registered company address is: Gewerbestrasse 11, 6330 Cham, Switzerland

Paper in this product is recyclable.

Foreword

The *Pacific Islands* of this book are home to people of three major Pacific cultures: Melanesia, Micronesia and Polynesia, ocean dwellers among the most celebrated of ancient navigators. Pacific people take seriously their place and responsibility for their ocean environment and resources—a vast ocean well acknowledged by science as now in evident peril in the face of severe global forces, serious and widespread pollution, and ever more dangerous climate change consequences.

Explorers of the sixteenth century had named it the peaceful seas, and there is much of our ocean that commends *pacific* imagery. Yet, in just the past century this region was an active theatre of globalized war. Island communities had no part in the causes but came to know *untold sorrow* and of the *scourge of war*. Worse, Pacific islands and environments had been the site of prolonged testing of weapons of mass destruction, conducted by three different colonial Governments over fifty years between 1946 and 1996, in ten different locations in the region. Today, there are ominous implications as Pacific communities watch the strategic "pivoting" of big power geopolitical competition for influence and influencing in the Indo-Pacific area.

As with my own country, Samoa, statehood and political independence for PICs was gained through an extended period of colonialism and the trusteeship under the United Nations system. All would have taken to heart the magnificent declarations in the preamble of the UN Charter, of international peace and security and of fundamental human rights, and more so the reaffirmation of the equal rights of nations large and small, and respect for international law. And so, there was a certain "naturalness" in working within the framework of the United Nations and its enlarging membership. For small communities, there is comfort in familiarity and the expected assurance of safety and strength in numbers.

Statehood determined that these countries, small as they are, take their place in the councils of the world community. It is the place to frame and to define relationships with others. In particular, there was purpose and encouraging aspiration in the UN Charter for seeking "social progress and better standards of life in larger freedom". Some, like Fiji, took early UN membership on gaining independence in 1970.

Samoa, the first of Pacific states to gain independence in 1962, waited fourteen years to join the UN, other PICs even longer. For many PICs there were more demanding priorities at home in national development and preparations to ensure their effective participation as independent states in the confronting challenges of an ever-modernizing international community.

I see this book in that general context. It is a work of masterful and substantial effort. While not a historical study as such, it does explore in historical perspective the general and combined experience of Pacific Island Countries (PICs) in their statehood, political and international development over the years following independence, with a particular focus on PIC interactions with the United Nations and the relationships developed with the UN system. In doing so, Professor Hassall creates more specific boundaries for study and assessment of a subject area that is vast and hugely diverse and complex and with issues of major practical importance not only to eighteen individual PICs but also for the Pacific region considered and functioning as a whole.

There will be volumes written of the PICs and their work with particular or selected UN agencies and other international organizations. But I agree with the view that there is comparatively little dealing specifically with PICs and their interactions with the UN and the UN system. In that respect, the material presented in this book, extensively covered and researched in admirable detail, provides an especially timely and superbly informed general overview of a relationship with the UN which, while in earlier times ubiquitous in a general sense, is now, in my view, essential in partnership and integral in the continuing development and sustainability of PICs in the modern world. It is a relationship which calls for proper and more focussed assessment and which needs to be of informed understanding, equality and of complete reciprocity in every measure. This book is a singular and substantial contribution in that direction.

The UN in the Pacific

The Canberra Agreement of 1947 establishing the South Pacific Commission (SPC) provided for the fullest possible cooperation with the UN and its agencies, and for Governments then participating in the SPC to consult the UN, in part to define the relationship and to ensure effective cooperation with the UN. By the time the Pacific Islands Forum (PIF) was established in 1971, there was already an established series of UN offices in regional capitals as well as engagement by a number of UN agencies with Pacific regional organizations on a variety of technical and development programmes. The relationship has of course expanded exponentially since then, both in subject-matter coverage and in depth of policy content, management structure and operational strategies and provisions. The effort, in true sense summed up in the *2030 Agenda for Sustainable Development* (with, among others, a dedicated UN programme of action for small island developing states at the heart of it), binds us all in a truly global partnership for peace and prosperity for people and the planet, now and into the future. By its terms, what happens in the Pacific region determines the success, or failure, of the SDGs. I would contend that with effective measures, of Pacific regional cooperation and in partnership with the UN, we have every prospect of meeting SDG development expectations.

Professor Hassall is particularly good at pointing to practical situations and raising queries. He notes, for example, that at least in the formative stages of the relationship the UN effort seemed more focussed on the basis of the primacy of individual nation-states rather than on the basis and strengths of Pacific regionalism. This is an informed perspective and, I believe, a point validly made. In my own time at the Forum, in reviews conducted of the then *Millennium Development Goals* (MDGs), I felt that the UN system seemed to be more concerned with the MDGs as such, and not the *Pacific Plan* determined by Pacific Leaders in 2005 for the strengthening of Pacific regional cooperation and integration—almost as if the MDGs were distinct and separate from the Pacific Plan in development purpose and standards. In the development of the SDGs, a much more encompassing and transparent process, considerable UN and international effort was given to the consideration of the essential role of regional institutions, and it seems clear there is now wide acceptance and recognition of the critical importance of regional institutions and their arrangements in the effective implementation of the 2030 Agenda.

The Pacific in the UN

Peace and security, protection of their homeland, the ocean and environment, and the development and improvement of their countries and populations are key determinants in the international policy and positions of the PICs. From the first meeting of Pacific Island Forum Leaders in 1971, the dangers of nuclear testing, concerns for the ocean and its resources, early advancements in education, communications and transportation, were on the Leaders' agenda. Many of these issues remain today on the Pacific Islands Forum agenda.

Nuclear testing has been a long-standing and serious Pacific concern. From 1973 when Australia and New Zealand first took issue with France before the International Court of Justice (ICJ), the judicial arm of the UN, countries of the Pacific region as a whole have made hugely important and sustained efforts on the nuclear issue. The *Treaty of Rarotonga* in 1985 creating the South Pacific Nuclear Free Zone (second only to the *Treaty of Tlatelolco*, for the prohibition of nuclear weapons in Latin America and the Caribbean) was a high-point initiative of Pacific leadership and declaration of international responsibility. PICs filed submissions before the ICJ in support of the proceedings by Australia and New Zealand in 1973 and again in proceedings by New Zealand in 1995 against the resumption of testing by France. Later in 1995, a number of PICs participated in the hearings before the ICJ on the referrals from the UN General Assembly and WHO on the issue of the *Illegality of Nuclear Weapons;* and in 2014, Marshall Islands filed applications in the ICJ against all nine states possessing nuclear weapons (China, DPRK, France, India, Israel, Pakistan, Russia, UK and the USA) based on what the Marshall Islands considered to be violations of international obligations under the UN *Non-Proliferation Treaty*. Combined, all this amounts to considerable effort by the region and the PICs in taking an emphatic stand on an issue of global concern and of centrality to the UN and the struggle for international peace and security. The same can be said of the PICs role and participation in other areas of the highest priority and concerns for the UN and the international community as a whole, especially with respect to the oceans, climate change and sustainable development.

While only Fiji (along with Australia and New Zealand from the region) had been able to participate in the UN Seabed Committee in the preparations for the negotiations at the Third UN Conference on the Law of the Sea (UNCLOS), other PICs were able to join in the process from 1974 and the commencement of the substantive UNCLOS sessions. Fiji's leadership and prominence in the UN work and developments under UNCLOS to this day and the service of distinguished Fijians (on the International Seabed Authority and as UN Special Envoy for the Ocean) are well recognized in the public record. Representatives of other PICs as well including my own country have been active in their engagement and in presiding over UN processes on the Law of the Sea. PICs are impressive in the quality of their contribution and performance when they work together in unity, as they did in the important UN conference (chaired by the highly regarded Fijian diplomat and Law of the Sea notable, the late Ambassador Satya Nandan) which produced the *Straddling Fish Stocks Agreement* of 1995.

Professor Hassall's book provides a range of other examples of the participation of PICs in the work of the UN and active roles being played by Pacific people and representatives in other key areas, and more particularly in strengthening the relationship with the UN system. There is now an established process for Pacific Leaders to meet in bilateral discussions with the UN Secretary General each year on the occasion of the opening of the UN General Assembly sessions. These are invaluable opportunities for direct exchanges on global issues as well as those pertaining more to the region—and personally to renew courtesies, noting in particular the significance of UN Secretary General António Guterres being able to visit the Pacific in person, as did his immediate predecessor SG Ban Ki-moon.

The public record is equally clear on the extensive and important contribution being made by PICs to the UN work on climate change. Pacific Leaders are taking on personal roles to lead UN negotiating processes or as advocates and lead speakers for their countries and the region; and PIC representatives have become skilled and effective climate negotiators and policy strategists, either for their countries or in their roles as chairpersons of the Alliance of Small Island Sates (AOSIS) (an office occupied by Vanuatu, Samoa and Nauru over the years) or as lead authors on technical studies and assessments by the Intergovernmental Panel on Climate Change (IPCC). With their known oceanic exposure and front-line vulnerabilities, PICs bring to the international debate an essential ethical dimension.

By comparison with other larger country experiences, and notwithstanding the constraints of small populations and resources, PICs and representatives of PIC countries have been able to participate and contribute to the work of the UN system and associated bodies in significant ways: high-level seniority in UN bureaucracy; as judges or deputy prosecutor of the international judicial system—at the International Criminal Court (ICC), International Criminal Tribunal for Rwanda (ICTR) and Special Court for Sierra Leone (SCSL); and UN peacekeeping. The record and professional performance of Fijian soldiers, women and men, in UN peacekeeping are long-standing and unmatched; and now supplemented by un-armed police contingencies from several contributing PICs.

Professor Graham Hassall has devoted his professional life to the study and understanding of the Pacific and its people, to research and teaching at Universities in the Pacific region and internationally. He has travelled extensively throughout the Pacific region, with close involvement and engagement with Pacific people at all levels, students, policymakers, professional colleagues, among others. These few words are more than merely to introduce. Rather, they serve also as warm tribute to scholarship of considerable measure in taking on a subject of such magnitude and importance, with a natural sense of empathy and understanding, at the same time with scholarly balance and objectivity offering informed insights and assessments.

Tuiloma Neroni Slade

Formerly Secretary General of the Pacific Islands Forum Secretariat, Suva, Fiji

Judge of the International Criminal Court, The Hague, The Netherlands

Samoa's Ambassador to the United Nations, New York, NY, USA

Preface

The *United Nations and the Pacific Islands* is a study in trans-national public administration. It offers an initial survey of the role that the United Nations has played in the emergence of the Pacific Island Countries and in their coming together as a region on the global stage. It also explores participation of the Pacific states within the UN—the General Assembly and the Economic and Social Council—and the UN's many specialized commissions, councils, agencies and programmes. It is not an historical work, but it does draw on historical records to the extent that they help us understand contemporary public policy and public administration settings. I've also taken this approach to assist with retention of institutional memory, since the rapidity of staff turnover can quickly remove an organization's knowledge and understanding of what has been previously experienced, achieved and learnt.

The larger subject concerns the extent to which the structure and operation of global multilateral organizations such as the United Nations provide pathways for effective and meaningful participation by small states, such as those from the Pacific Islands region, in global public policy and global governance. Small states may not have much prospect of influencing world affairs through economic or political power, but they can surely aspire to being "good states", even smart, intelligent states, in the serious collaborative projects "global public policy" and "global governance". The book is concerned with progress and politics, but the politics of virtue rather than power, and coordination rather than division. The quest is for states that are peaceful, governments that are just, and societies that are prospering. There is so much more to explore about these relationships and interactions than has been possible to achieve in this survey, but my hope is that other researchers will identify the most significant gaps and write additional studies.

My interest in the United Nations commenced as a teenager, fortunate to participate in a "model UN" conference for secondary school students, hosted in Canberra in 1975 by a branch of the United Nations Association of Australia. Each participant was assigned a country to represent and the one allocated to me was, ironically, Iran. My awareness of United Nations procedures, particularly in the field of human rights, increased rapidly after Iran's 1979 Islamic Revolution, when the regime's treatment of that country's Bahá'í community became too frequent a focus of the UN General Assembly and the Human Rights Commission.

The ubiquitous presence of UN agencies in Pacific Island countries came to my notice whilst undertaking PhD fieldwork in Melanesia in the mid-1980s, and later periods working and travelling in the region afforded numerous opportunities to experience UN–Pacific Islands interaction. When teaching at Landegg Academy in Switzerland (2000–2003), I included fieldtrips to such international organizations in Geneva as the United Nations, the World Trade organization and the InterParliamentary Union, in an undergraduate course "Development, Civil Society, and the United Nations System".

On returning to the Pacific, six years at the University of the South Pacific's Suva-based Pacific Institute of Advance Studies in Development and Governance (2004–2009) followed by ten years at the Victoria University of Wellington's School of Government (2010–2020) allowed for close observation of, and participation in, UN-sponsored or affiliated initiatives. I was fortunate to teach Governance at the University of the South Pacific when the Pacific's regional institutions were being revised and revitalized by the "Pacific Plan" and when UNDP and other UN programmes were seeking to increase their participation in the region. The Millennium Development Goals were the immediate focus, but within a broader framework of objectives promised under a host of "sustainability" pledges dating back to the 1992 Earth Summit in Rio de Janeiro. In Australia and New Zealand, opportunities to engage with UN studies came through participation on the boards of the New Zealand Centre for Global Studies and New Zealand's Council for International Development, and the United Nations Association committees for Victoria (1997–1999) and New Zealand (2012–2016).

These contexts provided opportunities to participate in numerous seminars, workshops and networks: WSSD Type II Initiatives; UNDP Pacific's Crisis Prevention and Recovery Consultative Group and its "Peace, Stability and Development" assessments; WHO's Working Group on Pacific Human Resources for Health Alliance; DPA's Regional Workshop on Conflict Prevention & Peace-building in the Pacific; UNESCO's "right to information" campaign; UNFPA's population and development seminars; UN-Habitat's Pacific Urban Forums; UNESCAP/ITU's Sub-regional conference on ICT for Development; UNDESA's "reinventing government" and "Dialogue in the Social Integration Process" projects; the SIDS III Pre-Conference in Apia; the Global Partnership for the Prevention of Armed Conflict; "track II" conflict prevention dialogue regarding squatter settlements in Fiji; and UNDP projects in constitutional reform, public service ethics and accountability, building trust in government, human development, parliamentary induction, digital government, peace education and security studies. They provided opportunities to associate with staff at UNDP, UNRISD, UNDESA, DPA, ESCAP, OCHA, ILO, UN-Habitat, FAO, WHO, UNFPA and UN Women. They also provided opportunities to visit UN multi-country offices in Fiji and Samoa, the UN Office for Papua New Guinea, the UNDP field office for Solomon Islands, UN headquarters in New York and Geneva, the ESCAP centre in Bangkok, the United Nations University for Peace in Costa Rica and the United Nations Centre in Vienna.

In the course of so much engagement with UN agencies and officials, I observed both admirable and perturbing actions: senior officials flying in from Bangkok or New York to give keynote addresses at regional meetings and immediately departing without stopping long enough to hear the views of local actors; mid-career officials treating their short-term appointment in the Pacific as merely a "stepping stone" towards their promotion to one or other UN head office; program directors cancelling projects initiated by their predecessors in order to commence their own, irrespective of resources already invested and results as yet incomplete; appointment of short-term international consultants with meagre understanding of local conditions rather than appointment of local experts with greater local knowledge and connections but less international experience; and capricious discontinuity of projects regardless of statements made that had professed deep and on-going commitment. This book does not dwell on the clientelism and patronage prevalent in UN procurement and project implementation, but I note these tendencies in this introduction to make the point that they invariably influence attitudes towards UN agencies and officials within Pacific Island elites.

This period at the University of the South Pacific brought me into contact, furthermore, with other collaborating organizations: the Pacific Islands Forum Secretariat, the Pacific Community, the Asian Development Bank, and the Pacific Islands Association of Non-Governmental Organizations (PIANGO), the London-based Commonwealth Secretariat and Commonwealth Local Government Forum, Transparency International, the East-West Center in Honolulu, the Manila-based Asian Development Bank, the Australian Agency for International Development (AusAID), New Zealand Aid and the European Community; and research centres, including the Australian National University and the University of Melbourne in Australia, and in Papua New Guinea the Institute for National Affairs and the University of Papua New Guinea.

In New Zealand, UN-related research projects were carried out with the Victoria University of Wellington, the University of Canterbury, the New Zealand Institute of International Affairs, the New Zealand Centre for Global Studies and the United Nations Association of New Zealand.

This book complements two others that I have written or edited in recent years: *Government and Public Policy in the Pacific Islands* (Hassall, 2021) provides a comprehensive survey of the emergence of the Pacific Island countries in the twentieth century, together with the challenges they face as "small island developing states" in the twenty-first century global order. *Achieving Sustainable E-Government in the Pacific Islands*, edited with Rowena Cullen (Cullen & Hassall, 2017), focuses on the use of digital technologies to transform the delivery of government services to rapidly expanding populations. Whilst both books refer to the United Nations and its diverse range of programs in the Pacific Islands context, most discussion of UN-related issues was directed into the current volume. Given the extensive coverage of the structure and operation of Pacific governments in *Government and Public Policy* and *Achieving Sustainable E-Government*, the current work, *The United Nations and the Pacific Islands*, focuses on interaction between the Pacific Island Countries and the UN system, and with UN-related activities within the region. It was conceived during study leave at the University of Gothenburg's School of Global Studies in 2016, for which I thank Jan Aart Scholte; and further developed during study leave at the University of Queensland's Law School in 2019, for which I thank Jennifer Corrin.

I want to thank the many individuals whose conversations contributed to the coproduction of knowledge about United Nations-Pacific Islands interactions and relationships. At the Pacific Islands Forum Secretariat, these include the late Secretary-General Greg Urwin, former Secretaries-General Dame Meg Taylor and Tuiloma Neroni Slade, former Deputy Secretaries-General Andie Fong Toy and Cristelle Pratt, Merewalesi Falemaka, Isabel Calvert, Joel Nilon, Paki Ormsby, Alifereti Soakai and Sione Tekiteki. At the Secretariat of the Pacific Community, I received invaluable assistance from Audrey Aumua, Alitia Nasila and Mohammed Khan. Within the UN System, several associates have been particularly helpful—in Suva: Iosefa Maiava at ESCAP, Garry Wiseman, Eugenia Piza Lopez, Jean-Luc Stalon and Charmaine Rodriquez at UNDP and Wame Baravilala at UNFPA; in Port Moresby: Roy Trivedi and Iwonna Spytkowski at UNDP; in Apia: Georgina Bonin at UNDP; in Geneva: Naren Prasad at ILO and Arthur Dahl, formerly with UNEP and founder of the International Environment Forum.

I have endeavoured to use as much primary material as possible both because it is "first-hand" documentation of thought and action and because it otherwise remains out of reach of most readers. Online resources are easier to access, and, while much less permanent than printed materials, are an increasingly important source for policy documents that librarians categorize as "grey". These I have distinguished from printed matter by referring to them in footnotes rather than as citations. For access to the more permanent documents, I am grateful to the staff at the National Library of New Zealand (Roger Swanson), Archives New Zealand, the Australian National Library, the National Parliament of Papua New Guinea Parliamentary Library (Elesallah Matatier), National Parliament of Solomon Islands Parliamentary Library (Boniface Supa), the Pacific Islands Forum Secretariat Library (Kesaia Vilisoni), the League of Nations Archives in Geneva (Jacques Oberson) and the Victoria University of Wellington Library (Corrina DeWolff).

Other colleagues who have shared insights on, or experiences with, UN-Pacific engagement that have influenced this book include the late Nandi Glassie, former Cook Islands Minister for Education then for Health; the late Litia Mawi, Fiji's "Ambassador at large"; John Momis, Bikeni Paeniu, Sharon BhagwanRolls, champion of UNSC resolution 1325 on Women and Peace and Security, and the Global Partnership for the Prevention of Armed Conflict; Mark Borg, who collaborated with me on a study of Pacific Parliamentary Websites for UNESCO; Filipo Masarura, now with Oxfam Pacific; Shamim Kazemi, who undertook fieldwork on health governance in the Pacific; Feue Tipu and Karibaiti Taoaba whose work on the Pacific Urban Agenda at the Commonwealth Local Government Forum drew us into the activities of UN-Habitat's Pacific Urban Forum; Michael Powles, New Zealand's Permanent Representative to the UN 1996–2001 and my predecessor as President of UNANZ; the late Peter Donigi, former Permanent Representative of Papua New Guinea to the United Nations who authored for the Governance

Programme at the University of the South Pacific a pilot course in diplomacy for the Pacific Islands; William Dihm, former Secretary for Foreign Affairs in Papua New Guinea, and Ambassador Peter Eafeare, my counterpart in a training program for Papua New Guinea's new foreign service officers. I am ever grateful to New Zealand colleagues for their interest in my UN research: Evan Berman, Neil Boister, Rowena Cullen, Joy Dunsheath, Girol Karacaoglu, John Hopkins and Pedram Pirnia. Special thanks to Kennedy Graham, Tuiloma Neroni Slade and Phillip Taula for their invaluable feedback on the draft text. To all, I say thank you for sharing your knowledge and experience so generously and stimulating the desire to write this book.

Wellington, New Zealand Graham Hassall

References

Cullen, R., & Hassall, G. (Eds.). (2017). *Achieving sustainable E-Government in Pacific Island States* (Vol. 27): Springer International Publishing.

Hassall, G. (2021). *Government and Public Policy in the Pacific Islands.* Bingley: Emerald Publishing Limited.

Chronology

- 1947

  South Pacific Commission established.

- 1962

  Samoa gains independence.

- 1965

  January: WHO opens office in Suva, Fiji.
  Cook Islands becomes self-governing territory with New Zealand.

- 1968

  Nauru gains independence.

- 1969

  West Papua becomes province (Irian Jaya) of Indonesia.

- 1970

  Fiji and Tonga gain independence.

- 1971

  South Pacific Forum established by Australia, New Zealand, Fiji, Western Samoa, Tonga, Cook Islands, and Nauru .

- 1972

  CCOP/SOPAC established as a United Nations Development Programme.

- 1973

  South Pacific Forum establishes the South Pacific Bureau for Economic Cooperation (SPEC).

- 1974

  Papua New Guinea becomes a member of the South Pacific Forum.
  Niue becomes self-governing territory with New Zealand.

- 1975

  September—Papua New Guinea gains independence.

- 1976

  Pacific Forum Line established.

- 1977

  Kiribati becomes a member of the South Pacific Forum
  South Pacific Regional Environment programme (SPREP) established
  Forum Fisheries Agency (FFA) established.
  UNESCO first convenes the Pacific Heads of Education System (PHES) network.

- 1978

  Solomon Islands and Tuvalu gain independence and become members of the South Pacific Forum.
  Commonwealth of the Northern Mariana becomes Commonwealth of the USA.

- 1979

  Federated States of Micronesia, Kiribati and Marshall Islands gain independence.
  Nauru (on behalf of AOSIS) submits first draft of Kyoto Protocol on Climate Change

- 1980

  Vanuatu gains independence and becomes a member of the South Pacific Forum.
  South Pacific Regional Trade and Economic Cooperation Agreement (SPARTECA) established.

- 1981

  Palau gains independence.

- 1982

  Energy Unit established at SPEC.
  United Nations Convention on the Law of the Sea (UNCLOSS).

- 1984

  South Pacific Telecommunications Development Programme commenced.

- 1985

  South Pacific Nuclear Zone Treaty.

- 1987

Federated States of Micronesia and Marshall Islands become members of the South Pacific Forum.

- 1988

SPEC name changed to South Pacific Forum Secretariat.

- 1989

Post Forum Dialogue process established.
South Pacific Organizations Coordinating Committee established (Name changed to Council of Regional Organizations in the Pacific in 1999).
Tarawa Declaration on Drift-Net fishing.

- 1990

South Pacific Forum Ministerial Committee on New Caledonia established.
China joins the Post Forum Dialogue Partners.

- 1991

PIC/Development Partners annual meetings initiated by Forum Secretariat.
South Pacific Forum Secretariat Agreement.
Forum recognizes the special development requirements of the Smaller Island States.

- 1992

Declaration on Regional Security Cooperation in the Pacific (Honiara Declaration).
January 16–17: Inaugural Summit of Smaller Island State Leaders held in Rarotonga, Cook Islands.
UNDP Regional Pacific Programme 5th cycle (1992–1996) begins.

- 1993

September: *Port Vila Declaration on Population and Sustainable Development.*

- 1994

South Pacific Forum gains Observer Status at the United Nations.
April 25–May 6: Global Conference on the Sustainable Development of SIDS in Barbados.
South Pacific Forum endorses *Suva Declaration on Sustainable Human Development in the Pacific.*

- 1995

Waigani Convention.
South Pacific Forum Vision Statement.
Plan of Action on the theme "Securing Development Beyond 2000".

Palau joins the South Pacific Forum.
Japan and Korea join Post Forum Dialogue Partners.

- 1996

Agreement on establishment of the Pacific Islands Centre—Japan, Tokyo.
April 9–12: Inaugural meeting of the Association of Pacific Islands Maritime Training Institution and Maritime Authorities.
Secretariat Technical Programmes phased out at the end of 1996.

- 1997

Aitutaki Declaration on Regional Security Cooperation.
July 11: First Forum Economic Ministers Meeting (FEMM)—Cairns, Australia.
Malaysia joins the Post Forum Dialogue Partners.
October: Inaugural South Pacific Japan Leaders' Summit held in Tokyo, Japan.

- 1999

New Caledonia gains Observer status at the Pacific Islands Forum.
South Pacific Forum Leaders agree on a new name for the Forum—"Pacific Islands Forum".
Pacific Islands Forum Leaders endorse in principle a Free Trade Area among Forum members.
Forum Vision for the Pacific Information Economy.
June 1–2: First Pacific Islands Forum Trade Ministers Meeting (FTMM) in Suva, Fiji.
Pacific Islands Forum Leaders agree to the establishment of a regional representative office in Geneva.

- 2000

PALM 2000 Summit between Japan the Forum Countries.
Biketawa Declaration.
March 27–30: Pacific Regional Conference on Governance for Parliamentarians.
September: Millennium Development Goals adopted by UNGA for the period 2000–2015.
October 30: New Agreement Establishing the Pacific Islands Forum. Secretariat.
Philippines joins Post Forum Dialogue Partners.

- 2001

Pacific Island Countries Trade Agreement (PICTA).
August 18: Pacific Agreement of Closer Economic Relations (PACER). PICTA & PACER open for signature.
First Forum Election Observer Mission—Solomon Islands.
Vanuatu initiates campaign to seek ICJ opinion on Climate Change
May 14–15: First Education Ministers Meeting—Auckland, New Zealand.

- 2002

  Nasonini Declaration on Regional Security.
  Pacific Islands Regional Oceans Policy.
  East Timor becomes Special Observer at Pacific Islands Forum.

- 2003

  Regional Assistance Mission to the Solomon Islands (RAMSI).
  Forum Principles of Good Leadership.
  Pacific Islands Air Services Agreement (PIASA).
  Thailand joins Post Forum Dialogue Partners.

- 2004

  April 6: *Auckland Declaration* and "Leaders Vision" adopted at Special Leaders Retreat.
  HIV/AIDS Regional Strategy.
  Pacific Regional Assistance to Nauru (PRAN).
  French Polynesia gains Observer status at Pacific Islands Forum.

- 2005

  Pacific Islands Forum Leaders adopt the Pacific Plan.
  Agreement establishing the Pacific Islands Forum. This includes a new policy regarding admission, criteria and entitlements for associate membership and observer status.
  Tokelau gains observer status at Pacific Islands Forum.
  Pacific Islands Framework for Action on Climate Change (2006–2015).

- 2006

  New Caledonia and French Polynesia become associate members of the Pacific Islands Forum.
  Wallis and Futuna, the United Nations, Commonwealth Secretariat and Asian Development Bank invited to become Pacific Islands Forum observers.
  India joins Post Forum Dialogue Partners.
  The United Nations gains observer status at the Pacific Islands Forum.
  November 22: Forum Leaders' *Declaration on Deep-Sea Bottom Trawling to Protect Biodiversity in the High Seas.*

- 2007

  Vava'u Declaration on Pacific Fisheries.
  Italy joins Post Forum Dialogue Partners.
  Pacific Aid Effectiveness Principles agreed by Forum Leaders.

- 2008

  Niue Declaration on Climate Change.
  First official participation by UN in a Pacific Islands Forum meeting (Executive Secretary for ECOSOC attended Forum Leader's meeting in Alofi, Niue).

- 2009

  Cairns Compact on Strengthening Development Coordination in the Pacific.
  May 2: Fiji suspended from Pacific Islands Forum.
  Pacific Regional Strategy on Disability.

- 2010

  Port Vila Declaration on Millennium Development Goals.
  Cairns Compact Peer Reviews.
  Pacific Leaders endorse *Framework for a Pacific Oceanscape.*
  Pacific Environment Community Fund (PEC FUND) established.

- 2011

  Waiheke Declaration on Sustainable Economic Development.
  Pacific Islands Forum Leaders Statement on Non-Communicable Diseases.
  First attendance at the Forum Leaders meeting by a Secretary-General of the UN – Mr Ban Ki Moon
  Joint Statement of Pacific Islands Forum Leaders & the Secretary-General of the United Nations.
  Secretary General Tuiloma Neroni Slade appointed as Pacific region's first Ocean Commissioner.

- 2012

  Pacific Islands Development Forum Established.
  Pacific Leaders Gender Equality Declaration.
  UNDAF for Papua New Guinea Commences (2012–2017).

- 2013

  August 20: Nauru Signs and Ratifies PICTA Trade in Services Protocol.
  Majuro Declaration for Climate Change.
  UNDAF for the Pacific commences (2013–17).

- 2014

  Dame Meg Taylor appointed first female Secretary General of the Pacific Islands Forum.
  Palau Declaration "The Ocean: Life & Future" Charting a Course to Sustainability.
  Tokelau becomes an Associate Member of the Pacific Islands Forum.
  International Organization for Migration (IOM) invited to become an Observer in the Forum.
  Framework for Pacific Regionalism (FPR) (replaces the Pacific Plan)
  Spain and Turkey become the Post Forum Dialogue Partners.
  September 1–4—SIDS III Conference in Apia
  November 17–19—Pacific Consultation on Progress on the Millennium Development Goals and the Post-215 Development Agenda.

- 2015

SSCR through the FPR identify five issues for Leaders' consideration: Fisheries, Climate Change, Information Communications Technologies, Cervical cancer, West Papua.
Pacific Islands Forum Leaders endorse *Hiri Declaration on Strengthening Connections to Enhance Pacific Regionalism.*
Regional Roadmap for Sustainable Fisheries.
First Pacific Civil Society Organizations held in the margins of the Forum Leaders meeting.
September: Sustainable Development Goals agreed to by UNGA for the period 2015–2030.

- 2016

Inaugural meeting of Joint Steering Committee for the Analysis of Governance and Financing.
First Standing Meeting of Pacific Islands Forum Foreign Ministers.
Federated States of Micronesia joins the Smaller Islands States (SIS) Grouping.
New Caledonia and French Polynesia join as Full Forum Members.
Germany joins as the 18th Post Forum Dialogue partner.

- 2017

June 30: RAMSI (Regional Assistance Mission to Solomon Islands) concludes (commenced on 24 July 2003).
June: Fiji hosts inaugural United Nations Ocean Conference.
"The Blue Pacific" Narrative as catalyst for deeper Pacific Regionalism.

- 2018

PIFS granted approval to establish Permanent Observer Office at the UN Office at Geneva (UNOG).
Regional Security Declaration (Boe Declaration).
UN Development Assistance Framework for 2018–2022.
High-level dialogue on climate at SPC Headquarters, Nouméa.

- 2019

FOC Sub-Committee on Prioritization and Resources established (succeeds Sub-Committee on Prioritization & Funding and Sub-Committee on Remuneration.
July: UN Secretary General Guterres attends Leader's Forum in Tuvalu
July: *Kainaki Declaration for Urgent Climate Change Action Now.*
July: Forum Leaders endorse *2020 Strategy for the Blue Pacific Continent.*

- 2021

October: Commission of Small Island States on Climate Change established at COP 26

- 2022

  Pacific Regional Culture Strategy.
  July: Kiribati withdraws from Pacific Islands Forum.
  September: UNSG Guterres addresses Pacific Islands Forum Leader's Meeting, Suva, Fiji.
  November: Pacific Islands Forum joins UNCTAD-led "eTrade for all initiative".
  November: PICs participate in COP 27 UN climate summit in Sharm el-Sheikh, Egypt.

Contents

About the Author

Graham Hassall has held teaching and visiting appointments traversing the fields of history, comparative constitutional law, governance, public policy, public administration, and global studies, at the University of Melbourne, the University of Papua New Guinea, Nagoya University in Japan, Landegg Academy in Switzerland, the University of the South Pacific in Fiji, the University of Gothenburg in Sweden, and the Victoria University of Wellington in New Zealand. He is a founding member of the New Zealand Centre for Global Studies, and a life-member of the United Nations Association of New Zealand, having served as National President 2012–2016. He was Academic Director of Papua New Guinea's Foreign Service Training Programme 2014–2017. He has undertaken consultancies for the Pacific Islands Forum Secretariat, UNDESA, UNDP, Commonwealth Local Government Forum, Global Integrity, the Commonwealth Secretariat, and the Papua New Guinea Department of Higher Education, Research, Science and Technology and has served on editorial and advisory boards including the New Zealand Association for Comparative Law; the Comparative Law Journal of the Pacific; the Commonwealth Journal of Local Governance; and the New Zealand Council for International Development.

Abbreviations and Glossary

ACP	Africa, Caribbean, and Pacific Group of States.
ADB	Asian Development Bank.
AIDAB	Australian International Development Assistance Bureau.
AOSIS	Alliance of Small Island States.
AusAID	Australian Development Agency.
BPoA	Brussels Programme of Action for the Least Developed Countries.
BPOA	Barbados Programme of Action agreed at the Global Conference on the Sustainable Development of SIDS held in Barbados 25 April–6 May 1994.
C24	Special Committee on Decolonization established by the General Assembly in 1961.
CC	Climate change.
CCA	Climate change adaptation.
CCOP	Committee for Coordination of Joint Prospecting for Mineral Resources in South Pacific Offshore Areas.
COP	Conference of the Parties
CROP	Council of Regional Organizations of the Pacific, established in 1988.
DM	Disaster management.
DRM	Disaster risk management.
DRR	Disaster risk reduction.
DSP	District Services Programme—later renamed District Services Improvement Programme (Papua New Guinea).
ECAFE	United Nations Economic Commission for Asia and the Far East—subsequently renamed ESCAP—United Nations Economic Commission for Asia and the Pacific.
EEZ	Exclusive Economic Zone.
EPA	Economic Partnership Agreement.

FAIDE	Framework for Actions for Development of ICT for the Pacific.
FAO	Food and Agriculture Organization.
FFA	Forum Fisheries Agency.
FIC	Forum Island Country.
FLNKS	Front de Liberation National Kanak et Socialiste—Pro-independence political party in New Caledonia.
FDC	Foundation for Development Cooperation.
Forum Secretariat	Known until 1988 as the South Pacific Bureau of Economic Cooperation and later as the Pacific Islands Forum Secretariat. It acts as the secretariat for the Pacific Forum, a regional organization comprising the leaders of independent Pacific Island Countries.
FRSC	Forum Regional Security Committee.
GNP	Gross National Product
HDI	Human Development Index.
HDR	Human Development Report.
IAEA	International Atomic Energy Agency.
ICAO	International Civil Aviation Organization.
ICC	International Criminal Court.
ICT	Information and Communication Technologies.
ICT4D	Information and Communication Technologies for Development.
IFAD	International Fund for Agricultural Development.
ILO	International Labour Organization.
IMF	International Monetary Fund.
IPCC	Intergovernmental Panel on Climate Change.
ITU	International Telecommunications Union.
IUCN	International Union for the Conservation of Nature.
LDC	Least Developed Country.
M4D	Mobile phones in Development.
MDGs	Millennium Development Goals.
MSG	Melanesian Spearhead Group
MSI	Mauritius Strategy for Implementation.
NSA	Non-state Actor—term used by PIFs in preference for "civil society organization".
NDMO	National Disaster Management Office.
New Hebrides	British/French condominium renamed Vanuatu at independence in 1980.
NZ	New Zealand.
OCHA	United Nations Office for the Coordination of Humanitarian Affairs.
OCO	Oceania Customs Organization.
ODA	Overseas Development Assistance.

OECD	Organization for Economic Cooperation and Development.
OPCW	Organization for the Prohibition of Chemical Weapons.
PACER	*Pacific Agreement* on Closer Economic Relations.
Pacific Islands Forum	Regional Organization led by leaders of independent Pacific Island Countries, previously known as the South Pacific Forum.
Pacific Islands Forum Secretariat	Secretariat established by agreement on 30 October 2000.
PASO	Pacific Aviation Safety Office.
PCCR	Pacific Climate Change Roundtable.
PCRAI	Pacific Catastrophe Risk Assessment and Financing Initiative.
PFTAC	Pacific Financial Technical Assistance Centre.
PIANGO	Pacific Islands Association of Non-Governmental Organizations.
PIC	Pacific Island Country—The term "country" is preferred because not all Pacific states are sovereign nations.
PICISOC	Pacific Internet Society.
PICT	Pacific Island countries and territories.
PICTA	Pacific Island Countries Trade Agreement.
PIDF	Pacific Islands Development Forum.
PIDP	Pacific Islands Development Programme.
PIFACC	*Pacific Islands Framework for Action on Climate Change.*
PIF	Pacific Islands Forum—an annual meeting of leaders of all Pacific Island countries.
PIFS	Pacific Islands Forum Secretariat—Suva-based secretariat established under the name SPEC, to service the policy agenda of the Pacific Island leaders, as agreed at their annual forum.
PIPSO	Pacific Islands Private Sector Organization.
PITA	Pacific Islands Telecommunications Association.
PM	Prime Minister.
PNG	Papua New Guinea.
PPA	Pacific Power Association.
PrepCom	Preparatory Committee Meeting.
PSC	Public Service Commission.
PUA	Pacific Urban Agenda.
RAMSI	Regional Assistance Mission to the Solomon Islands.
SAMOA	SIDS Accelerated Modalities of Action
SDGs	Sustainable Development Goals.
SIDS	Small Island Developing States.
SOPAC	South Pacific Geoscience Organization, formerly a CROP agency, later integrated into SPC.

South Pacific Forum	established by six Pacific Island leaders in 1971. The name was changed to the Pacific Islands Forum in 1999.
SPARTECA	South Pacific Regional Trade and Economic Cooperation Agreement.
SPBEA	South Pacific Board of Educational Assessment. Formerly a CROP agency, after 2008 integrated into SPC.
SPC	Secretariat for the Pacific Community—established in 1947 as the South Pacific Commission—known from 1998 as the Pacific Community.
SPEC	South Pacific Bureau for Economic Cooperation. Established in Fiji by the South Pacific Forum in 1973 and renamed "Forum Secretariat" in 1988.
SPF	South Pacific Forum.
SPOCC	South Pacific Organizations Coordinating Committee—established in 1989—precursor to CROP.
SPREP	Secretariat of the Pacific Regional Environment Programme, until 2004 known as the South Pacific Regional Environment Programme.
SPTO	South Pacific Tourism Organization.
SRDP	Strategy for Disaster and Climate Resilient Development in the Pacific
UN Women	United Nations Entity for Gender Equality and the Empowerment of Women established by UNGA in 2010.
UN-REDD	United Nations collaborative initiative on Reducing Emissions from Deforestation and Forest Degradation.
UNCAC	United Nations Convention Against Corruption.
UNCDF	UN Capital Development Fund.
UNCLOS	United Nations Convention on the Law of the Sea (1982).
UNCTAD	United Nations Conference on Trade and Development.
UNDAF	*United Nations Development Assistance Framework.*
UNDAT	*United Nations* Development Advisory Team.
UNICEF	United Nations Children's Fund.
UNIFEM	United Nations Development Fund for Women—in 2010 amalgamated with other UN agencies into UN Women.
UNDP	United Nations Development Programme.
UNFCCC	*United Nations Framework Convention on Climate Change.*
UNGA	United Nations General Assembly.
UNIDO	United Nations Industrial Development Organization.
UNISDR	United Nations Office for Disaster Risk Reduction.
UNOMB	United Nations Observer Mission in Bougainville
UNORHRLLS	United Nations Office of the High Representative for the Least Developed Countries, Landlocked Developing Countries, and Small Island Developing States.

UNPAN	United Nations Public Administration Network.
UNRISD	United Nations Research Institute for Social Development.
UNSC	United Nations Security Council.
UNSG	United Nations Secretary General.
UPR	Universal Periodic Review.
UPU	Universal Postal Union.
USD	United States Dollars.
USP	University of the South Pacific—regional university co-owned by 12 Pacific Island Countries.
Western Samoa	PIC renamed Samoa in 1997.
WMO	World Meteorological Organization.
WSIS	World Summit for the Information Society (Geneva 2003 and Tunis 2005).
WSSD	World Summit for Social Development (Copenhagen, 1995).
WSSD	World Summit on Sustainable Development (Johannesburg, 2002).

List of Figures

List of Tables

Chapter 1
Introduction: The Pacific Islands Region and the United Nations

This book explores relations between the Pacific Islands region and the United Nations Organization. It is not, however, a study of each Pacific country's individual experience with the UN—that would be a far larger enterprise. The term "Pacific Islands" distinguishes the countries in question from those covered by such terms as "Pacific Rim", "Asia Pacific", or "trans-Pacific"—all of which are too broad. Such a study is important for several reasons. First, since its establishment in 1945, the United Nations has come to hold a ubiquitous position in the lives of Pacific Islanders. By this I mean to say that every step of their progression from colonial status to independence, every step toward improved social and economic circumstance, and every stage of legal and infrastructural advance, has been directly or indirectly influenced by values promoted by, or programs initiated by, a UN organ or agency. The UN is not a "single-issue" organization, as are some other international organizations which are focused on security, or finance, or development. As will become clear in this book, the UN engaged in political affairs, in development at the most technical levels, in governance and public administration, and at global level, in coordination of policy articulation, and in the creation of international law. The human rights protections for citizens all flow from the *UN Declaration on Human Rights* and the subsequent International Bill of Rights. The Pacific Islands Forum made eligibility of new members into its ranks dependent on successful completion of the UN's self-determination processes.[1] United Nations standards and procedures

---

[1] The Forum also welcomed the expression of interest by the Government of the Federated States of Micronesia in membership of SPEC and the Forum. When the Federation of Micronesia sought Forum membership in 1979 after it a new constitutional arrangement with the UN had come into effect, the Forum Leaders responded that the Federated States of Micronesia could join SPEC but that Forum membership would have to wait until termination of the United Nations Trust Territory agreement in 1981. This early rebuff may have sown the seeds of later tensions between the Fiji-based Forum and the member states of the North Pacific. The New Hebrides was similarly advised in 1979 that it could become an observer, but that full membership was only possible after scheduled independence in 1980.

G. Hassall, *The United Nations and the Pacific Islands*, United Nations University Series on Regionalism 24, https://doi.org/10.1007/978-3-031-34155-7_1

have become essential ingredients in policy making, legislation, case law, treaty observance, and politics, from local to global levels.

From 1945 to the present, the UN has broadened and deepened its influence in Pacific Island Countries (hereafter PICs). A periodic report on cooperation between the UN and regions around the world tabled at the United Nations General Assembly (UNGA) in 2006, for instance, identified ten UN agencies assisting in Pacific regional projects in progress at that date, which ranged from disarmament, to the Pacific Plan, the Millennium Development Goals, food security, education, connectivity, and the development of local government (United Nations General Assembly, 2006). The Pacific Community's exercise to map its strategy for 2021–2030 commenced with seven key UN global campaigns (from the 2030 Agenda for sustainable development to Voluntary Guidelines for Securing Sustainable Small-scale Fisheries) (The Pacific Community, 2021).

I have written this book because I do not think that the extent of the UN's influence on the foundations and operation of PICs and Pacific regional architecture is sufficiently understood or appreciated. In many instances this is understandable, since much of the UN's contribution occurs deep inside the operating procedures of a specific sector, visible only to those familiar with that sector. In the field of shipping, for instance, a proposal in the 1970s to establish a Regional Shipping Council for the Pacific was accompanied by a reminder that the UN's second development decade included a recommendation that "Freight rates, conference practices and adequacy of shipping services and other matters of common interest to shippers and shipowners should be the subject of consultation ... Every effort should be made to encourage the institution and operation of shipper's councils." (South Pacific Forum Secretariat, 1973, 45) This is an example of UN standards informing a field far from the public eye. In the field of telecommunications, the International Telecommunications Union (ITU) has participated from the outset in the establishment of global standards, building infrastructure and encouraging Pacific Island Country (PIC) representation at global meetings such as the periodic World Administrative Radio Conference (from 1993 the World Radiocommunication Conference)—again, far from public notice. These and so many similar examples demonstrate the extent to which technical and administrative standards have been set in diverse sectors through UN-affiliated policy processes with very little public awareness or appreciation.

Positive contributions made by the UN do not detract from the fact that the region's development has not progressed as well as envisaged or as required, and that the entire model of development assistance needs re-evaluation. PICs interact with the United Nations organization bilaterally and multilaterally, and at national, regional, and international levels. Furthermore, there is interest in how these nations—some of the smallest nation-states in the world—are coping with the heightened expectations for delivering effective government in the twenty-first century among increasingly literate and mobile populations which are no longer satisfied with status quo conditions and are in search of better livelihoods and well-being at individual, family, and community levels. There is need to probe beyond rhetoric, beyond the frenetic pace of activity, and beyond calculation of the quantity of material resources transferred, to consider what qualities distinguish mature Pacific societies and people. Literacy, numeracy, and health are vital objectives, but well-being also requires integrity in public institutions, the ability of leaders to solve complex problems and to promote justice at both individual and societal levels, the increasing ability of individuals, families and communities to not be reliant on welfare, and the ability of nations to collaborate not only regionally but across the globe. The UN can provide assistance in these matters, but results are reliant on the efforts of Pacific Island communities and leaders.

It is also the case that the small island nations, including PICs, have helped define the work of the United Nations. Although they were not among the UN's founding members and did not have a voice in the shaping of its principal organs, their needs, expectations, and special interests contributed significantly to the operation and expansion of UN programmes and agencies since the latter decades of the twentieth century. The UN now acknowledges the contexts of its smallest, most vulnerable members, which must build capacity in the face of challenges of such magnitude as climate change and sea-level rise. It recognizes amongst its members and associate members 52 Small Island Developing States (38 sovereign and 14 dependent) which share these characteristics. The 14 of these 52 that are in the Pacific now constitute one of the largest voting blocs in the United Nations. Although American Samoa, the Cook Islands, French Polynesia, Guam, New Caledonia, Niue, and the Northern Mariana Islands are not full members of the UN, they are "associate members" through the United Nations Economic and Social Commission for Asia and the Pacific (ESCAP)—and this membership, in turn, has allowed some of these territories to become members of other Pacific regional organizations, such as the South Pacific Geoscience Organization (SOPAC)—and SOPAC successor arrangements.

Another purpose in writing this book has been to fill a gap in regional studies literature. When the world's regions are considered, the Pacific Islands has been the least represented, and the Pacific is the last of the world's region's to be formally recognized by global actors. This is true of academic studies as much as international policy studies. Alley, Graham, Powles, and Manoa are among the few scholars to have written about the United Nations in Pacific Islands context (Alley, 1998; Graham, 2008; Manoa, 2015; Powles, 2002).

Few Pacific Island scholars have published under UN academic auspices, such as the United Nations University and its affiliate centres, or through the United Nations Research Institute for Social Development (UNRISD) (Amosa, 2012; Prasad & Kausimae, 2012). Some Pacific Island studies have appeared through the United Nations University Press although, curiously, not since 2013. Two Pacific regional seminars and participation at the 7th World Conference on Reinventing Government produced seminar reports but no major Pacific-focused publications (Hassall, 2006; UNDESA, 2006; UNDESA, UNDP, & PIDP, 2004). Pacific Islands' experience was briefly noted in Graham and Felício's 2006 *Regional security and global governance* (Graham & Felício, 2006), but was not considered in the 2012 *Third world report on regional integration* (De Lombaerde et al., 2012). Thakur and Van Langenhove's survey of "global governance through regional integration" makes no mention of the Pacific:

> Regionalism has become an integral part of contemporary multi-layered and multi-actor governance. Neither states by themselves nor the UN as their universal collective forum can substitute for regional governance. Within Africa, the Americas, Asia, and Europe, countries share certain policy problems and approaches on a regional scale that they do not hold in common with all countries on a global scale. At the same time, however, regional governance cannot substitute for the UN, particularly in promoting security and development in the world. The task is therefore to build effective partnerships between regional and global agencies. (Thakur & Van Langenhove, 2006, p. 235)

In 1993 the UN Security Council invited regional organizations to study how they could strengthen their functions in peace and security and enhance cooperation with the United Nations. That initiative resulted in Secretary-General Boutros Boutros-Ghali's *An Agenda for Peace* (Boutros-Ghali, 1995), accompanied by six high-level meetings on regional-global security between 1994 and 2005 at which the Pacific region was not represented. The UN was seeking to work with regional agencies under UN Charter Chapter 8 but there was a view within the organization that there were too many "regional agencies" and this exercise sought to clarify which were the most "representative" of a region.

The 4th UN Conference on Least Developed Countries held in Istanbul in 2011 grouped the five Pacific country reports under the "Asian" category. The concept note for a High-level Thematic Debate on "Strengthening Cooperation between the United Nations and Regional and sub-regional organizations" held by the General Assembly in 2015 cited "evolving relationships" with the African Union (AU), European Union (EU), Caribbean Community (CARICOM), Association of South East Asian Nations (ASEAN); League of Arab States (LAS), Organization of Islamic Cooperation (OIC), Organization of American States (OAS), the Community of Latin American and Caribbean States (CELAC), the Organization for Security and Cooperation in Europe (OSCE), and the Collective Security Treaty Organization and the Shanghai Cooperation Organization—but not with any equivalent regional body from the Pacific Islands (United Nations. President of the General Assembly, 2015).

A report to the UN Secretary General on investing in development to achieve the Millennium Development Goals (MDGs) referred to investment priorities in Africa, East Asia, South Asia, Central Asia, Latin America, the Middle East and North Africa, and the Transition and Former Soviet (Commonwealth of Independent states—CIS) countries and Eastern Europe—but not to the Pacific (UN Millennium Project, 2005). The Organization for the Prohibition of Chemical Weapons (which is not a UN agency but works in close collaboration), includes its Pacific Island member states under "Asia." The Security Council included its deliberations on Papua New Guinea and Solomon Islands under "Asia."[2]

"It's unfortunate that despite all our efforts," an official wrote to colleagues when marshalling participants for an Asian Regional Conference in Tokyo in January 2003 leading up to the World Summit on the Information Society,

[2] https://www.un.org/securitycouncil/content/repertoire/asia

> ... the official title for the event remains the "Asian Regional Conference"—without reference to the Pacific. Be that as it may, our combined efforts have resulted in an excellent opportunity to put Pacific States on the map vis a vis other WSIS participants, and to use the opportunity to ensure that there is a strong Pacific contribution at the meeting and incorporation of Pacific interests and issues in the report that goes to PREPCOM II. (Gwynne, 2002)

All who have represented a Pacific country at a regional or global event can recount similar experiences of their marginalized Pacific. As recently as 2020 a study of *Group Politics in UN Multilateralism* lacked reference to the Pacific Islands region and institutions (Laatikainen, 2020). At the present time several of the Pacific's largest external partners, including the US, Australia, France, and Japan, engage the Pacific Islands in the context of their "Indo-Pacific" strategic and diplomatic programs (Wallis & Batley, 2020). Such examples demonstrate the challenge Pacific Island leaders faced as they sought to gain the world's attention for their concerns about atmospheric nuclear testing, incomplete decolonization, and, more recently, genuine responses from the world community on fishing rights, global warming and rising sea levels. Within the UN, regional groups exist for the African, Asian, Eastern European, Latin American and Caribbean, and Western European and other states—but till now there is no such regional group for the Pacific.

The relationship between the United Nations and Pacific Island Countries and region has evolved across at least three (albeit overlapping) phases:

1945–1994—self-determination and the establishment of new nations
1962–2021—independence, national consolidation, and regional institutions
1972–2021—international relations; global policy dialogue, and summitry

In the first phase, PICs were politically powerless societies administered by metropolitan powers, in which the UN implemented development programs on the instructions of colonial authorities. Post-independence, the activities of UN agencies and programmes continued, but through agreement with newly independent governments, which viewed the UN predominantly as a development agency and themselves as beneficiaries entitled to maximum support in nation-building. Over time, issues that were most important to Pacific leaders, such as the continued struggle for independence in New Caledonia and elsewhere, continued atmospheric testing of nuclear weapons in French Polynesia by France, the impact of nuclear tests on the people and environment of Marshall Islands, the drafting of the Law of the Sea, which had direct impact on Pacific nations' fisheries revenues and marine sustainability, followed by the making of global commitments on climate change—these issues and more—sharpened their political and policy skills, enticed their collaboration with others facing similar issues elsewhere on the planet, and thrust the small Pacific Island nations into the realms of global summitry and the collective project of securing global public goods.

In the context of the three phases set out above, this book investigates four fundamental questions:

- What has been the United Nations contribution to the development of the Pacific Island countries and region?

- Conversely, how have the Pacific Island countries and region contributed to the development of the United Nations Organization?
- And, based on this evidence, what does the relationship between the United Nations and Pacific Island countries say about the role of small island countries in global multilateral organizations?
- What reforms to the UN do small states such as those in the Pacific suggest would make it more effective?

These lead to numerous follow-up questions, such as when and how should PICs engage with the UN on political issues? When and how should the PICs engage the UN on a regional basis? In what ways do PICs engage in owning development aspirations, processes, and results? Is the UN more important to small states than to large ones because larger states have more venues available other than the UN within which to exert their influence? What framework can be applied to assess the role and effectiveness of UN-Pacific Islands relations? How can we overcome the "methodological nationalism" that privileges the study of any one country over all others to assess the contribution of Pacific nations as "good states", as ones that promotes the international values of the UN, that seeks to apply them domestically, and that look beyond their individual interests to contribute to policy dialogue on an ever-expanding global public policy agenda?

The answers to these questions are embedded in the discourses of history, politics, communications, development, public management, and global studies. Evidence is embedded in personal correspondence, institutional records, public statements, and academic essays. Each category has its anticipated range of recipients, from narrow and confidential (such as within an organization or institution) to wide and public for all to see (such as vision statements, declarations, plans, and completion reports). The intentions of such discourse are as important as their provenance and intended audiences. There are, for instance, communicative discourses that strive to understand the views of all parties involved in a specific issue, while there are, on the other hand, strategic discourses in which the views of a particular party are advanced with the intent of vanquishing all rival viewpoints.

Reading through the official records of the Pacific Islands Forum and its Secretariat, with a focus on their relationship with the United Nations, provides a genealogical account of what these organizations knew, what they wanted to know, what they wanted to achieve, and how they sought to do this. We can apply these tags of historical setting, political stance, and discursive intentionality, to each of the texts referred to in this study.

Development narratives, for instance, were crafted for audiences according to strategic intent. Official documents began to feature strategic plans, annual work programs, declarations, and vision-statements. The message to the international community was primarily "send more money"; the message from the international community was primarily "show more evidence of results!" Decision-making was consolidated in the hands of "the people's representatives." Chapters 6–12 demonstrate how these themes were performed for international audiences as well as for regional and national ones.

Finally, a word is required about the use of acronyms in this book. Approximately one-hundred acronyms are used as shorthand descriptions of organizations and programs in place of their full names. While this may be disconcerting to the novice reader, they are terms in wide use and already familiar to those experienced with the Pacific Islands and/or the United Nations Organization. Each term will be written in full when first used. Major policy statements, declaration, and treaties, are indicated in italics. The term "Pacific Island Countries" (abbreviated to "PICs") is used throughout the book for all territories irrespective of whether they are sovereign nations or dependent territories. Although Australia and New Zealand are members of numerous organizations in the Pacific Island region they are not considered as "Pacific Island Countries" in this study.

1.1 The Pacific Islands and People

Although now organized into approximately 22 dependent and independent states and territories, the peoples of the Pacific can also be described as 9 to 10 million people, of whom half live in urban locations and half in rural locations, on thousands of islands, in approximately 25,000 villages and 100 cities and towns. There is more cultural, ethnic, and linguistic diversity within some of these countries than there is across them. Since their "exclusive economic zones" are considerably larger than their landmasses, the ability to govern the seas and to protect the ocean's economic resources is as important as governing on land. Pacific leaders refer to the Pacific Ocean as the "Blue Pacific Continent." The "Blue Pacific vision" is:

> . . . a region of peace, harmony, security, social inclusion and increased prosperity so that all Pacific people are leading free, healthy and productive lives. Our Blue Pacific Identity reinforces the potential of our shared stewardship of the Pacific Ocean and reaffirms the connections of Pacific peoples with their natural resources, environment, cultures and livelihoods. We understand that this vision can only be achieved through regional collective action.[3]

The Pacific Islands region thus has some unique characteristics when compared to other regions of the world: small but fast-growing populations with low average incomes, living on small land areas spread over vast sea areas, cherishing their cultures and traditions but also anxious to partake of the best that globalization can offer, and hampered by the long distances to markets and high costs of transport and communications. The ratios between land and sea (especially territorial waters) is made clear in Fig. 1.1 (Hassall, 2021, pp. 4–5) as well as in Table 1.1.

---

[3] https://www.forumsec.org/vision-for-the-2050-strategy/

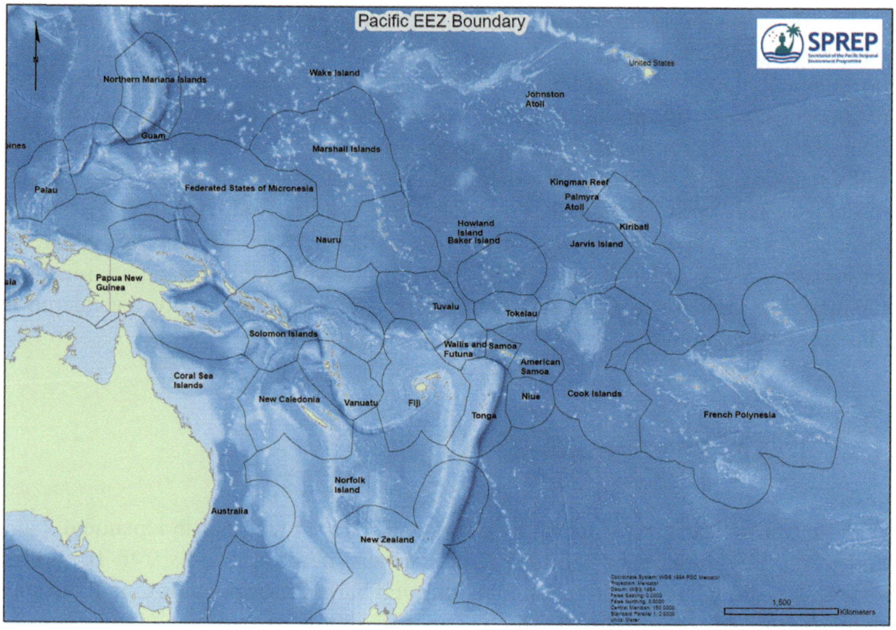

Fig. 1.1 Countries in the Pacific Islands region

1.1.1 Colonial Era Followed by Sovereign and Non-sovereign States

Several recurrent themes are associated with the emergence of Pacific nations and their identity as a region in the post-World War II era. First is an awareness of the extent to which their boundaries and privileges as nation-states have been shaped by European exploration and the colonial impositions from which they have emerged. Of the 22 Pacific territories, 12 enjoy full national sovereignty and the remainder have limited sovereignty or remain dependencies of a metropolitan power. Pitcairn, for instance, belongs to Great Britain; New Caledonia, French Polynesia, and Wallis and Futuna are overseas territories of France; American Samoa, Guam, and the Commonwealth of the Northern Mariana Islands are administered by the United States of America; and Cook Islands, Niue, and Tokelau are freely associated states of New Zealand. This leaves just 12 PICs with sovereign status and full voting rights in international organizations: Federated States of Micronesia, Fiji, Kiribati, Nauru, Palau, Papua New Guinea, the Republic of the Marshall Islands, Samoa, Solomon

Table 1.1 Population, land and sea areas of Pacific Island Countries

Region/country or territory	Population estimates (mid 2020)[a]	Land area (km²)	Exclusive Economic Zone[b]
MELANESIA		**540,030**	
Fiji	894,961	18,333	1,290,000
New Caledonia	273,015	18,576	1,740,000
Papua New Guinea	8,934,475	462,840	1,617,917
Solomon Islands	712,071	28,230	1,340,000
Vanuatu	294,688	12,281	680,000
MICRONESIA		**3156**	
Federated States of Micronesia	105,503	701	2,978,000
Guam	176,664	541	218,000
Kiribati	118,744	811	3,550,000
Marshall Islands	54,590	181	2,131,000
Nauru	11,690	21	320,000
Northern Mariana Islands	56,608	457	1,823,000
Palau	17,930	444	629,000
POLYNESIA		**8126**	
American Samoa	56,813	199	390,000
Cook Islands	15,281	237	1,830,000
French Polynesia	278,908	3521	5,030,000
Niue	1562	259	390,000
Pitcairn Island	50	47	759,287
Samoa	198,646	2934	120,000
Tokelau	1506	12	290,000
Tonga	99,780	749	700,000
Tuvalu	10,580	26	900,000
Wallis and Futuna	11,441	142	300,000
Total	12,325,506	551,312	

[a]https://Stats.pacificdata.org.spc
[b]http://www.marineregions.org/about.php

Islands, Tonga, Tuvalu, and Vanuatu. These PICs, their dates of independence, and their former status, are set out in Table 1.2.

Table 1.3 depicts four territories which have chosen dependent rather than independent status. However, as this chapter will explain, these constitutional acts have not been free of controversy and contention.

Although national sovereignty is highly treasured by the independent Pacific states, it sits beside a desire for Pacific unity that transcends notions of independent and dependent states and peoples. Sovereign status is not a prerequisite for membership in most Pacific regional organizations. Tokelau, for instance, participates fully in the Pacific Community, the South Pacific Regional Environment Programme, the Council of the University of the South Pacific, and the Forum Fisheries Agency, and as an observer at the Pacific Islands Forum. It is an associate

Table 1.2 Independent Pacific Island countries

Territory	Date	Former status
Samoa[a]	1962	U.N. Trusteeship (NZ)
Nauru	1968	U.N. Trusteeship (Australia, N.Z. and Britain)
Fiji	1970	British Crown Colony
Tonga	1970	British Protectorate
Papua New Guinea	1975	U.N. Trusteeship
Solomon Islands	1978	British Protectorate
Tuvalu	1978	British Colony (GEIC)
Kiribati	1979	British Colony (GEIC)
Marshall Islands	1979	Self-governing Republic in assoc. w. U.S.A.
Federated States of Micronesia	1979	Federation in compact of Free Assoc. w. U.S.A.
Vanuatu	1980	British/French Condominium
Palau	1981	Self-governing Republic in assoc. w. U.S.A.

[a]Gained independence as "Western Samoa" in 1962 and renamed "Samoa" in 1997

Table 1.3 Dependent Pacific Island countries

Territory	Date	Status
Cook Islands	1965	Self-governing territory in Free Assoc. w. N.Z.
Irian Jaya	1969	Province of Indonesia[a]
Niue	1974	Self-governing territory in Free Assoc. w. N.Z.
Northern Mariana Islands	1978	Commonwealth of the U.S.A.

[a]The situation of the Indonesian Province of Irian Jaya (West Papua) requires extensive re-examination by the international community. The people of West Papua are Melanesians who live within Indonesia's state boundaries as a legacy of Dutch colonization and a troubled decolonization process centred on a 1969 UN-sanctioned vote

member of the World Health Organization, the Food and Agriculture Organization and the United Nations Educational, Scientific, and Cultural Organization, and a full member of the Forum Fisheries Agency.

Associated with this awareness of the need to transcend their divisive colonial heritage has been the willingness of regional organizations—but especially the Pacific Islands Forum—to advocate on behalf of New Caledonia and West Papua, which it believes have not had genuine opportunities to vote on self-determination. There are, in addition to the PICs set out in Table 1.3, additional territories, set out in Table 1.4, which have not had the opportunity to express their preferences, some of which, nonetheless, participate in the processes of Pacific regionalism.

Although the origins and purposes of the two most prominent Pacific Regional organizations—the Pacific Islands Forum and the Pacific Community—will be introduced in subsequent chapters, some aspects of their functioning will be noted in this introduction to assist in understanding their basic pattern of operation. In the case of the Pacific Islands Forum, the leaders of the independent PICs have met annually as "the Forum" since 1971 and issued a communique detailing the topics consulted on and decisions reached. In 1973 the leaders established a "Bureau" to

Table 1.4 Pacific Island Countries that have not had acts of self-determination

Territory	Status
American Samoa[a]	Unincorporated Territory of the United States
French Polynesia[a]	Overseas Territory of France
Galapagos Islands	Province of Ecuador
Guam[a]	Unincorporated Territory of the United States[b]
Hawaii	Statehood in the American Federation
Lord Howe Island	dependency of the state of New South Wales, Australia
New Caledonia[a]	Overseas Territory of France
Norfolk Island	Australian external territory
Ogasawara	Territory of Japan
Pitcairn[a]	British dependency administered by New Zealand
Rapa Nui (Easter Island)	Territory of Chile
Tokelau[a]	Independent territory of New Zealand
Torres Strait Islands	Australian possession administered by the state of Queensland
Wallis and Futuna	French Overseas Territory

[a]Denotes currently on the UN C-24's list of non-self-governing territories
[b]In 1976 US Congress gave Guam the right to frame its own constitution, but the terms were restricted by congress such that the people rejected it at referendum

follow up on these decisions. The names of these two entities have evolved over time, and this book thus makes numerous references to "The Forum", the "Forum Leaders Meeting", the "South Pacific Forum", the "Pacific Islands Forum"—in reference to PIC leaders' actions; and SPEC (South Pacific Bureau for Economic Cooperation), "the Bureau", "the Pacific Islands Forum Secretariat", and the Bureau/Secretariat's Director/Secretary General—in reference to the actions of their administrative and research office and leadership.

The second major regional organization, the Pacific Community, commenced in 1947 as the South Pacific Commission, is led by a Secretary-General. From 1951 the Commission's work program was agreed at triennial meetings of member-states known as the "South Pacific Conference." In the 1980s these meetings became annual, and the Commission changed its name to "The Pacific Community", with its bureaucracy retaining the term "Secretariat" thus allowing continued use of the familiar acronym "SPC." The "South Pacific Conference", now known as the 26-member "Committee of Representatives of Governments and Administrations" (CRGA), held its 52nd session in November 2022.[4]

[4]https://www.spc.int/crga

1.2 The United Nations

Having introduced the Pacific Island Countries and the region in brief, our focus will now shift to the United Nations. The UN was established in 1945 with the four purposes of ensuring peace and security, human rights, social and economic development, and policy coordination. These are set out in the first article of its Charter:

1. To maintain international peace and security, and to that end: to take effective collective measures for the prevention and removal of threats to the peace, and for the suppression of acts of aggression or other breaches of the peace, and to bring about by peaceful means, and in conformity with the principles of justice and international law, adjustment or settlement of international disputes or situations which might lead to a breach of the peace;
2. To develop friendly relations among nations based on respect for the principle of equal rights and self-determination of peoples, and to take other appropriate measures to strengthen universal peace;
3. To achieve international co-operation in solving international problems of an economic, social, cultural, or humanitarian character, and in promoting and encouraging respect for human rights and for fundamental freedoms for all without distinction as to race, sex, language, or religion; and
4. To be a centre for harmonizing the actions of nations in the attainment of these common ends.

Understanding the scope of the UN set out in these four purposes is essential to understanding the interactions examined in this book. To some, the UN is a political organization, to others it is a security organization, while yet to others it is primarily a development organization; but viewed more widely, its work has each of these dimensions, in addition to another that seeks to "harmonise" the actions of all nations. Some refer to this as global governance but because the global dimension remains under-developed I prefer, for the present, the term "international cooperation."[5] An excellent overview of the contribution made by UN ideas to global challenges is provided in publications of the United Nations Intellectual History Project (Emmerij et al., 2001).

To work toward achieving its four purposes the UN Charter establishes six main "organs", each with specific powers and structure. Their operation is discussed in more detail in the chapters that follow but briefly put, these six organs are the *General Assembly*, which all member-states belong to and participate in; the *Security Council*, which has five permanent members and an additional 10 members elected for two-year terms; the *Economic and Social Council*, whose 54 members are elected for 3 year terms; the *Trusteeship Council*, which no longer functions as its work is deemed to have finished; the *International Court of Justice*, whose 15 judges are

[5] A more complete overview of the UN is not required here. However, for those who are interested, this paper by Szasz provides a useful introduction: (Szasz, 1999).

elected to a 9 year term by the General Assembly and the Security Council; and the *Secretariat*, whose many departments are overseen by the Secretary General.

In the course of their work these six organs have established a considerable number of committees, agencies, commissions, high-level panels, and programmes, to assist in the implementation of their work, each accommodating the participation of member-states in ways appropriate to the needs of those tasks. These principal organs, as well as their major subsidiary entities, as at 2021, are set out in Fig. 1.2. The General Assembly is the only body that all member countries are automatically part of; all other engagements within the UN are optional and subject to decision by a member state's government. As will be shown, United Nations organs continue to refer to the organization's four founding purposes when passing resolutions or establishing new initiatives.

The United Nations came to the Pacific region with a 'bang' in the form of UN visiting Missions investigating the progress of Pacific Island Trusteeships toward social and economic development and towards self-government. After several PICs achieved independence and joined the United Nations Organization, UN programmes rapidly emerged across the region. Endeavours under United Nations auspices led the process of articulating the meaning of development, fostering global knowledge of development policy, fostering development programming and implementation, aligning national and global development planning, and measuring development progress.

Weiss has suggested that the United Nations is best understood as a collection of "three UNs" (Weiss, 2010): (1) the Member states—currently 193; (2) the UN's employees—approximately 37,000; and (3) civil society—approximately 141 organizations in general consultative status, 4483 in special consultative status and 967 on the Roster,[6] together with all other non-state actor support for the UN's principles, ideas, and programmes. Pacific Island personalities participate in all three UNs: political leaders such as Prime Ministers and Ministers for Foreign Affairs and other portfolios; Permanent Representatives to the UN together with other officials and diplomats, and those employed as "international public servants" in junior and senior positions in UN organs and agencies; and the many Pacific NGOs, activists, academics, experts, intellectuals, and others who engage with the UN formally or informally, and who very often provide the "voice of conscience" in UN dialogue.

[6] https://en.wikipedia.org/wiki/List_of_organizations_with_consultative_status_to_the_United_Nations_Economic_and_Social_Council#:~:text=There%20are%20141%20organizations%20in,pdf%20format%2C%20is%20also%20available

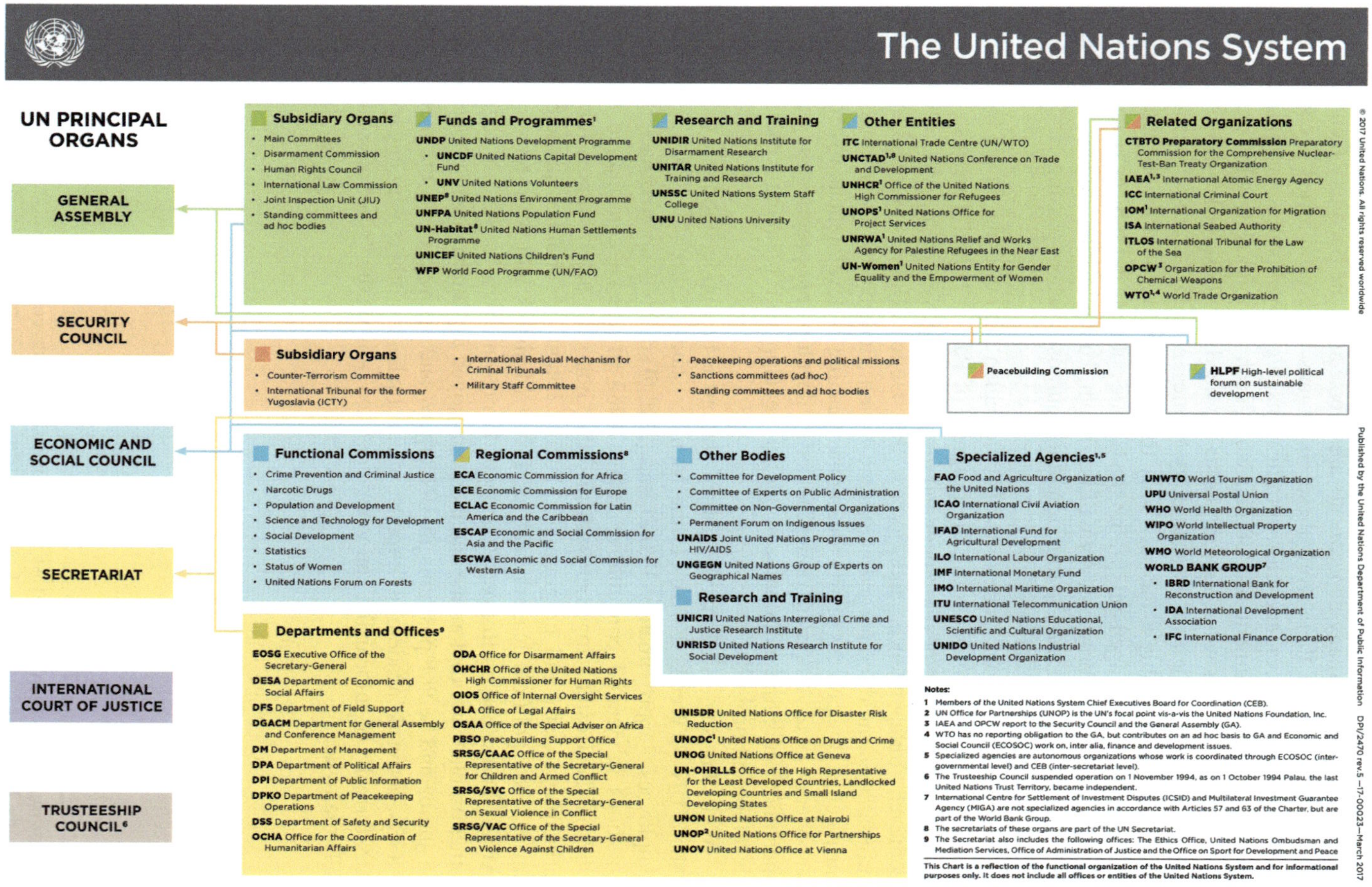

Fig. 1.2 The United Nations System Chart (2021)

1.2.1 Contemporary Challenges for the United Nations: Globally, and in the Pacific Region

The United Nations faces major challenges in both its structure and its operation. In the broad context, Kaul has identified major changes in how the UN has functioned, and how it has been perceived by its members, since its inception until recent times. Whereas it is true that no major worldwide conflagration has broken out, the UN could not prevent the "cold war" between the liberal/western and communist states, nor a considerable number of related proxy wars, nor has it been able to reduce stockpiles of either nuclear or non-nuclear weapons. One significant inflection point was 1990, when the collapse of communism brought an end to the cold war dynamics that had inhibited cooperation within the United Nations for four decades. Post-1990, however, other inter-state dynamics have come to the fore. The larger, predominantly western states, which had shaped the UN and dominated its proceedings up to the 1960s, are now outnumbered by more numerous but less economically powerful post-colonial states, predominantly from the global south, who have their own views about how their economies and societies should be shaped. Commitment to the United Nations appears to have declined amongst western nations no longer able to impose their development preferences on this "developing" world and declined also amongst non-western nations no longer willing to be told how best to pursue social and economic development.

Additionally, the UN is no longer the only intergovernmental organization addressing global policy issues. The reality that sovereignty is "distributed" rather than exclusively possessed and exercised by nation-states is evidenced in the rise of multi-level, multi-agency organizations, platforms, and networks focused on the attainment of targeted global public goods. Yet other groupings seek to keep control of global policy agendas out of the reach of the new UN majorities. As Kaul explains:

> The post-1990 era has brought relatively few additional gains for any country group. From the perspective of developing countries, it may even have meant some reversals. This is because of the increased intrusiveness of UN policies and the suggested revision of sovereignty, which mainly concern developing countries, but in its current formulation do not apply to global threats emanating from industrial countries like pollution or financial contagion effects.
>
> The current UN is highly divided, with industrial countries having chosen to address many of the major policy issues in non-UN fora; and developing countries, too, often 'exiting'—giving more attention to regional fora or just 'lying low' and avoiding decision-making. (Kaul, 2010, 331)

Structurally, the UN system has not evolved in response to changing circumstances despite numerous official and unofficial reform initiatives. The organization remains trapped in bureaucratic proceduralism and in disputes about sovereignty and territorial boundaries. The Security Council remains dominated by the interests of its five veto-wielding permanent members, and, despite its mandate and extensive powers, has been unable to resolve conflict in the Middle East, or to protect civilians in Syria,

Iraq, Somalia, Afghanistan, and elsewhere (Hassall & Partow, 2020). The Security Council was impotent in the face of Russian aggression against Ukraine from 2014 onwards, and particularly from early 2022: no significant progress has been made toward global collective security. The International Court of Justice still lacks compulsory jurisdiction over states-parties that appear before it (i.e., it has no power to enforce its judgements), and whereas an International Criminal Court has been established, the United States is among the UN members that refuse to recognize its jurisdiction. Global inequalities between the richer and poorer nations continue to grow rather than decrease.

The UN has overseen several spectacular operational failures in the Pacific, including widespread dissatisfaction with the situation in West Papua, the unnecessary violence that accompanied the establishment of independence in Timor Leste, inability to successfully conclude acts of self-determination for New Caledonia and French Polynesia, and inability to intervene effectively over the Bougainville autonomy/separatist issue. The UN brought no relief from the effects of nuclear weapons testing by two permanent members of the UN Security Council, at Bikini Atoll in the Marshall Islands, and at Mururoa Atoll in French Polynesia. It contributed marginally at best to resolution of Solomon Islands civil unrest 2000–3, or to Fiji's crises of 1987, 2000 and 2006, in which the military or others under military influence overthrew democratically elected governments—in the very country which hosts the main Pacific regional organizations as well as United Nations offices. The United Nations has facilitated large aid flows into the Pacific, but analysis is divided as to whether this has resulted in development or continued dependency.

At the current time much effort is being expended within the UN system monitoring global carbon emissions, sea-level rise, and the impacts of climate change, but in Pacific Islands' perspective little rhetoric, let alone analysis or commentary, has been translated into meaningful action, despite the pledges and agreements at global climate change conferences. The low-lying island societies face existential threat and a level of uncertainty that brings a sense of futility to all other social, economic, and infrastructural development activities.

And yet, the UN has delivered—and continues to deliver—considerable benefits to Pacific Island countries and to the region more broadly. These challenges and accomplishments are examined in this book in three parts. Part I examines the UN's contribution to the emergence and development of the Pacific Island Countries, several of which were Trusteeships from soon after the establishment of the United Nations in 1945 and came to independence under the UN Trusteeship Council's stewardship. Colonies that were non-UN trusteeships were also assisted through the process of self-determination. Subsequently, UN agencies established working relationships with the new nations of the Pacific and collaborated in pursuit of a wide range of social and economic development objectives. Similar relationships were established with the Pacific's regional organizations, principally the South Pacific Commission and later the Pacific Islands Forum.

Chapters 2 and 3 draw on the discourses of self-determination. Actors included: (a) subject peoples (whose entitlements to protection could only be considered as "rights" at a most basic level); (b) international public servants (employees of the

League of Nations and the United Nations, propelled by personal and institutional imperatives); (c) the metropolitan powers (justifying the status quo and intent on resisting change); (d) their peer nations (judging each other by a formula of self-interest mixed with public interest); (e) the print media, and (f) the interested public (numerically small but strategic advocacy groups eventually known as "civil society organizations"). Power was not spread uniformly between these actors: the voices of the metropolitan powers, the international institutions, and even the metropolitan non-state actors, were far more influential than those of the subject populations whose interests were purportedly being served. The "public sphere" was confined to village and clan settings. There was no "indigenous elite"; individuals who demonstrated ambitions for progress were labelled as troublemakers (Hassall, 1991; O'Brien, 2017).

There are significant issues of agency, accountability, and democratic voice. These matters have long been settled at the level of the democratic nation state, at least theoretically, through the accountability of governments to "the will of the people" at periodic general elections. But when projects are agreed at regional or international level the lines of accountability are less clear: how does the public express its preferences? How is accountability for results to be applied? Whose agendas are being progressed through multi-actor programmes and platforms? Are current models for pursuit of national, regional, and global development meeting their targets? And if not, what is to be done? The most fundamental question to be asked about the relationship between the UN and the Pacific region is who is influencing whom, for what reasons, and with what results? Some clues to how Pacific Islands leaders are thinking may be gleaned from a close reading of their "Leader's Vision for 2050" statement issued in 2022 (Pacific Islands Forum, 2022). This statement, which acknowledges the need to engage with regional, multilateral and global partners without referencing specific organizations such as the United Nations, or specific global campaigns such as the SDGs, will be considered in more detail in later chapters.

Part II changes the focus to PIC engagement in the principal organs of the UN as well as its many other agencies and programs. It examines the extent to which they have engaged with the UN's principal organs and the extent to which they have adopted and promoted UN values, such as through acceding to and implementing treaties. Commitment to UN processes is demanding of human and material resources. Participation is expensive, and there are more options for engagement in UN agencies and initiatives than small states can exercise. The activities they do engage in at the UN gain little public notice domestically, and this results in a considerable communication deficit between political leaders and their constituents. Chapters 4–6, which investigate the ongoing quest for self-determination, and nation-building for those countries that had achieved that goal, the discourse continues to feature strategic communication, but by actors whose power-relations are undergoing change. In the new nations "subjects" were now "citizens", whose rights had expanded—but only to the boundaries of legal frameworks as they existed at the time of independence and improving at the pace their parliaments could achieve in fulfilment of nationalist promises about the lifting of oppression. The discourses of nation-building engaged domestic and international publics, donor nations and institutions, and peer nations both large and small.

Part III focuses on the extent to which the UN facilitates the participation of small states in global policy dialogue. World conferences are no longer the settings for mere articulation of ideals, but for pledging of commitments to act on these ideals through effective implementation at regional, national and sub-national levels, and for judgement of pledged implementation. The PICs were building capability at national level during an era when the major policy challenges evolved from national to global in scope. Care for the environment, the quest for prosperity in an era of increasing global disparities in the wealth of nations, and the urgency of addressing rapid climate change, impelled the Pacific states to find their voice and approach the United Nations as not merely a development agency but as a key portal through which to influence world opinion on issues of fundamental concern. Public policy problems that are global in scope rather than national—such as environmental protection, economic and financial stability, and suppression of global aggression, criminality and terrorism—require global agreement and action rather than cumbersome exchanges between 193 nation states. These and other issues will be touched on in subsequent chapters and answered at least in part.

References

Alley, R. (1998). *The United Nations in Southeast Asia and the South Pacific*. Macmillan.

Amosa, D. U. (2012). *Social policies in Samoa* (Vol. 6). UNRISD.

Boutros-Ghali, B. (1995). *An agenda for peace, 1995*. United Nations.

De Lombaerde, P., Baert, F., & Felício, T. (Eds.). (2012). *The United Nations and the Regions: Third World Report on Regional Integration*. Springer.

Emmerij, L., Richard, J., & George, W. T. (2001). *Ahead of the curve? UN ideas and global challenges*. Indiana University Press.

Graham, K. (Ed.). (2008). *Models of regional governance for Pacific Island States: Sovereignty and the future architecture of Pacific Regionalism*. University of Canterbury Press.

Graham, K., & Felício, T. (2006). *Regional security and global governance: A study of interaction between regional agencies and the UN Security Council with a proposal for a regional-global security mechanism*. VUB University Press.

Gwynne, B. (2002). WSIS - Asia regional meeting: Pacific initiatives. *SPC records*.

Hassall, G. (1991). The Failure of the Tommy Kabu Movement: a reassessment of the evidence. *Pacific Studies, 14*(2).

Hassall, G. (2006). *Establishing social trust through transparent governance*. Paper presented at the Reinventing Government in the Pacific Islands Region: Exchange and Transfer of Innovation for Transparent Governance & State Capacity, Nadi, Fiji.

Hassall, G., & Partow, N. (Eds.). (2020). *A Seat at the Table: New Zealand and the United Nations Security Council, 2015–2016*. Massey University Press.

Hassall, G. (2021). *Government and Public Policy in the Pacific Islands*. Emerald Publishing.

Kaul, I. (2010). The changing role of the United Nations: Lessons for multi-level governance. In H. Enderlein, S. Wälti, & M. Zürn (Eds.), *Handbook on multi-level governance*. Edward Elgar.

Laatikainen, K. V. (2020). In K. Smith & K. V. Laatikainen (Eds.), *Group politics at the UN: Conceptual considerations* (pp. 20–44). Brill | Nijhoff.

Manoa, F. (2015). The New Pacific Diplomacy at the United Nations: The rise of the PSIDS. In G. Fry & S. Tarte (Eds.), *The New Pacific Diplomacy*. ANU Press.
O'Brien, P. (2017). *Tautai: Sāmoa, world history, and the life of Ta'isi O. F. Nelson*. University of Hawaii Press.
Pacific Community. (2021). Pacific Community Strategic Plan 2021+ Planning for our Future our SPC Our Blue Pacific.
Pacific Islands Forum. (2022). *2050 Strategy for the Blue Pacific Continent*. Pacific Islands Forum Secretariat.
Powles, M. (2002). Making waves in the Big Lagoon: The influence of Pacific Island Forum countries in the United Nations. *Juridique Polynesienne, 2*, 59–76.
Prasad, B. C., & Kausimae, P. (2012). *Social policies in Solomon Islands and Vanuatu* (Vol. 7). UNRISD.
South Pacific Forum Secretariat. (1973). The establishment of a South Pacific Regional Shipping Line. *SPEC, 73*, 17.
Szasz, P. C. (1999). The complexification of the United Nations system. *Max Planck Yearbook of United Nations Law Online, 3*(1), xxi–57. https://doi.org/10.1163/187574199X00018
Thakur, R., & Van Langenhove, L. (2006). Enhancing global governance through regional integration. *Global Governance, 12*(3), 233–240.
UN Millennium Project. (2005). *Investing in development: A practical plan to achieve the millennium development goals*. Earthscan.
UNDESA. (2006). *Reinventing government in the Pacific Islands Region: Exchange and transfer of innovations for transparent governance and state capacity*, Nadi, Fiji Islands.
UNDESA, UNDP, & PIDP. (2004). *Report on the Findings of the Regional Forum on Reinventing Government in the Pacific Islands*. UNDESA and UNDP.
United Nations General Assembly. (2006). *Cooperation between the United Nations and regional and other organizations: Report of the Secretary-General*. UN.
United Nations. President of the General Assembly. (2015). Concept note: High-level thematic debate of the general assembly on "Strengthening Cooperation between the United Nations and Regional and sub-regional organizations" 4 May. New York.
Wallis, J., & Batley, J. (2020). How does the 'Pacific' fit into the 'Indo-Pacific'? The changing geopolitics of the Pacific Islands. *Security Challenges, 16*(1), 2–10.
Weiss, T. G. (2010). How United Nations ideas change history. *Review of International Studies, 36*, 3–23.

Part I
Self-determination and the Development of New Nations

Whether the colonial experience was oppressive or benevolent it was, either way, a condition of subordination. It was a condition of injustice. One presumes that all subsequent developments—by these colonies as they trod the path of self-determination, and by the international, intergovernmental agencies established to assist them—have served the purpose of building more just societies; that their interactions have removed sources of oppression as Freire would have it (Freire, 1993); and have been "emancipatory" as Habermas has envisioned (McCormick, 2007). But who, in the dialogic interaction between the nation-state and the United Nations organization, best understands what is oppressive, and what actions are emancipatory? Who is the "principal", and who is the "agent"? Or does a different conception such as "communicative action" more adequately describe the interactions between national and international actors? Could both be influential, one upon the other? Is "interdependence" a better term for the political, economic, and social relationships that now exist between small states and international intergovernmental organizations such as the United Nations?

This chapter explores these and other questions. It commences by reviewing the League of Nations Mandates era, and then the United Nations Trusteeships under which the UN became an ally of colonized peoples to the extent that the Trusteeship Council's oversight of their governance was oriented toward not merely "well-being" (invariably a euphemism for the minimal social and economic development undertaken during the League's Mandates) but more active engagement in improving health, education, economy, and infrastructure—and political advance. Whereas metropolitan countries invariably claimed full credit for bringing Fiji, Samoa, Papua New Guinea, Solomon Islands, Kiribati, Tuvalu, and other small nations to independence, there can be no doubt that the speed of decolonization was much influenced by the probing reports of United Nations Trusteeship Council Visiting Missions.

Whereas self-determination was an essential entry point to admission into the community of nations, relationships with the United Nations differed significantly depending on whether a nation chose sovereign independence or some degree of association with their former colonizer, and this was because the United Nations was founded on the Westphalian philosophy of state, according to which only fully sovereign states could gain admission as voting members of the General Assembly. This distinction has required the small states to weigh the merits of choosing sovereign independence against the merits of continued dependency. This primacy of the nation-state limited the UN's capacity to formally recognize the Pacific region, people, and organizations. This constitutes one of this book's major themes. For the first several decades of the UN's existence and operation, the concerns of Pacific peoples could only be viewed within the UN system through the lens of individual states, and formal dialogue with the Pacific's key regional institutions only commenced later.

By the 1980s the Pacific nations were much focused on nation-building and on maximising access to resources on offer from the UN's various agencies and programmes for their individual country development programmes. But also, by the 1980s, additional developments emerged to challenge the primacy and effectiveness of international arrangements based on agreements between Westphalian states and their bureaucracies. One such development was the rise of non-state actors: non-government organizations, transnational corporations, academics, media, and expert consultants and professional bodies, all possessed insights and perspectives useful to decision-making by governments. New governance arrangements were required to accommodate this increased number and range of voices: networks, working groups, advisory groups, and other collaborative platforms and initiatives. This progression in status of the Pacific territories from colonial subordination to small developing nation-states, regionally networked, making full use of new governance arrangements to access the resource and development opportunities offered by intergovernmental organizations such as the UN, is the subject matter of Part I.

References

Freire, P. (1993). *Pedagogy of the oppressed.* Continuum.

McCormick, J. P. (2007). *Weber, Habermas, and transformations of the European state: Constitutional, social and supranational democracy.* Cambridge University Press.

Chapter 2
Self-Determination

2.1 League of Nations Mandates

League of Nations mandates in the Pacific Islands may appear to some as little more than a "prequel" to the main themes of this study but they are essential to understanding the context. After WWI the victors established the *League of Nations* to prevent the outbreak of a similar conflict in the future. That project failed: insightful studies by Bobbit (2003) and Lopez-Claros et al. (2020) document the League's structural weaknesses, especially its inadequate collective security measures. But the League made advances in other areas of international public administration, including promoting intellectual cooperation and the systematic collection of social and economic data. It also instituted the *Permanent Mandates Commission* to oversee the administration of dependent colonies and territories. The relevant article of the League's Covenant reads:

> ARTICLE 22.
>
> To those colonies and territories which as a consequence of the late war have ceased to be under the sovereignty of the States which formerly governed them and which are inhabited by peoples not yet able to stand by themselves under the strenuous conditions of the modern world, there should be applied the principle that the well-being and development of such peoples form a sacred trust of civilisation and that securities for the performance of this trust should be embodied in this Covenant.
>
> The best method of giving practical effect to this principle is that the tutelage of such peoples should be entrusted to advanced nations who by reason of their resources, their experience or their geographical position can best undertake this responsibility, and who are willing to accept it, and that this tutelage should be exercised by them as Mandatories on behalf of the League.
>
> The character of the mandate must differ according to the stage of the development of the people, the geographical situation of the territory, its economic conditions and other similar circumstances.
>
> There are territories, such as South-West Africa and certain of the South Pacific Islands, which, owing to the sparseness of their population, or their small size, or their remoteness from the centres of civilisation, or their geographical contiguity to the territory of the Mandatory, and other circumstances, can be best administered under the laws of the

G. Hassall, *The United Nations and the Pacific Islands*, United Nations University Series on Regionalism 24, https://doi.org/10.1007/978-3-031-34155-7_2

> Mandatory as integral portions of its territory, subject to the safeguards above mentioned in the interests of the indigenous population.
>
> In every case of mandate, the Mandatory shall render to the Council an annual report in reference to the territory committed to its charge.
>
> The degree of authority, control, or administration to be exercised by the Mandatory shall, if not previously agreed upon by the Members of the League, be explicitly defined in each case by the Council.
>
> A permanent Commission shall be constituted to receive and examine the annual reports of the Mandatories and to advise the Council on all matters relating to the observance of the mandates.

The League's Covenant outlined a three-tiered classification of Mandates:

- Category A. Mesopotamia and Palestine, allotted to Great Britain, Syria (including Lebanon), allotted to France.
- Category B: Portions of East Africa, Togo and the Cameroons, allotted to Britain, France and Belgium.
- Category C: The former German possession in the Pacific, south of the Equator, except for Nauru and Samoa, allotted to Australia. The former German possessions north of the Equator, allotted to Japan.[1] Samoa, allotted to New Zealand; Nauru, allotted to the British Empire, Southwest-Africa, allotted to South Africa.

"C" category mandates were regarded as the least developed, and least likely to become developed. Of interest to the current study were the four "Category C" mandates in the Pacific Islands:

1. South Seas Islands—administered by Japan from 1920 (until, formally, 1946).
2. New Guinea—administered by Australia from 1920 to 1946.
3. Nauru—Administered by Australia from 1923 to 1946; and
4. Western Samoa—administered by New Zealand from 1920 to 1947.

The intentions behind the establishment of "C" class League of Nations Mandates remain contested. The League's view was that such mandates were legitimate expression of the international community's desire to assist the progress of these peoples and territories (Pedersen, 2015). An alternate view is that these were "veiled annexations", implemented without any consultation with the peoples affected, (Evans, 1927) "subtle and complex" instrumentalization of law to "give effect to colonisation." (Heathcote, 2022, p. 62) In any case, the PICs had no say in the establishment and conduct of mandates (just as they had no say in the later establishment of Trusteeships under the United Nations). In 1921 the Australian parliament passed a law recognizing its authority over New Guinea. New Zealand passed a similar law for Western Samoa, and Britain for Nauru. If early essays by anthropologist Felix Keesing are any guide, the metropolitan powers viewed mandated territories as colonies suited to exploitation rather than development:

> . . .at the present time there appears to be emerging into the practice of government in many if not most dependencies a new sense of responsibility toward the dependent peoples. This

[1] The USA protested the allocation of Yap to Japan.

> administration of the native into the scheme of things seems evidently to be no simple matter... many Pacific dependencies are already vital economic reservoirs for raw materials: rubber, sugar, metals, fruits, copra, guano (fertilizer), and shell. If the food, population and industrial problems of the Pacific are to be understood, the facts concerning present exploitation (sources of capital, ownership of natural resources, labor, transportation, markets and the like), and also concerning their future possibilities, should be known. (Keesing, 1930, p. 450).

Although widely considered as the "colonies" of Australia, Japan, or New Zealand, the quality of administration in these mandated territories was as a matter of international law supervised by the League of Nations' Permanent Mandates Commission. The "Mandatory Authorities" were obliged to report their efforts to improve the conditions of labour, health, education, public finance, district administration, law and justice, infrastructure, agriculture, and other vital sectors (League of Nations. Permanent Mandates Commission, 1930b, 1930–32, 1935–38). The details of these reports were queried by the Commission at its annual sessions in Geneva.

Subjects from the mandated territories had the right of petition, which provided an early means for the expression of grievances about their government (League of Nations. Permanent Mandates Commission, 1928, 1930a). New Zealand's administration of Samoa for instance, was critiqued closely by the Commission throughout the 1920s and 1930s (Hassall, 2017). Table 2.1 depicts the numbers of petitions lodged by residents of mandated territories and those for which decisions were reported (Pedersen, 2012, p. 237). The small numbers of petitions from the Pacific mandates reflects the smaller size of their populations but may also reflect the restricted access that their residents had to means of communication with the international body.

The experience in the mandates is perplexing. On the one hand they were established non-consensually by the First World War victors while on the other hand they brought the mandated territories into the orbit of international administration, principally through annual reports, visiting missions, and the right to petition—although the Commission tended to give more weight to the explanations and justifications put to them by the Mandatory powers. For instance, the Commission was confounded by reports reaching it from individual petitioners as well as from press articles about military build-up and tight control of movement in Japan's South Seas Mandate—as compared with the Japanese government's reports which maintained that no such developments were taking place. Year upon year, Japan's annual report to the PMC recorded, under the chapter "Military clauses", a complete absence of military activity in the Territory. The report for 1937, not received by the PMC until 9th September 1938, stated:

> Simultaneously with the establishment of the South Seas Bureau, in April 1922, the South Seas Defense Corps was abolished, and the entire naval contingent stationed in the Islands was withdrawn. Since then, absolutely no military or naval forces have been stationed there. No military or naval basis nor any fortification has ever been maintained or newly built within the Territory. Again, the natives have never been subjected to military training. (Government of Japan, 1937, p. 102)

Table 2.1 Petitions received and considered by the Mandates Commission (Pedersen, 2012)

Mandated territory	All petitions and communications (Van Ginnken)	Petitions in section register, 1828–40	PMC Reports on petitions, 1925–39
Syria/Lebanon	1322	743	145
Palestine/Transjordan	1237	428	88
Iraq	69	53	15
(All A mandates)	*(2628)*	*(1224)*	*(248)*
Tanganyika	66	62	12
Ruanda/Urundi	6	3	1
Cameroon (French)	40	42	13
Cameroon (British)	2	1	0
Cameroon (both)	2	9	0
Togo (French)	71	43	11
Togo (British)	1	4	0
Togo (both)	3	0	0
(All B mandates)	*(191)*	*(155)*	*(37)*
South West Africa	146	86	25
Western Samoa	35	26	7
New Guinea	10	3	2
Japanese islands	9	9	4
Nauru	3	0	0
(All C mandates)	*(202)*	*(124)*	*(38)*
General/multiple	23	6	2
Total	3044	1509	325

Sources: Figures for column 1 (all petitions and communications) are from Van Ginneken, Volkenbondsvoogdij, 211–17. Figures for column 2 are calculated from the Mandates Section's Register of Petitions, in LNA, S1681. Figures for column three (PMC reports) are my count of reports listed in the comprehensive published index of PMC documents: Reno, *League of Nations Documents, 1919–1946*

By 1938 the League was receiving press articles as well as letters from individuals reporting the military activities that appeared to contradict Japan's formal statements. But because Japan withdrew credentials from its representative in Geneva, the Commission had no one to formally interview. After taking legal advice, the Commission proceeded with an assessment of the report on the basis that both Japan and the Commission still had legal duties under the League of Nations Covenant (League of Nations. Permanent Mandates Commission, 1938).

In the case of Samoa, the Commission was aware of civil unrest in the 1920s, but reliant for the most part on information provided by New Zealand's representatives, who were protagonists in the very affairs they were commenting on. A group known as Mau (The "Samoan League") declared its objective to be:

> The advancement of Samoa, and to present to the Administration and the Government of New Zealand from time to time subjects concerning the Government of Western Samoa which may be considered by the members of the League essential for the promotion of the peace, order, good government, and the general welfare of the territory (Davidson, 1967, p. 119).

The New Zealand administration regarded the Mau and such of its leaders as Olaf Nelson as agitators. After police fired on a peaceful Mau street march, a Commission of Inquiry conducted by New Zealand and published on 10th December 1927, exonerated New Zealand's Administrator For Western Samoa, Sir G.S. Richardson.[2] Although reports of New Zealand's harsh treatment of Samoans continued to reach Geneva, the Commission continued to privilege the justifications provided by New Zealand, and to dismiss the petitions submitted by and on behalf of the Samoan people. "The Mandates Commission's dismissal of Samoan grievances was based on three assumptions," Pederson explains,

> …first, that Samoans' involvement in the May was based on their suggestible and wilful character and if deprived of Nelson's leadership would vanish overnight; second, that Nelson was seeking power and personal gain and had no deep convictions; and third, that the Administration could, by a policy of 'firmness', defeat the protest movement. Over the course of the next few years, to the bewilderment of the Mandates Commission and the New Zealand government, every single one of those assumptions was proven wrong." (Pedersen, 2015, p. 185)

An international regime, ostensibly established to assist a people learn about governance from their European tutors, demonstrated race prejudice and notions of cultural supremacy: the seeds were sown for Samoan response to a second round of foreign domination. From the mid-1930s the League of Nations collapsed, the second world war unfolded, and Japan launched incursions into the Pacific Islands from bases covertly developed on islands across Micronesia. Across the Pacific, the Mandated Territories experienced military incursion or military administration, at the least. One era of colonial imposition was at an end and another about to begin.

[2]Unfortunately for the Samoans, in 1928 New Zealand appointed Major General Sir George Richardson as one of its representatives to the Permanent Mandates Commission.

2.2 Self-determination and Independence: UN Trusteeships

After the establishment of the United Nations in 1945, Pacific territories that were formerly under League of Nations authority were constituted, once again with no local consultation, under the terms of the UN Charter's article 73 concerning "non-self-governing territories." One major difference was that Japanese control over Micronesia was replaced by American. In contrast to League of Nations' Article 22, the UN Charter's article 73 anticipated not mere "tutelage" by more "civilized" nations, but political, economic, social and education advancement, and even self-government. The term "territories inhabited by peoples not yet able to stand by themselves under the strenuous conditions of the modern world . . ." was replaced by "territories whose peoples have not yet attained a full measure of self-government" (United Nations. Secretary General, 1946, p. 2). Keesing's 1947 article noted the influence of UN thinking on self-determination:

> Today, instead of two small territories, the United States holds a great zone of islands scattered across the Pacific. These, taken from Japan, are at present also being controlled by our Navy under a system of "military government" as prescribed for occupied enemy territories under the international law of war...the United States took over Guam and Samoa in days when expansion of empire and colonial monopoly were the rule, and backward peoples "the white man's burden." Today, as we come to fix the status of such holdings, it is against the backdrop of United Nations principles, of growing international standards of trusteeship and welfare, and of the burgeoning desire of colonial peoples for self-government and even "independence. (Keesing, 1946, p. 61)

The three Pacific territories and their administering authorities overseen by the UN Trusteeship Council were: Nauru (Australia—1947-1968); New Guinea (Australia—1946–1975); and Western Samoa (New Zealand—1945–1962). A fourth territory, the Trust Territory of the Pacific Islands, was set up as a "strategic trust" under the United States of America (1947–1994). Prince has argued that this latter arrangement entailed a contradiction in terms, with implications for the legal processes by which the Trusteeship should have, at a later date, been terminated:

> The Charter does not provide in detail what the differences between a strategic and non-strategic trust should be, but the provisions of the Trusteeship Agreement for Micronesia do establish some differences. Among some of the special provisions is article 5 which allows the United States to set up military facilities in the Trust Territory. Also, article 13 of the Agreement grants the United States the power to close off certain areas of the Trust Territory from even Security Council supervision. (Prince, 1989, p. 23)

The geographic extent of these Trusteeships is set out in Fig. 2.1, the United Nation's 1947 map of "Southwest Pacific Trust Territories."

2.2.1 Annual Reports

Whatever their deficiencies, the Trusteeships brought Pacific Islands' administration and political development under unprecedented levels of international scrutiny, and

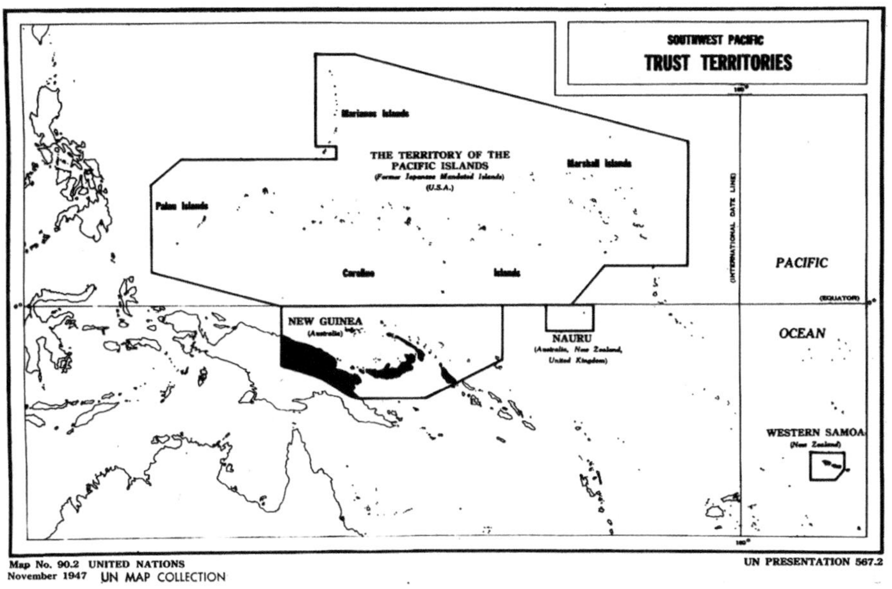

Fig. 2.1 UN Southwest Pacific Trust Territories

this pushed the administering states (Australia in New Guinea and Nauru, and New Zealand in Samoa) toward self-government at much faster rates than would have occurred otherwise. In Micronesia the United States came under similar pressure (Meller & Meller, 1969). Annual reports lodged with the Council and inspected by its members invariably resulted in the mandatory power fielding questions. Australia, for instance, was asked 34 questions following lodgement of its 1946–47 report for Papua New Guinea, which ranged from naïve to constructive: for what reason were the territories of Papua and New Guinea combined into a single administration? Were their peoples of the same race and language? Why wasn't the International Labour Organization included in the list of general and multilateral international agreements? What was the relationship between the Trust Territory of New Guinea and the South Pacific Commission? Was Papua New Guinea included in the plan for the Fiji Medical School? Other questions probed economic conditions, demographics, financial arrangements, labour conditions, and plans for infrastructure development (United Nations. Trusteeship Council, 1948). These examinations continued at three-yearly intervals, beginning in 1951, until the Trusteeship territory attained self-determination and was promoted beyond the Trusteeship Council's purview.

2.2.2 *Visiting Missions*

A second source of information for the Council were visiting missions, which generally comprised three to four members of the Council, none of whom had very much familiarity with the Pacific region. Visiting missions were instructed to give attention to issues that had been raised in annual reports and petitions, to accept or receive petitions while traveling and to investigate petitions "on the spot" when they felt this necessary, and to transmit their full report to the Trusteeship Council as soon as possible (United Nations. Trusteeship Council, 1950c, p. 2). Given the vast distances involved and the speed and range of air travel in the post war years, a visit from New York to all four Pacific Trust Territories in 1950 was expected to take 110 days (United Nations. Trusteeship Council, 1949a). Mander suggested in 1953 that assessments by UN visiting missions to the Pacific were "complex and unsatisfactory" due to different levels of understanding of the issues by their members, depending on whether they were from administering or non-administering countries—with those of the latter "have been less cognizant of anthropological and culture-contact factors and are apt to judge from formal categories." (Mander, 1953, p. 145)

Between 1947 and 1993 the Trusteeship Council undertook no less than 36 missions to the four Pacific islands Trusteeships, as indicated in Table 2.2. Samoa received the least number of visiting missions prior to the attainment of independence in 1962, followed by Nauru whose independence was achieved in 1968, then Papua New Guinea (which had been administratively attached to Australia's adjacent colony Papua and granted independence in 1975). Missions to the Trust Territory of the Pacific Islands (i.e., Micronesia), were more complex, since they were established as a "strategic trust" under the UN Security Council rather than under the Trusteeship Council, and were administered by the US Navy rather than by civilian authorities. According to Meller, the efforts of new nations within the UN, coupled with the voice of the communist bloc, placed US' administration of the Trust Territory on a defensive footing. The report of the 1961 visiting mission, for instance, stressed the need to transform the Council of Micronesia into a legislating body (Meller & Meller, 1969)—an idea much against the plans and interests of the United States at that time.

The use of the term "trust" was ironic: in 1995 the Marshall Islands Ambassador to the UN relayed to UNGA newly disclosed information to the effect that at the UN Secretary General had on 10th May 1954 agreed to a US request that a petition from the Marshall Islands pleading for an end to nuclear tests in the Territory be withheld until completion of the test series. "In this petition", stated Ambassador Muller,

> . . .the people of the Marshall Islands, through their leaders, had expressed their genuine fears and concerns to the only world institution they believed could help them. In return, however, the high executives in this institution not only failed to consider this a matter of any significant importance, but actually engaged in an act most unbefitting the integrity of this august body. (United Nations General Assembly, 1995, p. 3)

Table 2.2 Trusteeship Council visiting missions to Pacific Islands Trusteeships

	Samoa	Nauru	New Guinea	Trust Territory Pacific Islands
1947	√			
1951	√	√	√	√
1953	√	√	√	√
1956	√	√	√	√
1959	√	√	√	√
1961				√
1962		√	√	
1964				√
1965		√	√	
1967				√
1968		√	√	
1970				√
1971			√	
1972				
1973				√
1976				√
1980				√
1982				√
1985				√
1989				√ (Palau)
1992				√ (Palau
1993				√ (Palau

During the cold war decades, the Pacific Trusteeships were inevitably drawn into the politics of big-power animosities (just as they continue to be in the present day) and visiting missions on occasion became sites for the expression of rivalry between the "western" and "communist" states. Yet the willingness of mission reports to expose the inadequacies of one or other administrative regime, whatever the motive, assisted Pacific Islanders in their quest to achieve self-determination as rapidly as possible. It may even be the case that the United States' willingness to withdraw from Micronesia in the 1990s was hastened by the end of the cold war and the end of a perceived imperative to deny the Soviet Union access to the North Pacific.

2.2.3 *Petitions*

Petitions to the Trusteeship Council from individuals and communities living under trusteeship were a third source of information for the Trusteeship Council about the administering authority's quality of governance. From Nauru, for instance, came appeals about the inadequacy of war compensation, salaries, and health and housing conditions. In 1953 more than 100 individuals from five districts petitioned the

Trusteeship Council to recognize their claims for war damage. Europeans had been compensated for damage but not Nauruans, and the petitioners understood that in Papua New Guinea the Australian government compensated Papua and New Guinea inhabitants, even to the extent of bush-knives and blankets lost (United Nations. Trusteeship Council, 1953c). Danang had lost an arm during an allied naval raid and was unable to fend for himself (United Nations. Trusteeship Council, 1956b), Fritz Gaunibwe sought compensation for a car that had been sunk in a Japanese attack (United Nations. Trusteeship Council, 1956c).

In response to such claims the Australian government replied that it had no legal liability to individual Nauruans for war damage compensation, that "reasonable assessment of individual claims was quite impracticable", and that any money the government might make available would be better spent on the "general community betterment scheme, such as housing and provision of furniture. . ." (United Nations. Trusteeship Council, 1953f) Housing was another pressing concern. J. Aroi, superintendent of native primary schools, petitioned the Trusteeship Council for 3 consecutive years seeking compensation for the loss of his house and car, having failed to obtain relief from the Australian administration (United Nations. Trusteeship Council, 1953b). Aggi H. of Anetan district, appealed for help with housing, as he had not been allotted one and was living in a "poor hut (like a lean-to)." (United Nations. Trusteeship Council, 1956d) Australia's standard response to claims for war compensation was that a "substantial" housing scheme had been implemented at a cost of £304,000, with a fund of £15,000 available to provide furniture. In 1956 it was still formulating a Nauru Social Services Ordinance which when completed would allow for consideration of individual claims (United Nations. Trusteeship Council, 1956f).

Then there were appeals about healthcare. Apadinuwe, a victim of polio, received no more assistance than a wheelchair (United Nations. Trusteeship Council, 1953a), but conceded to the 1956 visiting mission that Nauruans were ". . .better off at present as a result of the last Mission representations to higher authority". Apadinuwe also appealed about employment opportunities and for an investigation into royalties owed him for phosphate mining on Ocean Island (United Nations. Trusteeship Council, 1956a).

This matter of remuneration was always present. Thoma Q., chief of the District of Aiwo, petitioned because he had not received remuneration as had all other chiefs (to which the Australian Minister for Territories replied that Thoma received a salary for his work as a medical orderly and did not need additional support for his responsibilities as a chief) (United Nations. Trusteeship Council, 1953d). Mr. Victor sought an increase in salary for his work as laboratory assistant at Nauru hospital from £17 to £25 per month on the basis of holding superior qualifications, appealing first in 1953 (United Nations. Trusteeship Council, 1953e) and again 3 years later (United Nations. Trusteeship Council, 1956e).

The Australian government examined such petitions and invariably provided the Trusteeship Council with supplementary information that was not disclosed by the petitioners, to help justify its decisions. Where the merit of a case was accepted, administrative action was sped up (United Nations. Trusteeship Council, 1956f).

Australia gave little ground on matters of political and administrative education. When the Nauru Local Government Council suggested that some indigenous representatives be attached to Australia's delegation to the UN Trusteeship Council sessions, the Australian government replied:

> . . .the Administering Authority is in full accord with any proposal that gives indigenous inhabitants of Trust Territories opportunity to develop their preparation in public affairs. It does not consider, however, in relation to the Trust Territory of Nauru, that such an objective would be advanced by the acceptance at this stage of the proposal submitted by the Nauru Local Government Council. (United Nations. Trusteeship Council, 1953f, p. 3)

In New Guinea, visiting missions and people's appeals had the similar effect of subjecting Australia's administration to international scrutiny. The Territory of Papua and New Guinea first came under joint administration with passage of the *Papua New Guinea Act 1949* by the Australian parliament. The Trusteeship Agreement approved by the United Nations General Assembly in December 1946 was incorporated as its fourth schedule. Based on this Act and subsequent amendments to it, a modified Westminster model of government was gradually introduced to the Territory. The first local government agencies appeared with the *Native Village Councils Ordinance* (1949) and a comprehensive *Local Government Ordinance* was passed in 1964.

Between 1951 and 1971 the Trusteeship Council sent eight visiting missions to New Guinea, which provided local residents with opportunities to air their grievances and relate their circumstances. The Trusteeship Council received petitions from New Guinea subjects as well as from their supporters in other countries. Many concerning race-based discrimination: the Papua and New Guinea Native Ex Service Association, for instance requested that they be allowed to obtain loans of equal size to those granted to European ex-servicemen (United Nations. Trusteeship Council, 1967b). Tsang Sang, who had lived in Kavieng New Ireland for 52 years, complained about discrimination against the Chinese population:

> All coloured people, such as the Chinese and the Japanese who live in New Guinea are discriminated. At first, the coloured people were restricted to apply for farm land and were banned to go into the public hospital for treatment. Though these bans were lifted up by the Trustee authorities after my repeated and strong protest. However, when the coloured people out of the territory come to help them in the development of industry and commercial management as well as cultivation, they were turned down by the Trustee Authorities while the white people, such as the British, the German, the Italian and the Russian are allowed to enter the territory without limitation.

Sang said that 7000 Chinese lived in New Guinea, including 700 in Kavieng, but that his application to bring Chinese teachers to a school he built at his own expense had been denied, which he asserted contravened United Nations Charter, article 76:

> The basic objectives of the trusteeship system, in accordance with the Purposes of the United Nations laid down in Article 1 of the present charter, shall be: . . .(c) to encourage respect for human rights and for fundamental freedoms for all without distinction as to race, sex, language, or religion, and to encourage recognition of the interdependence of the peoples of the world, and (d) to ensure equal treatment in social, economic, and commercial matters for all members of the United Nations and their nationals, and also equal treatment for the

> latter in the administration of justice, without prejudice to the attainment of the foregoing objectives and subject to the provisions of Article 80. (United Nations. Trusteeship Council, 1966a)

The Australian government's reply sought to set the record straight. It could find no record of formal application for entry by teachers from the Republic of China. In any case "Such an application would not accord with the underlying principle concerning entry to Papua and New Guinea for both permanent and temporary residents. That principle is that the racial composition of the population should avoid being made more complicated than it already is." Besides, the government replied, the petitioner's figures were inaccurate: there were just 3000 Chinese nationals in the Territory, including 160 in Kavieng. With regard to entry to public hospitals, "...certain Chinese who so desired were admitted to hospitals without cost and on the same basis as any person, regardless of race, who could not afford hospital fees." (United Nations. Trusteeship Council, 1966b) On reading such petitions and responses the Trusteeship Council may have been justified in thinking that much of the story was left out by both parties.

Other petitions were from communities and organizations rather than individuals, and covered a raft of issues rather than a single grievance. In late 1966 the Tolai of New Britain addressed an 8-page letter to the UN Secretary-General that commenced:

> We urgently need your help, as we two people have been picked confidentially by our people to tell you about our needs and concern for our country, in the hope that you may be able to change the way things are going ...you may tell us that we should send this letter to the Australian Government or the Territory Administration, but we would like to point out that this would do no good as several native people have written letters about minor complaints and Injustices to these people before and have lost their jobs, so it is apparent that it is not worthwhile to place a letter of this sort in their hands. (United Nations. Trusteeship Council, 1967c, p. 1)

The petitioners had heard that a UN mission might be coming to New Guinea and hoped they could meet with them on the Island of New Britain. The letter then expressed concerns about native wages, land rights, plantations, local councils, foreign labour, native business opportunities, training opportunities, insurance and investments, use of public money, taxation, schools, the racist attitudes of administrators, and representation in the House of Assembly and in the United Nations: "as explained before we are told nothing of Organizations such as United Nations and how they work. we would like to know if our island has a United Nations Representative apart from New Guinea?" (United Nations. Trusteeship Council, 1967c, p. 7).

The 6000-member Public Service Association wanted the Administering Authority to investigate the existing wage structure for Local Officers, with a view to establishing equal wages for local and expatriate public servants, and also to allow married women equal employment opportunities and status, and to investigate the current and projected housing needs of public servants (United Nations. Trusteeship Council, 1968a). Australia responded promptly, advising that such matters as equal pay for males and females and the possibility of permanent employment for married females, were being investigated. It rejected the idea of an inquiry into housing on the basis that a Housing Commission had recently been established (United Nations. Trusteeship Council, 1968b).

Other petitions sought support for the development of coffee, copra and other industries, whilst yet others were more akin to appeals from legal proceedings, which were more appropriately left to judicial processes. In 1969 the National Union of Australian University Students petitioned the UN Secretary General in protest against the actions of mining company Conzinc Riotinto, which had used tear gas and batons against landowners opposed to the establishment of the copper mine at Panguna. (United Nations. Trusteeship Council, 1969). In April 1970 Australia responded to five petitions concerning the Panguna mine that the Trusteeship Council had received from concerned citizens, students, and Branches of the Australian Communist Party, explaining to the Trusteeship Council that the terms of agreement with landowners had been reached on 16th January 1970 (United Nations. Trusteeship Council, 1970). Land compensation petitions sought remedies to inequitable land transactions made under German administration (i.e., prior to 1914) on the island of New Britain's Gazelle Peninsula and forced alienation of land by the Australian Administration Arawa on the island of Bougainville.

From the Strategic Trust in the North Pacific, the Trusteeship Council received petitions urging that traditional customs and institutions remain in place to facilitate self-government at local level. In May 1950 18 Iroij (chiefs) petitioned the Trusteeship Council about the impact of newly introduced Municipal Councils on their rights, and on the customs by which the Marshallese resolved conflicts amongst themselves (United Nations. Trusteeship Council, 1950b). These Iroij were most likely unaware that the Trusteeship Council already held the view that traditional governance arrangements were inhibiting the introduction of modern representative government:

> The Trusteeship Council noting that some of the 137 municipal governments established in the Trust Territory enjoy a real measure of autonomy, welcomes the efforts of the Administering Authority to establish democratic organs of purely local government and hopes that further steps may be taken to ensure that the will of the people, rather than hereditary considerations, prevail in the election or appointment of such bodies. (United Nations. Trusteeship Council, 1949b, p. 9)

In Western Samoa, relations with the United Nations did not get off to a good start. Soon after the UN was established New Zealand consulted Great Britain and Australia on a draft trusteeship agreement concerning Western Samoa before sharing it with Samoan leaders. When the country's traditional leaders heard that a draft had been submitted to the UN Trusteeship Council in such a manner, they immediately petitioned the Council stating in emphatic terms that they did not want continued New Zealand control as the draft envisaged. They wanted self-government, and much more rapidly than New Zealand intended. The petition resulted in the Trusteeship Council sending its first visiting mission to the Pacific region—in the words of one commentator, "the first example of Trusteeship Council operations and which may set a precedent for its future activities." (Finkelstein, 1948) "A direct result" of this 1947 mission, wrote Samoan Prime Minister Tofilau Eti Alesana in 1995, "was the Samoa Amendment Act 1947 which created the new Government system". (Tauafafu, 1995 foreward)

Western Samoa was fortunate to have sympathetic allies in positions of authority in the New Zealand government during the transition to independence. After the conflict of the 1920s and 1930s when autonomy movements were suppressed, more collaborative relations applied under Peter Fraser (Prime Minister from 1940 to 1949), Sir Guy Powles (High Commissioner from 1940 to 1960), and Professor Jim Davidson (advisor to the Samoan government from 1949). Peter Fraser had chaired the Trusteeship Committee at the San Francisco conference at which the UN's Charter was negotiated, and his only concern about Western Samoa attaining independence was its ability, as a small state, to act independently on the world stage. In 1952 High Commissioner Guy Powles published a small booklet to educate the Samoan public about the terms of the Trusteeship arrangements. (*The United Nations and Western Samoa (Being the Universal Declaration of Human Rights, with Preface, together with the more important of those Resolutions etc. of the Trusteeship Council and General Assembly that most directly concern Western Samoa, with commentary)* 1952)

Notwithstanding the comparatively sympathetic approach shown by New Zealand's political leaders in the 1940s, another challenge was cultural. Samoa's traditional structure was hierarchical rather than egalitarian, and authority was possessed by chiefs. Everyone had a voice, but within the appropriate settings. Given this context, it was somewhat ironic that the route to Samoa's political freedom required an act of self-determination using a universal voting system that went against traditional Samoan values. One of Samoa's key leaders, Malietoa Tanumafili II, addressed the Samoan people by radio on 6 April 1959 to explain the situation:

> Today and tomorrow members of the Legislative Assembly will be explaining to their districts why it will be necessary to hold a plebiscite among the people of Western Samoa before the United Nations Agrees to terminate the Trusteeship Agreement. And the ending of Trusteeship is, of course, the final act which will give our country independence.
>
> ...The position is this. When all the details of our new constitution have been prepared and agreed to by a Constitutional Convention, the New Zealand government will ask the United Nations, on our behalf, to bring the Trusteeship agreement to an end. ... the General Assembly is required by the Charter of the United Nations to satisfy itself that such a request is in accordance "with the freely expressed wishes of the people", that is, in our case, that the people of Western Samoa really do want to be independent, rather than to continue under Trusteeship. Only if at least two thirds of the nation are completely satisfied that the people do want to have independence will the request be agreed to. Now, over the last two or three years the United Nations has had to consider a number of requests for independence made to it on behalf of Trust Territories in Africa. As a result, it has become clear what sort of evidence the General Assembly will accept as proving that the request does have the backing of the people. The evidence they want is that a plebiscite shall have been held in which all persons over 21, both men and women, both titled and untitled, shall have taken part.
>
> ...this procedure is, of course, novel to us and is not in accordance with our own custom, by which the Matai speaks for his family. But the reason for it is a very special one. The United Nations contains nations which have many different forms of custom. When a request is made that a Trust Territory should be given its independence, a procedure must be used that will satisfy all these different nations that the people of that territory are really

> behind the request. All the nations will agree that the people have expressed their approval of the request if every man and woman has had a chance to say "yes" or "no." that is why such a plebiscite is necessary.
>
> There is one more point I must make clear. This plebiscite has nothing to do with the type of government we shall have in Western Samoa. It does not mean that every man or woman should have a vote in our elections. . . .it is the "gateway to freedom"; if we pass through it, the future is our own. If we fail to pass through it—that is if we refuse to have a plebiscite—then the Trusteeship Agreement would probably continue in existence; and the great object of so many years of effort—SAMOA MO SAMOA—would still be unattained. . . . (Malietoa Tanumafili II, 1959)

On 18th December 1960 the UN adopted a resolution recommending:

1. that New Zealand should take steps, in consultation with a United Nations Plebiscite Commissioner, to organize under UN super vision a Plebiscite in order to ascertain the wishes of the inhabitants of Western Samoa concerning their future;
2. that the questions to be put should be:

   "1. Do you agree with the Constitution adopted by the Constitutional Convention on 28 October, 1960?"

   "2. Do you agree that on 1 January 1962 Western Samoa should become an independent State on the basis of that Constitution?"

3. that the Plebiscite should be conducted on the basis of universal suffrage with all adult citizens of Western Samoa being entitled to vote

In April–May 1961 a 13-member UN mission headed by Ambassador Najmuddine Rifai of the United Arab Republic spent 3 weeks in Western Samoa, culminating on May 9th with the plebiscite based on universal suffrage. The vote on the first question was 31,426 for and 4909 against, with 1562 informal votes. The vote on the second question was 29,882 for and 5108 against, with 2907 informal votes (McKay, 1961, p. 354). The outcome was accepted by the Trusteeship Council and the General Assembly and independence was achieved on January 1st, 1962. The Samoan people did not again vote until universal suffrage was introduced in 1990.

Did the UN Trusteeship in Western Samoa and the other Pacific territories aid or hinder their progress toward self-determination? "In nearly 47 years of New Zealand rule. . ." an assessment made in 1962 said:

> . . . the most fruitful administrative periods have been confined to two bursts, one in the 1920s and one in the 1950s; they total scarcely 15 years. There are clear historical reasons for this: nevertheless, it was not entirely a coincidence that these were also the two periods when the League of Nations and the United Nations respectively showed their keenest interest in the Territory. The trusteeship system has been criticised as leading to excessive vulnerability to criticism, over-anxiousness to please, and hasty window-dressing to win international praise. These defects were noted by Keesing even under the mandate. But the Samoans themselves have been sceptical of the effectiveness of New Zealand's good intentions when not prodded by an international body. It is usually said in Samoa today, with a certain amount of truth if not of legal accuracy, that 'the United Nations is giving us independence' (Corner, 1962, pp. 138–139)

2.2.3.1 Nauru

Nauru's path to self-determination might have been as simple as that experienced by Samoa. It is a single island with a population by the 1960s of little more than 3000. The one complication was that Nauru was rich in phosphate, which had been mined profitably for nearly a century by German then by British and Australian companies, using Chinese indentured labour. In the wake of WWI the Nauru Island Agreement of 1919 established a Board of Commissioners (later the British Phosphate Commission), whose duty was (Art.9): "to dispose of the phosphates for the purpose of the agricultural requirements of the United Kingdom, Australia and New Zealand." (Storr, 2020, p. 150) The Nauruan people held property rights in land and trees, but not sovereignty—an arrangement that served colonial interests nicely, and was written into the Trusteeship Agreement drafted by Australia and accepted by the General Assembly:

> In answer to the administrative obligation, reformulated in Article 76(b) of the UN Charter, to promote the 'progressive development towards self-government or independence', Article t of the draft Trusteeship Agreement . . . .referred neither to self-government nor independence, instead placing on the Administering Authority the obligation only to 'assure the inhabitants of the Territory, as may be appropriate to the particular circumstances of the Territory and its peoples, a progressively increasing share in the administrative and other services of the Territory'. (Storr, 2020, p. 199)

Whereas the Trusteeship Council persisted with efforts to have the intransigent Australian administration chart Nauru's path in the spirit of the 1960 *Declaration on the Granting of Independence to Colonial Countries and People toward self-determination*, there was also general acceptance within the Council—and even, for a time, amongst Nauruan leaders—of the Australian proposition that the future of the Nauruan people included resettlement elsewhere once phosphate extraction made the island uninhabitable. However, once Western Samoa had successfully attained independence in 1962, and following the Trusteeship Council's visiting mission to Nauru in 1965, and, furthermore, upon the failure of resettlement talks between Australia and Nauru's Local Government Council, the focus of negotiations shifted to the terms of Nauruans gaining independence on their own island. The key issue remained, of course, ownership of Nauruan phosphate, with Australia attempting to distinguish future national sovereignty of the Nauruans from legal ownership of their main natural resource—the diminishing but still profitable phosphate. The Australian parliament's *Nauru Act* of 1965 established legislative and executive councils in a bid to defer independence. The Nauruans, however, placed their hopes in the judgement of the Trusteeship Council, which, upon reviewing the Australian negotiations, recommended to UNGA that a date for full political independence for Nauru and full transfer of control over the phosphate industry be set for not later than 31 January 1968. In 1967 agreement was reached on future phosphate royalties, a constitution for the Republic of Nauru was drafted, the Australian parliament passed the *Nauru Independence* Act, and independence was achieved on the final date stipulated. Here was an instance in which the Trusteeship Council's interventions made a decisive contribution.

2.2.3.2 Papua New Guinea

By the early 1970s the Australian government was being pressured both by those who sought self-government for Papua New Guinea at the earliest opportunity, and by those who sought to delay it for as long as possible. "We the Native people have enough with the colonial administration policies in our country", wrote the Mataungan Association (United Nations. Trusteeship Council, 1971a). The Australian Minister for External Territories stated that full internal self-government would occur sometime between 1972 and 1976, and that the date for independence was for "the wishes of the elected leaders of the time." (United Nations. Trusteeship Council, 1971b)

One issue presented to the Trusteeship Council was separatist sentiment across the Trusteeship, notably in Papua, New Britain, and Bougainville. "First of all...", read a Tolai petition of 1966,

> ... if Papua New Guinea get self government which seems very likely, we do not wish to come under their control as we are NOT Papuan or New Guinean. A certain Dr. Gunther of the University of Port Moresby, recently made a statement in Australia that we of these islands altogether, wished to become the seventh State of Australia. We are still trying to find out where he got this opinion from, because it very certainly is not the wish of the people here. This, in our opinion, has been decided for us without us having been consulted first. (United Nations. Trusteeship Council, 1967c)

The circumstances surrounding Bougainville's failed effort to seek UN assistance in gaining independence from Papua New Guinea in 1974–75 contrast with the emergence of Kiribati and Tuvalu as small sovereign states from the Gilbert and Ellice Islands Colony (Bougainville has far greater natural and human resources than either the Gilbert Islands or Ellice Islands). Bougainvillean MPs Fr John Momis and John Teosin petitioned the UNSG in August 1975 explaining Bougainville's disappointment at the structure of Papua New Guinea's proposed independence constitution, which they believe ignored the recommendations of the Constitutional Planning Committee. The petition read, in part:

> 4 - On 16 September 1975, the Government of Australia, with the approval of the United Nations, intends to make a grant to full sovereign and independent status to the Government and Territory of Papua New Guinea.
>
> 5- It is the declared and considered view of the people of Bougainville that any such grant of independence, and consequential termination of the Trusteeship Agreement, would involve the United Nations and the Administering Authority in a clear breach of the basic objectives of the Trusteeship System as defined and set out in Article 76 of the Charter of the United Nations, and the Administering Authority in breach of Articles 3 and 8 of the Trusteeship Agreement.
>
> 6- The People of Bougainville believe and assert that the United Nations and the Administering Authority are contravening both the word and the spirit of the Trusteeship System in permitting the untimely creation of a sovereign and independent State of Papua New Guinea under a Government which, quite clearly, is incapable of commanding the respect, assent and allegiance of a significant number of its citizens, and in the context of widespread dissatisfaction with the constitutional arrangements which it is intended shall accompany the proposed new State into nationhood.

> 7- In the light of these unhappy circumstances, the people of Bougainville firmly believe that the proposed grant of sovereign and independent status to the Government of Papua New Guinea is both premature and a gross betrayal of their interests and status as inhabitants of the Trust Territory. ("Petition from Father John Momis and Mr. John Teosin concerning the Trust Territory of New Guinea," 1975, p. 3)

Notwithstanding the warnings contained in this petition, Papua New Guinea's independence proceeded as scheduled on 16 September 1975 with Bougainville refusing to take part until a 1976 constitutional amendment introduced a provincial government system as an interim solution. In the late 1980s widespread dissatisfaction with the arrangement culminated in a decade-long civil war that concluded with the Bougainville Peace Agreement brokered by New Zealand, but with no resolution of the fundamental issues in conflict. The Endeavour Accords provided for establishment of an autonomous province and a moratorium prior to a referendum. That referendum, held in 2019, favoured independence. But the terms of the peace agreement left matters of timing and final decision to the Papua New Guinea national parliament and talks between the Government of Papua New and the Autonomous Province of Bougainville have continued, under several Prime Ministers, from that time to the present.

2.2.3.3 Trust Territory of the Pacific Islands

The final Pacific territory to achieve self-determination, the Trust Territory of the Pacific Islands, also happened to be the last of the Trusteeship Council's responsibilities. The delay was due to two factors: first, a desire by the various peoples administered in the Trust Territory to emerge as separate nations, and second, the time taken for each new nation to negotiate its compact agreement with the United States.

Fellow Pacific Islanders did what they could to speed up the process. The Ambassadors for Papua New Guinea and Western Samoa, Lohia and Slade, participated in the Trusteeship Council's Visiting Missions to Palau, where eight different plebiscites were held to gain approval of a Compact Agreement and then a constitution. In 1986 representatives of the Micronesian people and all Forum States members of the United Nations addressed the Trusteeship Council urging termination of the Trusteeship Agreement, and on 28 May the Council agreed on the basis that the Federated States of Micronesia, the Marshall Islands, Palau, and the Northern Marianas, had each exercised their right to self-determination in plebiscites observed by the United Nations.

These four countries chose to exercise differing degrees of autonomy and sovereignty: for the Marshall Islands, Federated States of Micronesia and Palau—free association with the US (which provides budgetary support in exchange for US rights to certain military rights); and for CMNI—Commonwealth Status with the United States of America (which keeps it under US federal jurisdiction). Prince has argued that since the Trust Territory was established as a Strategic Trust under the Security Council rather than as a Trust under the Trusteeship Council, its termination procedures should have taken place under Security Council supervision. That this step did not occur, Prince speculates, may have been due to reluctance on the part of the United States to have its administrative record in the Territory further scrutinized by other Security Council members, especially Russia (Prince, 1989). Notwithstanding such legal intricacies, the role played by the UN in assisting the Micronesian states gain their independence has not been forgotten. When closing his address to a High-Level Plenary Meeting of the 60th Session of the United Nations General Assembly in 2005, Federated States of Micronesia's President Urusemal was moved to add:

> Ever since the earliest days of the Trust Territory of the Pacific Islands, the people of Micronesia have placed our faith in the United Nations Charter. As we leave this critical Summit that faith is refreshed and strengthened. (Federated States of Micronesia, 2005)

2.3 Non-UN Colonies

The United Nations also assisted self-determination processes in colonies that were not under its jurisdiction. These were under British, French, American, Australian, New Zealand, or Chilean control. As Table 2.3 indicates, the UN despatched 14 visiting missions to 8 non-trusteeship territories in the Pacific region between 1965 and 2018. In the Gilbert and Ellice Islands, a colony to the north of Solomon Islands which had been ruled jointly by Britain since 1916, its two distinct ethnic communities expressed a desire to gain independence as separate nations. In 1973 a British commissioner appointed to examine the situation recommended their administrative separation and the British government agreed on condition that the Ellice Islanders declare their wishes by referendum. This came to pass in August–September 1974 when a plebiscite overseen by the United Nations Committee on Decolonization decided in favour by 3799 votes to 293. In 1975 the Ellice Islands

Table 2.3 UN Visiting missions to non-Trusteeship Pacific Islands

	American Samoa	Cook Islands	GEIC	Guam	New Caledonia	New Hebrides	Niue	Tokelau
1965		√						
1972							√	
1974			√				√	
1976								√
1979						√		
1981	√							√
1986								√
1994								√
2006								√
2007								√
2014					√			
2018					√			

became the British colony of Tuvalu, and in 1978 became an independent nation. In 1979 the Gilbert Islands attained independence, as Kiribati.

This outcome is remarkable given that the disintegration of existing states into smaller ones gains little support from international law—although Trinidad points out that exemptions are given to "non-contentious cases, where fragmentation is justified on the basis of the consent of the affected colonial populations and/or concerns of international order and stability" (Trinidad, 2018, p. 18). By defining "peoples" as the inhabitants of whole territories, and not as minority groups within such territories, instruments of international law such as the United Nations Charter and subsequent U.N. resolutions, as well as the Covenant on Civil and Political Rights, circumvent the assertion of independent status by ethnic minorities within states, except in such rare circumstances. Also, perhaps, the fact that GEIC was not a UN trusteeship may have in this instance provided the flexibility to achieve this mutually agreed outcome.

The Cook Islands is another non-Trusteeship colony whose self-determination referendum was overseen by the UN. In February 1965 the General Assembly authorized UN supervision of a plebiscite, which took place on 20 April 1965. The election supervisor's report was accepted by UNGA's Fourth Committee in December's twentieth session. An excerpt from UNGA's Fourth Committee consultation is included below to demonstrate the degree of oversight given to the self-determination process by the international community at that time:

> 3 - Notes that the Constitution of the Cook Islands came into force on 4 August 1965, from which date the Cook Islanders have had control of their international affairs and of their future.
>
> Ghana's proposal to delete the words: "from which date the Cook Islanders have had control of their international affairs and of their future" was rejected 29 votes to 28, with 43 abstentions. However, Liberia's proposal to add as paragraph 7:
>
> 7 - Expresses the hope that the United Nations Development Program and the specialized agencies will endeavour to contribute in every way possible to the development and strengthening of the economy of the Cook Islands",
>
> was adopted 78 votes to one, with 20 abstentions, so that the resolution's final wording was:
>
> . . .that the General Assembly:
>
> 1 – Approves the chapters of the reports of the Special Committee on the Situation with regard to the Implementation of the Declaration on the Granting of Independence to Colonial Countries and Peoples concerning the Cook Islands;
>
> 2 – Notes the findings and conclusions of the United Nations Representative for the Supervision of Elections in the Cook Islands and expresses its high appreciation to the Representative and his staff;
>
> 3 – Expresses its appreciation of the cooperation extended to the United Nations by the New Zealand Government in the study of the question of the Cook Islands;
>
> 4 – Notes that the Constitution of the Cook Islands came into force on 4 August 1965, from which date the Cook Islanders have had control of their internal affairs and of their future;
>
> 5 – Considers that since the Cook Islands have attained full internal self-government, the transmission of information in respect of the Cook Islands under Article 73 e of the Charter of the United Nations is no longer necessary;

> 6 – Reaffirms the responsibility of the United Nations under General Resolution 1514 (XV) to assist the people of the Cook Islands in the eventual achievement of full independence, if they so wish, at a future date;
>
> 7 – Expresses the hope that the United Nations Development Programme and the specialized agencies will endeavour to contribute in every way possible to the development and strengthening of the economy of the Cook Islands. (United Nations General Assembly, 1965, p. 7)

Although in 1965 a majority of Cook Islanders opted for self-governance in association with New Zealand, their leaders have periodically expressed interest in becoming a full member of the United Nations—although this is not a step supported by New Zealand, which holds constitutional responsibility for Cook Islands foreign affairs and defence. In 2015 New Zealand PM John Key suggested that New Zealand was open to reviewing the relationship but pointed out that if Cook Islanders opted for full national sovereignty they would lose the (considerable) benefits of New Zealand citizenship—such as access to more comprehensive health, education, employment, pensions, and welfare services—than would otherwise be available to them ("NZ PM Won't Discuss UN Membership With Cook Islands PM," 2015).

A third Pacific colony which gained independence with the UN's assistance is the New Hebrides. This archipelago, which had been ruled jointly as a Condominium by France and Great Britain, came to the attention of the Committee of 24 in 1978. France's change in policy, which opened up the prospect of independence, provoked both nationalist and separatist sentiments. Nagriamel, for instance, a separatist movement on the northern island of Santo, appealed unsuccessfully to the United Nations in 1971 to recognize that island's independence (Larmour, 1984). After considerable political tumult the New Hebrides attained independence in 1980 under its new name, Vanuatu.

Niue is a fourth non-Trusteeship PIC that exercised self-determination under UN supervision. The island had become a British protectorate in 1900 and was annexed by New Zealand in 1901. In October 1974 the UN supervised a vote at which Niueans chose to exercise self-government in free association with New Zealand. The consequent constitution provides the Niuean Government with full executive powers but leaves New Zealand responsibility for Niue's defence. Niueans hold New Zealand citizenship and have open access to enter, live and work in New Zealand (and therefore also Australia). Approximately 90 per cent of Niue's population lives in New Zealand.

Some have asked whether independence was granted too swiftly to the UN Trusteeships in the Pacific region, and cite later political and social instability due to immature and fragile institutions unable to handle complex situations. Colonial administrators had maintained order with minimum use of force, this line of reasoning suggests, and while their institutions were neither representative nor democratic, nor were they incompetent or corrupt. In the case of Papua and New Guinea, such reservations about the dangers of rapid change were voiced by Papua New Guinean elites, such as Mathias To Liman, who's resolution passed by the House of Assembly was conveyed to the President of the Trusteeship Council:

> We the elected representatives of the people of Papua and New Guinea desire to convey to the Parliament of the Commonwealth of Australia, the Trusteeship Council and the General Assembly of the United Nations organization, the expressed wish of the people that they, the people, and they alone, be allowed to decide when the time is ripe for self-government in Papua and New Guinea, and the form that such government will take and the people's further firm conviction that the road to self-government can best be travelled with one guide—and that guide the Administering Authority, and that undue pressure from without can lead only to that disruption, chaos and bloodshed which the people have observed with great alarm in certain newly independent countries. (United Nations. Trusteeship Council, 1964)

A similar sentiment reached the Trusteeship Council in 1967 through petition from the Tolai of New Britain, "We would like to make it clear that we do not think that we are ready for Self Government and that we can get on without European assistance, but want only to be spoken to, and treated as human beings, and encouraged to learn and help ourselves." (United Nations. Trusteeship Council, 1967c) "I really believe that Australia itself is going too fast" Catholic Father Bernard Jakubco offered helpfully in a petition in late December 1966, "...They are accelerating the pace; and some indigenous are holding positions for which they are not competent (Note that the number of indigenous clergy for the Roman Catholic Church is small, very small. We demand competence...)" (United Nations. Trusteeship Council, 1967a).

From the Northern Mariana Islands, petitions reached the Trusteeship Council as early as 1950 expressing a desire to remain as closely aligned with the United States as possible: "It is our fervent hope", wrote the leaders of the House of Council and House of Commissioners,

> ...that all of the islands of the Northern Marianas be incorporated into the United States of America either as a possession or as a territory, preferably as a territory. it is our desire that someday these islands may be considered a part of the United States and its people attain American citizenship..." (United Nations. Trusteeship Council, 1950a, p. 2)

Choice of government system, and leadership integrity, are certainly topics worth further consideration. During the process of self-determination most PICs chose to replicate in their new states the political and constitutional systems that had oppressed them throughout the colonial period. Few of the post-independence frameworks, whether Westminster or Presidential, granted much recognition—other than ceremonial—to traditional authorities and customary law, and this situation continued after independence. Executive power in the new states was to be attained through adversarial competition between political parties which had no roots in Pacific societies, traditional modes of consultation were side-lined; and bureaucracies were established which simultaneously expanded public expenditure and reduced the scope of community-level decision-making and responsibility. For better or for worse, the modern state imposed its rules and requirements on the lifeworld of Pacific Island communities and individuals. The fortunate few with an education found positions and security within the state apparatus, while for the majority, life continued much as it had prior to independence.

2.4 The Continuing Decolonization Agenda

Although the Trusteeship Council suspended its operations on 1st November 1994, 1 month after Palau gained independence, and once the four Trust Territories in the Pacific region had completed their self-determination journeys, the statuses of New Caledonia, French Polynesia, Guam, American Samoa, Norfolk Island, Pitcairn, and West Papua, remain contested. Pacific Island leaders continue to express concern about self-determination in their annual leaders' communiques, and the topic of decolonization remains very much alive.

In December 1960 UNGA agreed to the *Declaration on the Granting of Independence to Colonial Countries and Peoples* and the following year established a "Special Committee on the Situation with regard to the Implementation of the Declaration on the Granting of Independence to Colonial Countries and Peoples" (also known as the Special Committee on decolonization or C24), to monitor implementation. New Caledonia was on the UN's Decolonization list in 1946 but the following year was unilaterally removed from it by France (South Pacific Forum Secretariat, 1981). (In somewhat similar fashion Hawaii was removed from the UN's Decolonization list in 1959 after the United States made it a state of the Union).

In May 1990 the C24 convened a regional seminar in Port Vila, Vanuatu. The following year, which marked the commencement of the UN International Decade for the Eradication of Colonialism, Forum Leaders called on the UNSG or his specially designated representative of the Secretary-General, to visit each of the territories which remained on the United Nations list of non-self-governing territories, and that "...an extensive study of the remaining non self-governing territories be undertaken by the United Nations and the results widely distributed, and that countries be requested to consider adopting national legislation to promote and safeguard the human rights of peoples living under colonialism." (South Pacific Forum, 1991, pp. 6–7). Subsequent Pacific regional seminars have been held in Port Moresby (1993 and 1996), Nadi (1998, 2002 and 2014)), Majuro (2000), Madang (2004), Bandung (2008), Nouméa (2010), Quito (2012), Managua (2016), and St George's Grenada (2018). The committee is currently observing 17 non-self-governing territories, six of which (approximately one-third) are in the Pacific: American Samoa, French Polynesia, Guam, New Caledonia, Pitcairn and Tokelau.[3] The status of each of these six territories is considered below.

2.4.1 *The American Territories: American Samoa and Guam*

American Samoa is an unincorporated and unorganized Territory under United States law, administered by the Office of Insular Affairs of the United States Department of the Interior. As such, the United States government does not welcome

[3] http://www.un.org/en/decolonization/specialcommittee.shtml

or accept United Nations interest in what it regards as US domestic arrangements. From the UN's perspective, however, American Samoa remains on the C24's list because its people have not yet had an opportunity to vote on the issue of self-determination. The C24 has managed just one visiting mission to American Samoa, in 1981. Each year it places a resolution to the General Assembly recommending it remain on the list, which the General Assembly accepts, but which the United States does not respond to. A second matter is whether the "two Samoas" might at some time in the future become reunited. In 1947 Western Samoa had petitioned the Trusteeship Council to urge the UN to end the "unnatural" division between Western Samoa and American Samoa agreed amongst Great Britain, Germany, and the United States of America in 1898, but without response, or success (Keesing, 1947). In the decades since, American Samoans have considered the prospects of re-uniting with their independent (but comparatively poor) relations to the west, but remain attracted to the benefits of association with the United States. In 2016 the Governor of American Samoa established an Office of Political Status, Constitutional Review and Federal Regulations, to pursue greater self-governance and economic development, but with no with indication that this signalled a desire for change of political status.

A second US possession on the C24's list is the island of Guam, home to the Chamoro. Guam was ruled by Spain from 1535 and by the United States from 1898. The *Guam Organic Act* of 1950 established Guam as an unincorporated organized territory of the United States. Since that time dialogue within the UN has featured US assurances that Guam's status is an internal issue on the one hand, and a range of Guamanian initiatives to break free, on the other. Given that Guam has been the site of a significant US air force base, talks under UN auspices to achieve self-determination have been ineffective. The UN's only visiting mission was in 1979. In 1997 the Guam Legislature established a Commission on Self-Determination to commence public education on self-determination options. This is now titled the Commission on Decolonization, and although numerous presentations have been made at the General Assembly, and at the C24's Pacific and Caribbean regional seminars, and although Guam's governor Lou Leon Guerrero has attempted to legislate to hold a non-binding plebiscite on self-determination, such efforts have made little progress, and Guam sits with West Papua among the world's "unrepresented nations and peoples".[4]

2.4.2 *The French Territories: French Polynesia and New Caledonia*

There are two French territories on the C24's list, French Polynesia, and New Caledonia. Notwithstanding the French view, similar to that of the United States

---

[4]https://unpo.org

concerning American Samoa and Guam, that the status of French Polynesia and New Caledonia is an internal matter, the Pacific's leaders have expended more effort on the New Caledonian case than on any other campaign for independence. The South Pacific Forum leaders issued their first statement following their 1981 meeting. Vanuatu's Prime Minister, for one, feared that the French government would take no concrete steps toward self-determination and that the violence that had begun to break out would worsen. Confrontation between the FLNKS (Front de Liberation National Kanak et Socialiste—the Pro-independence political party seeking Kanak sovereignty) and the RPCR (Rassemblement pour la Calédonie dans la Republique/ Rally for Caledonia in the Republic—which sought to remain within French sovereignty) continued until the Matignon agreements of 1988, which set out a pathway for constitutional dialogue and for decision-making through election and referendum: "For all its limitations in terms of social divisiveness, the formal conflict of elections here emerges as not merely preferable to but also more effective than the informal conflict of civil disruption, political extremism, and violence." (Clark, 1987, 92)

However, at their 1985 meeting in Rarotonga, the Forum Leaders failed to agree that seeking New Caledonia's inscription on the Committee of 24's decolonization list would help end the violence or shift the country closer to independence, and included a more general statement in their communique:

> The Forum discussed in some detail the question of possible involvement of the United Nations in the situation in New Caledonia. It was widely acknowledged that keeping the question before the international community was important, and Forum members felt that speeches by Forum members in the United Nations General Assembly's general debate might take up the objectives of the Forum in relation to New Caledonia. However, the Forum reaffirmed that it had a continuing primary role, as a regional body, to continue its consideration of developments in the territory and to respond as and where appropriate. While noting the arguments on the question of reinscription of New Caledonia on the United Nations List of Non-Self-Governing Territories, and without accepting re-inscription at this juncture, the Forum would seek information from the appropriate organs of the U.N. on the applicability of the U.N. Charter and the 1960 Declaration on the Granting of Independence to Colonial Countries and peoples. It agreed to give further consideration to this question at its 17th meeting." (SPEC Cooperation, 1985, p. 21)

Having completed these steps, which included mentioning the subject in their statements to UNGA in 1985 and writing to the UN Secretary General, the leaders established a "Standing committee of officials on the decolonization of New Caledonia", which first met in Suva in November 1985. When the UNSG refused (on the basis of standing) to circulate the Forum Director's letter concerning New Caledonia to the Committee on Decolonization, PICs were advised to write separately to the UNSG with a request that the letter be so forwarded. Papua New Guinea Ambassador to the UN Renagi Lohia wrote to the UN Secretary General requesting that the Committee of 24 "...consider the applicability of the UN Charter and the 1960 Declaration and prepare an appropriate report." (Pacific Islands Forum. Standing Committee of Officials on the Decolonisation of New Caledonia, 1985, p. 3). Ambassador Lohia did write that letter, dated 2nd October 1985, but on behalf of the Government of Papua New Guinea and without reference to the Pacific Islands Forum (United Nations General Assembly, 1985b).

Tensions were high in New Caledonia at this time. Prior to an impending referendum, France had enlarged its military presence in those islands to twice the size of the Kanak population, and had given soldiers the right to vote in local elections. In July 1985 French agents sank the Greenpeace vessel *Rainbow Warrior* at a wharf in Auckland, New Zealand, to prevent it participating in anti-colonial campaigns in the region. On 29th September the FLNKS won control of Councils in three of four provinces at general elections in the territory (although the pro-independence vote was just 38.42%).

In August 1986 Forum Leaders surmounted their previous differences and decided at their 17th meeting to request re-inscription of New Caledonia on the C24 list. Through Fiji, they placed the issue before the Decolonization Committee, and Vanuatu lobbied at the annual Conference of the Non-Aligned Movement. "France is now the only major remaining colonial Power in our Region," Papua New Guinea's Ambassador Vagi informed UNGA in October 1986,

> . . .it is unashamedly and insensitively so. The Government of France seems to regard the South Pacific as little more than a useful site for testing weapons and a backdrop for the expression of its residual imperial aspirations. . .France is not a South Pacific country. New Caledonia is not a part of France. To suggest that either is the case, as the French government does, is a delusion. (United Nations General Assembly, 1985a, pp. 24–25).

The UN threatened losing credibility, Ambassador Vagi advised, unless the colony be reinscribed on the UN list of non-self-governing territories.

In May 1987 the Forum called on the United Nations Special Committee on Decolonization to recommend to UNGA a UN-sponsored referendum 'consistent with universally accepted principles of self-determination and independence" (South Pacific Forum, 1987). In August 1987 the Committee voted 18 to 1 in favour of New Caledonia's re-inscription (with Cote D'ivoire voting against, and five abstentions) This was a major diplomatic victory. However, concerns that the referendum in New Caledonia on 13 September 1987 did not meet UN standards were tabled confidentially at its 1988 meeting in Nuku'alofa:

> The Referendum offered two stark options: independence or continued integration in France. As an act of political consultation, the Referendum was flawed in this, as well as in other respects: cooperation with the UN was refused; there was no political education regarding the options presented; there was no dialogue between various parties; punitive consequences were attached to the independence option and the French Government campaigned hard for its preferred option. The FLNKS resolved to exercise a passive boycott, urging its supporters not to participate in the Referendum process. In the event 59.1% of the eligible population voted on 13 September 1987, the overwhelming majority of whom (98%) voted for continued association with France. (Pacific Islands Forum Secretariat, 1988, pp. 1–2 Attachment: New Caledonia)

At the UN, the Forum UN countries worked with France to seek a consensus approach to the New Caledonia issue. A draft text for UNGA 44 provided for continued UN examination of the question, noted the positive measures pursued by France, and urged all parties to "...maintain their dialogue, to promote reconciliation and to refrain from violence. It also invited all parties to "...continue promoting a framework for the peaceful progress of the territory to self-determination." (Pre-Forum Officials Committee, 1989, p. 5 New Caledonia).

Under the Nouméa Accord of 21 April 1998 between the French government, the FLNKS and RPCR—ten years after the Matignon Accord of 1988—France agreed to transfer some powers to the Kanak population over a period of 20 years. Subsequently, the territory graduated in status under French sovereignty: in 2003 as a 'Collectivité d'outre-mer' and in 2004, as 'Pays d'outre-mer'. A visit by a C-24 UN delegation in March 2014 was approved by France, welcomed by the pro-independence FLNKS, but strongly criticised by anti-independence leaders.

Three referenda were held under the Noumea Accord's provisions, but the results continued to indicate deep division amongst New Caledonia's ethnic groups. The November 2018 result was 56.7% against and 43.35% in favour of independence (based on 81% of 174,185 registered voters participating), and the October 2020 result of 53.3% against and 46.7% in favour (based on 85% of 180,800 voters turning out) (Fisher, 2021). However, the December 2021 result in which 96.5% voted against and just 3.5% in favour came from a 43% turnout of 184,400 registered voters—indicating a boycott of the ballot once again by the Kanak population.

French Polynesia, much to the east of New Caledonia, was re-inscribed on the C24's list in 2013. A C24 draft resolution on the Question of French Polynesia tabled at the General Assembly on 17 June 2021 regretted that France had not since that time responded to its request to submit information about the Territory's re-inscription (United Nations. General Assembly, 2021).

2.4.3 Tokelau

With a population of just 1500, Tokelau is a non-self-governing territory for the purposes of the principle of self-determination of peoples enshrined in the United Nations Charter. New Zealand is obliged under Article 73 of the United Nations Charter to develop self-government. The 1964 "all Tokelau" *general fono* rejected all three self-determination options proposed by the UN (independence, self-government, and autonomy) and sought "continued 'direct association with New Zealand,'" a position that was put to subsequent UN visiting missions (Hoëm, 1999, p. 284) in 1976, 1981, 1986, 1994, 2002, 2006 and 2007. Since 1996, Tokelau's General Fono has had statutory authority to make rules for the peace, order, and good government of Tokelau. A New Zealand-appointed Administrator has constitutional power to overrule decisions of the General Fono but has never done so.

In 1969 Aikman succinctly summarized the status of New Zealand's Pacific possessions:

> The peoples of the four New Zealand territories have not been equally interested in a greater measure of self-government. The Samoans were the most insistent; the Cook Islanders responded to New Zealand stimulation; the Niueans have been reluctant to accept more

> responsibility for their own affairs; and it is clear that self-government is not a realistic objective for the Tokelau Islands. (Aikman, 1969, p. 309)

In 2006 and 2007 Tokelau held referenda supported by New Zealand and the C24 (Hooper, 2006). Both referenda failed to reach the two thirds threshold for a change in status and Tokelau has therefore retained its non-self-governing status. New Zealand and Tokelau have agreed for the medium term to focus on sustainable development rather than constitutional change. The UN Special Committee, for its part, has acknowledged the development that has taken place in Tokelau, but decided to continue to retain it on the committee's list.

2.4.4 West Papua

A final territory in the Pacific region on the C24's agenda, which is not a former Trusteeship, is the Indonesian Province of Irian Jaya, the western half of the island of New Guinea known as "West Papua." This is a delicate issue for Pacific leaders. West Papua was considered as being part of the Pacific from 1947, when the Netherlands participated as a member of the SPC on its behalf. The colony only came under Indonesian control in 1962 when the Dutch withdrew from this their last remaining foothold in what had been the "Netherlands East Indies." West Papuan Nationalists claimed independence on 14 December 1961 but failed to win international support. The Indonesian view (and the situation formally recognized by the UN) is that West Papua's status was settled by a UN-sanctioned 1969 act of self-determination. In the subsequent decades, however, the Free Papua Movement (Organisasi Papua Merdeka—OPM) has challenged the legitimacy of that event and has tried to keep the issue on the international agenda.

Indonesia bases the legitimacy of its authority over Irian Jaya on the "Act of Free Choice" it administered in 1969. A UN team appointed to oversee the 1969 vote arrived in West Papua in 1968 to find that the Indonesian government had already decided on an electoral procedure. The "act of Free choice" involved just 1026 handpicked members of eight regional councils, who were asked to state their preference using public *musyawarah* rather than a "one person one vote" secret ballot (*musyawarah* was a Javanese a term for consultation and consensus). All 1026 delegates were reported as having favoured integration with Indonesia and despite the misgivings of the UN team this outcome was conveyed to the UN General Assembly. The results were accepted by General Assembly Resolution 2504(XXIV) on 19 November 1969, (Musgrave, 2015) but did not quell unrest either inside or beyond the province.

Sympathy for the West Papuan position grew throughout the 1970s, 80s, and 90s. The Pacific Islands Forum, the Melanesian Spearhead Group, and Vanuatu, have been among the principal supporters of reconsideration of West Papua's integration with Indonesia, and this sentiment is also widespread in Pacific civil society. As early as June 1962 the Trusteeship Council received a petition from a "representative meeting of Kavieng public including 120 natives" seeking assistance in "resisting invasion of Dutch New Guinea by Indonesian forces." (United Nations. Trusteeship Council, 1962) The status of West Papua has remained on the Pacific leaders' agenda since 2000. It was elevated further in 2015 when civil society groups made the most of an invitation for public submissions on issues that the Forum Leaders should address to further the *Framework for Pacific Regionalism* by voting the West Papua issue into the top five most requested policy issues.

Papua New Guinea's situation in all this is delicate because it shares a land border with Indonesia, and its official position accords with a *Papua New Guinea-Indonesia Treaty of Friendship*. Nonetheless, in 2016 Papua New Guinea's Prime Minister Peter O'Neill wrote directly to Indonesian President Joko Widodo to convey the Forum Leaders' concerns about alleged human rights violations in West Papua and expressed their desire for a fact-finding mission. Although this did not occur, Indonesia allowed the Forum to send observers to provincial elections in 2017, after which time diplomatic engagement between the Forum and the Indonesian government has continued.

2.5 Discussion

The United Nations Trusteeship Council oversaw three Trusteeships (Western Samoa, New Guinea, and Nauru), and the Security Council nominally overlooked a fourth (Pacific Islands). Six independent nations and one dependency emerged from these Trusteeships (Western Samoa, Papua New Guinea, Nauru, Palau, Federated States of Micronesia and the Marshall Islands—the dependent country being the Commonwealth of the Northern Mariana Islands). In each of these cases the supervisory activities of the Trusteeship Council played a decisive role in guiding the self-determination process and timetable. In addition to these Trusteeships, other colonies in the Pacific were assisted in their self-determination processes by the General Assembly's Special Committee on Decolonization (Cook Islands, Vanuatu and Tokelau). Attitudes toward independence, and even self-governance, are in some cases determined by realistic appraisal of the resources required to maintain micro-states in the world of nation-states, internationalism, and globalization, and Hoëm has pointed to the irony that the UN's decolonization agenda can an imposition rather than moment of emancipation:

> The assumptions underlying the 'decolonisation scheme'—such as those that can be found in the 'Declaration on the granting of independence to colonial countries and peoples', namely the UN's recognition of 'the passionate yearning for freedom in all dependent peoples and the decisive role of such peoples in the attainment of their independence' and their recognition 'that the peoples of the world ardently desire the end of colonialism in all its manifestations'—is clearly more easily applicable to some other representatives in the Committee of Twenty-Four than to the particular case provided by atoll societies. (Hoëm, 1999, p. 285)

Political movement has only been possible in circumstances where the colonial power cooperated with the process. Where this political will was lacking, such as for New Caledonia, French Polynesia, West Papua, Guam, and American Samoa, the Committee's official proceedings bear "zombie-like" characteristics brought on by political intransigence—in the sense that they show movement but are without life.

References

Aikman, C. C. (1969). Constitutional development. In New Zealand's Island territories and in Western Samoa. In A. Ross (Ed.), *New Zealand's record in the Pacific* (pp. 308–341). Longman Paul for the New Zealand Institute of International Affairs.

Bobbitt, P. (2003). *The shield of Achilles: War, peace, and the course of history*. Anchor Books.

Clark, A. (1987). Conflict formal and informal elections in New Caledonia 1984-1986. *10*(3).

Corner, F. H. (1962). New Zealand and the South Pacific. In T. C. Larkin (Ed.), *New Zealand's external relations* (pp. 130–152). New Zealand Institute of Public Administration & Oxford University Press.

Davidson, J. W. (1967). *Samoa mo Samoa: The emergence of the Independant State of Western Samoa*. Oxford University Press.

Evans, L. H. (1927). Are "C" mandates veiled annexations? *The Southwestern Political and Social Science Quarterly, 7*(4), 381–400.

Federated States of Micronesia. (2005). *Address by H.E. Joseph J. Urusemal, President of the Federated States of Micronesia before the High-Level Plenary Meeting of the 60th Session of the United Nations General Assembly, Friday 16 September.*

Finkelstein, L. S. (1948). Trusteeship in action: The United Nations Mission to Western Samoa. *International Organization, 2*(02), 268–282. https://doi.org/10.1017/S0020818300019688

Fisher, D. (2021). *Uncertainties as New Caledonia Preparies for its final Independence referendum* (Vol. 3). Australian National University. Department of Pacific Affairs.

Freire, P. (1993). *Pedagogy of the oppressed*. Continuum.

Government of Japan. (1937). *Annual report to the Council of the League of nations on the Administration of the South sea Islands under Japanese Mandate for the year 1935*. Government of Japan.

Hassall, G. (2017). New Zealand, the league of nations, and the mandate over Western Samoa: The early years. *Comparative Law Journal of the Pacific, 23*, 21–44.

Heathcote, S. (2022). Legal models and methods of Western colonisation of the South Pacific. *Journal of the History of International Law, 24*, 62–101.

Hoëm, I. (1999). Processes of identification and the incipient national level. A Tokelau case. *Social Anthropology, 7*(3), 279–295.

Hooper, A. (2006). Against the wind: Tokelau 2001-2006. *The Journal of Pacific Studies, 29*(2), 157–194.

Keesing, F. M. (1930). Government of Pacific dependencies. *Pacific Affairs, 3*(5), 448–459.

Keesing, F. M. (1946). Administration in the Pacific Islands. *Far Eastern Survey, 16*(6), 61–65.

Keesing, F. M. (1947). Administration in Pacific Islands. *Far Eastern Survey, 16*(5), 61–65.

Larmour, P. (1984). Alienated land and Independance in Melanesia. *Pacific Studies, 8*(1), 1–47.

League of Nations. Permanent Mandates Commission. (1928). *New Zealand mandate over Samoa. Petition of April 24, 1928 from the Mau, requesting transfer of the mandate on self-government*. League of Nations.

League of Nations. Permanent Mandates Commission. (1930a). *New Zealand mandate over Samoa. Petition from A. Hall Skelton & E. Stevens, Auckland, May 5, 1930*. League of Nations.

League of Nations. Permanent Mandates Commission. (1930b). *New Zealand mandate over Samoa. Report on administration for the year ending March 31st 1930*. League of Nations.

League of Nations. Permanent Mandates Commission. (1930–32). *New Guinea under Australian Mandate. Report on the Administration 1st July 1930 to 30th June 1931*. Geneva: League of Nations.

League of Nations. Permanent Mandates Commission. (1935-38). South Sea islands under Japanese mandate. Report on the administration 1934. Geneva: League of Nations.

League of Nations. Permanent Mandates Commission. (1938). Permanent Mandates Commission. Thirty-Fifth Session. Islands Under Japanese Mandate. Draft Observation. League of Nations. Permanent Mandates Commission. R 4129 6A 35229 3192.

Lopez-Claros, A., Dahl, A., & Groff, M. (2020). *Global governance and the emergence of global institutions for the 21st century*. Cambridge University Press.

Malietoa Tanumafili II. (1959). *Termination of the trusteeship agreement*. James Wightman Davidson papers, (5105 Box 43). Australian National Library, Canberra.

Mander, L. A. (1953). New trends in South Pacific administration. *Far Eastern Survey, 22*(11), 141–145.

McCormick, J. P. (2007). *Weber, Habermas, and transformations of the European state: Constitutional, social and supranational democracy*. Cambridge University Press.

McKay, C. G. R. (1961). Western Samoa's plebiscite on Independence. *The Journal of the Polynesian Society, 70*(3), 107–110.

Meller, N., & Meller, T. (1969). *The congress of Micronesia; development of the legislative process in the trust territory of the Pacific Islands*. University of Hawaii Press.

Musgrave, T. D. (2015). An analysis of the 1969 Act of Free Choice in West Papua. In (pp. 209–228): Cambridge University Press.

NZ PM Won't Discuss UN Membership With Cook Islands PM. (2015). *Cook Islands news*.

Pacific Islands Forum Secretariat. (1988). Agenda item 6(1): Decolonisation - New Caledonia. In Nineteenth South Pacific Forum, Nukualofa, Tonga 20–21 September 1988. Meeting papers. Folder 1 (Suva: (88) MP. 3.

Pacific Islands Forum. Standing Committee of Officials on the Decolonisation of New Caledonia. (1985). Standing Committee of Officials on the Decolonisation of New Caledonia Inaugural Meeting, SPEC Headquarters, Suva 6–8 November 1985. Meeting papers. (85) MP. 12 ((85) MP. 12).

Pedersen, S. (2012). Samoa on the world stage: Petitions and peoples before the mandates Commission of the League of nations. *The Journal of Imperial and Commonwealth History, 40*(2), 231–261. https://doi.org/10.1080/03086534.2012.697612

Pedersen, S. (2015). *The guardians: The league of nations and the crisis of empire*. Oxford University Press.

Petition from Father John Momis and Mr. John Teosin concerning the trust territory of New Guinea. (1975). New York: UN.

Prince, H. C. (1989). The United States, the United Nations, and Micronesia: Questions of procedure, substance, and faith. *Michigan Journal of International Law, 11*, 11.

South Pacific Bureau for Economic Cooperation. (1985). Sixteenth South Pacific Forum. Rarotonga, Cook Islands, 5–6 August 1985. Summary Record. SPEC(85)REP(SPEC(85)REP).

South Pacific Forum. (1987). *Eighteenth South Pacific Forum, Apia, Western Samoa 29–30 July 1987*. Annotated Agenda. (87) MP. 2.

South Pacific Forum. (1991). *FSM Communiqué*.

South Pacific Forum Secretariat. (1981). New Caledonia: Right to self-determination calls for sympathetic consideration and greater solidarity within the forum to ensure maintenance of stability in the region - Vanuatu paper. Twelfth South Pacific forum, Port Vila, Vanuatu 10-11 August 1981: [meeting papers]. 81(MP)7(81(MP)7).

South Pacific Forum Secretariat. Pre-Forum Officials Committee, M. (1989). Pre-Forum Officials Committee Meeting, Tarawa, Kiribaiti 3–5 July 1989: [meeting papers]. (89) MP. 1.

Storr, C. (2020). *International status in the shadow of empire: Nauru and the histories of international law*. Cambridge University Press.

Tauafafu, F. (1995). *The United Nations in Western Samoa*. United Nations Development Programme.
Trinidad, J. (2018). *Self-determination in disputed colonial territories*. Cambridge University Press.
The United Nations and Western Samoa (Being the Universal Declaration of Human Rights, with Preface, together with the more important of those Resolutions etc. of the Trusteeship Council and General Assembly that most directly concern Western Samoa, with commentary) (1952). Apia.
United Nations General Assembly. (1965). *Report of the United Nations representative for the supervision of the elections in The Cook Islands*. UN.
United Nations General Assembly. (1985a). *Provisional verbatim record of the 7th meeting: Held on Tuesday 24 September 1985, New York, General Assembly, 40th session*. UN.
United Nations General Assembly. (1985b). *Special committee on the situation with regard to the implementation of the declaration on the granting of Independence to colonial countries and peoples*. A/AC.109/847: United Nations.
United Nations General Assembly. (1995). *General assembly official records, 50th session: 29th plenary meeting, Wednesday, 11 October 1995*. UN.
United Nations. General Assembly. (2021). *Question of French Polynesia: Draft resolution*. UN.
United Nations. Secretary General. (1946). *Non-self-governing territories transmission of information by members under Article 73(e) of the Charter*. New York: United Nations General Assembly.
United Nations. Trusteeship Council. (1948). Second session. Questions by members of the trusteeship council relating to the 1946-47 report on the Administration of the Territory of New Guinea. T/119.
United Nations. Trusteeship Council. (1949a). *Arrangements for a Visiting Mission to the Trust Territories in the Pacific in 1950*. T/366.
United Nations. Trusteeship Council. (1949b). *Examination of annual reports. Pacific Islands: First Report*. T/378.
United Nations. Trusteeship Council. (1950a). *Petition from the house of council and the house of commissioners, Saipan, concerning the Pacific Islands*. T/PET.10/5.
United Nations. Trusteeship Council. (1950b). *Petition from the Iroijs of The Marshall Islands concerning the Pacific Islands*. T/PET.10/7.
United Nations. Trusteeship Council. (1950c). *Terms of reference for the United Nations visiting mission to the trust territories in the Pacific*. T/458.
United Nations. Trusteeship Council. (1953a). *Petition from Mr Apadinuwe Concerning Nauru*. T/PET.9/11.
United Nations. Trusteeship Council. (1953b). *Petition from Mr J. Aroi Concerning Nauru*. T/PET.9/12.
United Nations. Trusteeship Council. (1953c). *Petition from Mr Roy Degorobore, on behalf of the people of Nauru*. T/PET.9/10.
United Nations. Trusteeship Council. (1953d). *Petition from Mr Thoma Q. Concerning Nauru*. T/PET.9/13.
United Nations. Trusteeship Council. (1953e). *Petition from Mr Victor Concerning Nauru*. T/PET.9/14.
United Nations. Trusteeship Council. (1953f). *Petitions Concerning the Trust Territory of Nauru. Observations of the Government of the Australian Government as Administering Authority*. T/OBS.9/1.
United Nations. Trusteeship Council. (1956a). *Petition from Mr Apadinuwe concerning Nauru*. T/PET.9/19.
United Nations. Trusteeship Council. (1956b). *Petition from Mr Danang concerning Nauru*. T/PET.9/15.
United Nations. Trusteeship Council. (1956c). *Petition from Mr Fritz Gaunibwe concerning Nauru*. T/PET.9/16.

United Nations. Trusteeship Council. (1956d). *Petition from Mr H. Aggi Concerning Nauru*. T/PET.9/17.

United Nations. Trusteeship Council. (1956e). *Petition from Mr Victor Eoaio concerning Nauru*. T/PET.9/18.

United Nations. Trusteeship Council. (1956f). *Petitions concerning the trust territory of Nauru*. T/OBS.9/2.

United Nations. Trusteeship Council. (1962). *Petition from the Kavieng community concerning New Guinea*. T/PET.8/L.9.

United Nations. Trusteeship Council. (1964). *Communication from Mr. H.L.R. Niall Concerning the Territory of Papua New Guinea*. T/COM.8/L.10 21.

United Nations. Trusteeship Council. (1966a). *Petition from Mr. Tsang sang concerning the trust territory of New Guinea*. T/PET.8/21.

United Nations. Trusteeship Council. (1966b). *Petition from Mr. Tsang Sang Concerning the Trust Territory of New Guinea* T/PET.8/21. T/OBS.8/11.

United Nations. Trusteeship Council. (1967a). *Petition from father Bernard Jakubo, M.S.C., of the Catholic Mission in Kavieng concerning the trust territory of New Guinea*. T/PET.8/L.11.

United Nations. Trusteeship Council. (1967b). *Petition from Papua New Guinea native ex-Service Association in Rabaul Concerning the trust territory of New Guinea*. T/PET.8/22.

United Nations. Trusteeship Council. (1967c). *Petition from the Tolai people of New Britain concerning the trust territory of New Guinea*. T/OBS.8/L.10.

United Nations. Trusteeship Council. (1968a). *Petition from the public service Association of Papua and New Guinea concerning the trust territory of New Guinea*. T/PET.8/24.

United Nations. Trusteeship Council. (1968b). *Petition from the Public Service Assocation of Papua and New Guinea Concerning the Trust Territory of New Guinea* T/PET.8/24. T/OBS.8/15.

United Nations. Trusteeship Council. (1969). *Petition from the National Union of Australian University students and other organizations concerning the trust territory of New Guinea*. T/OBS.8/29.

United Nations. Trusteeship Council. (1970). *Petition from Mr. Paul Marsh and Others (T/PET.8/28); Petition from National Union of Australian University Students and Other Organizations (T/PET.8/29); Petition from miss Fiona O'Beirne (T/PET/8.30); Petition from the South Coast District Committee, Communist Party of Australia (T/PET.8/32) Concerning the Trust Territory of New Guinea*. T/OBS.8/18.

United Nations. Trusteeship Council. (1971a). *Petition from Mr. Damien Kereku, Mataungan Association, Concerning the Trust Territory of New Guinea*. T/PET.8/36.

United Nations. Trusteeship Council. (1971b). *Petition from Mr. Uwe Lilje Concerning the Trust Territory of New Guinea* (T/PET.8/33). T/OBS.8/21.

Chapter 3
National Development Planning, Implementation, and Assessment

Chapter 3 returns to consideration of relations between the UN and the newly independent Pacific nations. This is essential context for Chap. 4, which examines development planning processes and dynamics at regional level. The principal focus of the relationship for the first two to three decades was social and economic development. Development plans initiated in some countries during the colonial era set out objectives and priorities along the path to self-determination and, presumably, independence. The consolidation of PIC capability – including legal and constitutional competence, and a public sector able to plan and implement social and economic development, to protect and promote human rights, and to participate in coordination networks from local to global levels - was at the core of engagement with the United Nations.

Although large agencies such as UNDP expressed concern that national development plans should set out national priorities irrespective of donor assistance, the programme nonetheless became increasingly involved in planning processes as the years progressed. The quality of development plans depended greatly on the expertise and resources within planning offices and may also have been influenced by progress observed by the UN Trusteeship Council. Post-independence development plans set out national aspirations in a more forthright manner, and progressively tailored priorities and activities to match the availability of development funds. The first sectors that UN agencies assisted were health, agriculture, telecommunications, trade, tourism, transport, and energy, with additional assistance later for public sector development, institutional strengthening and even constitutional reform.

Countries differ in the extent to which their development planning remained independent from the influence of international partners. Fiji's Ratu Mara, for instance, emphasised the desire for autonomy:

Although we were glad to receive aid, I did not feel we should simply accept it because it was on offer. In framing Fiji's program with the United Nations Development Programme, and with offers from other countries, I decided we should study our Development Plan VI and see what parts of it could be taken over by others. In this way, staff we had provided for in the estimates would not be taken away to be

G. Hassall, *The United Nations and the Pacific Islands*, United Nations University Series on Regionalism 24, https://doi.org/10.1007/978-3-031-34155-7_3

"counterparts" – more jargon - to some project never in our thinking. I further decided that we would expect those agreements we signed to be consonant with our position as an independent sovereign nation and not to impose conditions in conflict with this. (Mara, 1997, pp. 118–119)

It is also the case, on the other hand, that the PICs contributed significantly to the various UN frameworks - including Agenda 21, BPOA, MDGs, and SDGs - that came to influence their national plans. Viewed in this way, the adoption of UN development frameworks does not necessarily imply imposition or domination by the UN but rather the acceptance of development frameworks which have been framed, at global level, with PIC participation. Over time, UN Development frameworks became the foundation of most national planning practices. "The starting point for meeting our long-term vision is the National Sustainable Development Plan…" Cook Islands Prime Minister Marurai wrote in the forward to the Cook Islands plan for 2007–2010:

> …While comprehensive, "Te Kaveinga Nui" re-affirms our commitment to our international and regional partners through the World Summit on Sustainable Development (WSSD), Mauritius Declaration, Millennium Development Goals (MDGs), CEDAW, Multilateral Agreements (MEAs), Pacific Plan, PICTA, PACER amongst other Conventions. This ensures our development efforts are balanced across the three pillars of sustainable development: economic, environment and social growth" (Government of Cook Islands, 2007, p. 2).

A voluntary report tabled by Solomon Islands at a High-level political forum on sustainable development convened by the Economic and Social Council in 2020 indicates the extent to which its National Development Strategy for 2016–2035 was based on the UN's global and small state development agendas:

> …the 2030 Agenda for Sustainable Development, the SIDS Accelerated Modalities of Action (SAMOA) Pathway, the Programme of Action for the Least Developed Countries for the Decade 2011–2020 and other relevant international and regional frameworks have been integrated into the Strategy, through its five long-term objectives of inclusive economic growth; poverty reduction; access to quality health and education; resilient and environmentally sustainable development; and effective governance. The targets and benchmarks for achieving the Sustainable Development Goals at the national level are also set out in the Strategy (United Nations. Economic and Social Council. High-Level Political Forum on Sustainable Development, 2020, p. 64).

National planning processes have thus become intertwined with UN plans, campaigns, assessments, technical advice, and funding. Assistance has also been provided to the devising of sub-national development plans. In the late 1980s, for example, a UNDP "Integrated Atoll Development Project assisted Kiribati's Rural Planning Unit and Abaiang Island Council develop its a development plan for island for the period 1991–1995 (Government of Kiribati. Ministry of Home Affairs and Rural Development, n.d.).[1] The Island of Eua in Tonga has similarly received

[1]The plan was "prepared by the people of Abaiang", the plan's front page noted, alongside its acknowledgement of the government's, UNDP's Equitable and Sustainable Human Development Program.

assistance in formulating a development plan (Kingdom of Tonga & UNESCAP, 2006). In 2008 the Solomon Islands devised a Provincial Governance Strengthening Programme in collaboration with UNDP and UNCDF. (Government of the Solomon Islands, United Nations Development Progamme, & United Nations Capital Development Fund, 2008).

3.1 Common Country Assessments and Development Assistance Frameworks

UN procedures for development planning, implementation, and evaluation, established the planning blueprint for all recipient countries. Until 1971 UNDP made decisions about projects on a sector-by-sector basis. Thereafter it shifted to Country Programs in support of each country's development plan. These country programs generally ranged from 3 to 7 years and required the preparation of Indicative Planning Figures (IPF) for approval by the UNDP's New York-based Governing Council. Assistance was then delivered through an integrated package of staff, training, equipment, and capital inputs. *Standard Basic Agreements* set out the types of assistance to be provided to each PIC, the levels of mutual information sharing expected, the distribution of costs, privileges, and immunities of UNDP staff, and procedures for settlement of disputes.

A new phase in collaborative planning followed reforms to UN functioning initiated by SG Kofi Annan in 1997, which required UN agencies at national level to develop Common Assistance Frameworks. For PICs with Standard Basic Agreements in place, UN Country Teams (UNCT) prepared these *Common Country Assessments* during 2001–2002 in consultation with national and regional governmental and non-governmental stakeholders. Although NGOs were involved, these consultative processes received minimal press coverage and hence resulted in little increase in public awareness. *Development Assistance Frameworks* (UNDAFs) subsequently developed by UN agencies in-country, and again in collaboration with government departments and agencies, synthesized the priorities of the UN Secretary-General, Pacific Forum Leaders, and individual country programming. They also set out common objectives, strategies for development assistance, timeframes, and the UN's programme resources and business plan. Each PIC had to take responsibility for how it integrated UN assistance with that given by its other bilateral and multilateral development partners.

Several decades of practice under these arrangements raised fundamental questions of governance and accountability. Critiques from within the UN, from member countries, and from other stakeholders, expressed concerns about duplication of effort, redundant administrative procedures, wastage of resources, and lack of alignment between UN and member-country priorities. Notwithstanding PIC input, plans tended to privilege UN priorities and initiatives. Across the Pacific, "public policy" came to be widely perceived as being driven by foreign rather than domestic influences and interests (Aiafi, 2017). More recent approaches, under the term "One UN", place greater emphasis on integrating UN priorities into national priorities and development plans and processes.

Although some countries experienced remarkable continuity in their development planning, phases of certainty and disruption were also possibilities, depending on political events in a country, unpredictable natural disasters, and the performance of finance and planning departments. More investigation is required if we are to understand the circumstances that resulted in significant gaps occurring in formal development planning within individual countries. Table 3.1 constitutes an initial inventory of 120 such plans and U.N. agreements covering 16 PICs. The table does not include the numerous sectoral plans that flow from them, which if calculated at a (conservative) average of ten sectors per plan, amounts to somewhere between 1200 and 1500 in number across the region since independence. Nor does the chart include regional plans, or country-level agreements with the World Bank, IMF, ADB, and other major development partners). The Pacific, it can be said, is a region replete with development plans.

Development planning in the Pacific has been more of a technocratic than democratic exercise. Its main actors have been PIC public servants and politicians plus development agency and donor partner officials. There has been minimal engagement with local communities apart from occasional input from selected civil society organizations. This limited scope of development discourse has had the advantage of creating efficiencies through all actors possessing the requisite technical knowledge and development discourse etiquette. In the longer term, however, lack of broader stake-holder involvement may have alienated local communities from their own development aspirations and processes. Lopeti Senituli made this point when addressing the 1994 regional NGO meeting preparing a report for that year's Conference on Population and Development in Cairo:

> The Port Vila Declaration calls for partnership between governments and non-governmental organizations in the pursuit of population and sustainable development. This is most welcome. But Pacific Island governments today see development as something to be done for the people rather than by the people. This is evident in the prevalent top-down development philosophies which encourage structural adjustment programmes and economic pricing of public services and goods. . . .Similarly Pacific Island governments do not have a long history of working with NGOs. I acknowledge that there are some NGOs in this conference who can claim otherwise. But I am referring to those NGOs who dare to question the inequitable distribution and wealth in Pacific Island countries (Lopeti Senituli "Port Vila Declaration: An NGO Perspective" Pacific Islands Association of Non-Governmental Organizations, 1994, p. 41).

As will be shown later, UNDP worked assiduously in later years to remedy this power imbalance between the voices of state and non-state actors – although authority for decision-making in matters of public policy obviously remained with government, and much of the power to start or halt specific projects dependent on UN agency preferences.

Table 3.1 Pacific Island Country National Development Plans and United Nations bi-lateral development agreements

Country	National Development Plans[a]	UN bilateral Development Agreements[b]
Cook Islands	• 1982–1985 First Development plan (1983–1987 First Development Plan?) • 1988–1992 Second Development Plan • 2007–2010 National Sustainable Development Plan: Te Kaveinga Nui – Pathway for Sustainable Development • 2011–2015 National Sustainable Development Plan • 2016–2020 National Sustainable Development Plan • 2021 – Cook Islands Economic Development Strategy 2030	• 1963 Standard Basic Agreement (with New Zealand, covering Cook Islands, Niue and Tokelau) • 2003–2007 Multi-Country Programme (Cook Islands, Niue and Tokelau) • 2008–2012 Multi-country Program (Cook Islands, Niue, Samoa and Tokelau)
Federated States of Micronesia	• 1985–1989 1st National Development Plan • 1992–1996 2nd National Development Plan • 1997– • 2004–2023 Strategic Development Plan	• 2008 Standard Basic Agreement
Fiji	• 1946 Ten-Year Development and Welfare Plan • 1949–1958 Revised Ten-Year Plan • 1961–1965 Development Plan • 1966–1970 5th Development Plan • 1971–1975 6th Development Plan • 1976–1980 7th Development Plan • 1981–1985 8th Development Plan • 1986–1990 9th Development Plan • 2001–2020 Strategic Development Plan (50/50 by 2020) • 2003–2005 Strategic Development Plan • 2007–2011 Strategic Development Plan • 2017–2036 Twenty Year Development Plan • 2017–2021 Five Year Development Plan	• 1970 Standard Basic Agreement • 2019 Standard Basic Agreement
Guam	• 1951 Reconstruction and Development Programme • 1980 Comprehensive Development Plan • 2020 review • Guåhan 2050 Sustainability Plan	
Kiribati	• 1970–1972 1st National Development Plan • 1971–1973 2nd National	• 1987 Standard Basic Agreement • 2002 Common Country Assessment

(continued)

Table 3.1 (continued)

Country	National Development Plans[a]	UN bilateral Development Agreements[b]
	Development Plan • 1973–1978 3rd National Development Plan • 1979–1982 4th National Development Plan • 1983–1986 5th National Development Plan • 1987–1991 6th National Development Plan • 1992–1995 7th National Development Plan • 1996–1999 National Development Strategy: kauataboa te rikirake • 2000–2003 National Development Strategy • 2004–2007 National Development Strategy • 2008–2011 National development Plan • 2012–2015 Kiribati Development Plan • 2016–2019 Kiribati Development Plan • 2020–2023 Kiribati Development Plan	• 2003–2007 Development Assistance Framework • 2008–2012 UNDAF Kiribati Implementation Plan • 2022 Common Country Assessment
Marshall Islands	• 1985–1989 1st Five year development plan • 1986–1989 rephased initial phase of 15-year development program • 1991–1996 2nd national development plan • 2003–18 Vision 2018 "Jepilpilin Ke Ejukaan": Strategic Economic Development Plan • 2020–2030 National Strategic Plan	• 1994 Standard Basic Agreement
Nauru	• 2005–2025 National Sustainable Development Strategy	• 2012 Standard Basic Agreement
Niue	• 1949–1953 Four Year Development Plan • 1980–1985 National development plan: mata ki mua • Niue strategic development plan • 2003–08 Integrated strategic plan • 2016–2026 National Strategic Plan	• 1963 Standard Basic Agreement (with New Zealand, covering Cook Islands, Niue and Tokelau) • 2003–2007 Multi-Country Programme (Cook Islands, Niue and Tokelau) • 2008–2012 Multi-country Program (Cook Islands, Niue, Samoa and Tokelau)
Palau	• 1987–1991 Republic of Palau first national development plan • 1996 Palau national master development plan: Issues, options & strategies for Palau's development: Final	• 2008 Standard Basic Agreement

(continued)

Table 3.1 (continued)

Country	National Development Plans[a]	UN bilateral Development Agreements[b]
	report: The foundation for development/ prepared for the republic of Palau • 2005–2010 Medium term development strategy • 2009–2014 Actions for Palau's Future – Medium-Term Development Strategy • 2020 National Master Development Plan	
Papua New Guinea	• 1963–1967 First development plan • 1968–1972 Four-Year Comprehensive Development Programme • 1973–1975 eight-point improvement plan • 1976–1985 national development strategy • 1986–1990 National Development Plan • 1989–1997 Development Plan • 1997–2002 Medium Term Development Strategy • 2003–2007 Medium Term Development Strategy • 2005–2010 Medium Term Development Strategy • Vision 2050 • 2010–2030 PNG Development Strategic Plan • 2011–2015 Medium Term Development Plan 1 • 2016–2017 Medium Term Development Plan 2 • 2018–2022 Medium Term Development Plan 3	• 2001 Common Country Assessment • 2003–2007 Development Assistance Framework • 2008–2012 Country Programme Papua New Guinea: A Partnership for Nation-Building • 2012–2015 Development Assistance Framework • 2016 Common County Analysis • 2018–2022 Development Assistance Framework
Samoa	• 1966–1970 1st Development Plan • 1971–1975 2nd Five-year development plan • 1975–1979 3rd Five-year plan for the national development of Western Samoa • 1980–1984 4th Five-year development plan • 1985–1991 5th • 2002–2004 Strategy for the development of Samoa: 'Opportunities for all' • 2005–2007 Strategy for the development of Samoa • 2008–2012 Strategy for the development of Samoa[c] • 2012–2016 Strategy for the	• 2002 Common Country Assessment • 2003–2007 Development Assistance Framework • 2008 Standard Basic Agreement • 2008–2012 Multi-country Program (Cook Islands, Niue, Samoa and Tokelau)

(continued)

Table 3.1 (continued)

Country	National Development Plans[a]	UN bilateral Development Agreements[b]
	Development of Samoa • 2014–2018 Medium Term Development Plan • 2016/17–2019/20 Strategy for the Development of Samoa • 2021 Samoa 2040	
Solomon Islands	• 1955–1960 First Development Plan • 1960–1964 Second Development Plan • 1963–1966 Third Development Plan • 1965–1968 Fourth Development Plan • 1968–1970 Fifth Development Plan • 1971–73 Sixth Development Plan • 1975–1979 1st national development plan • 1980–1984 2nd national development plan • 1990–1994 National Development Plan • 2003–2006 National Economic Recovery, Reform and Development Plan • 2011–2020 National Development Strategy • 2016–2035 Solomon Islands National Development Strategy	• 1986 Standard Basic Agreement • 2002 Common Country Assessment • 2003–2007 Development Assistance Framework
Tokelau	• 2016–2020 National Strategic Plan • 2021–2026 National Strategic Plan	• 1963 Standard Basic Agreement (with New Zealand, covering Cook Islands, Niue and Tokelau) • 2003–2007 Multi-Country Programme (Cook Islands, Niue and Tokelau) • 2008–2012 Multi-country Program (Cook Islands, Niue, Samoa and Tokelau) • 2021 Common Country Analysis
Tonga	• 1965–1970 1st Development Plan • 1970–1975 2nd Development plan • 1975–1980 3rd Development Plan • 1981–1985 4th Development Plan • 1986–1990 5th Development Plan • 1991–1995 6th Development Plan • 2001–2004 Strategic Development Plan 7 • 2006–2009 Strategic Development Plan 8: Looking to the Future Building on the Past • 2011–2014 Tonga Strategic Development Framework	• 2013 Standard Basic Agreement

(continued)

Table 3.1 (continued)

Country	National Development Plans[a]	UN bilateral Development Agreements[b]
	• 2015–2025 Tonga Strategic Development Framework	
Tuvalu	• 1978–1980 1st Development Plan • 1980–1983 2nd Development Plan • 1984–1987 3rd Development plan • 1988–1991 Tuvalu national development plan IV • 1995–1998 National development strategy • 2005–2015 National Strategy for Sustainable Development • 2016–2020 National Strategy for Sustainable Development • 2021–2030 Te Keke – National Development Strategy	• 1979 Standard Basic Agreement • 2002 Common Country Assessment • 2003–07 Development Assistance Framework
Vanuatu	• 1978–1980 Transitional Development Plan • 1982–1986 1st National development plan • 1987–1991 2nd National development plan • 1992–1996 3rd National development plan • 2003–2008 Priorities & action agenda: supporting & sustaining development • 2006–2015 Priorities and Action Agenda • 2016–2030 National Sustainable Development plan (Vanuatu 2030 The People's Plan)	• 1984 Standard Basic Agreement • 2002 Common Country Assessment • 2003–2007 Development Assistance Framework

[a]69 National Development Plans were complemented by sectoral development plans, which have not been mapped here
[b]Sources: mptf.undp.org
[c](Samoa. Ministry of Finance, 2008). This plan included a review of the previous plan. Check whether the plan for post 2012 had a review of the period 2008–12

3.1.1 Case Study: Papua New Guinea

Relations between the United Nations and the Government of Papua New Guinea provide a case study in the evolution of development assistance planning. PNG is the most populous of all PICs and has the largest pool of human and natural resources. A Standard Basic agreement was established in 1981 (Government of Papua New Guinea & UNDP, 1981), after which UN agencies made separate agreements with government departments by sector. In 2001 a *Common Country Assessment* improved the integration of government and UN thinking about where UN assistance could have the most impact (UNDP & Government of Papua New Guinea,

2001), which was updated in 2016. Feedback on results was provided in the UN Resident Coordinator's annual reports and individual UN agency reports, but UN activities remained fragmented, and prone to duplication in programming, a lack of focus on outcomes, multiple funding streams, and multiple financial reporting requirements. In practice, this meant that each government department was obligated to prepare financial statements at different dates for different agencies. There was little operational coherence with more than ten UN agencies (not to mention many other non-UN governmental and non-governmental development partners). Within the health sector, for instance, WHO, UNFPA and UNICEF consulted individually with the Papua New Guinea Health Ministry and possibly also implemented their activities in isolation from each other.

In the 2000's the UN revised its operations to "Delivering as One." Upon introduction of the Resident Coordinator system after 2007, these UN agencies formed a UN Health Task Team prior to consulting with the National Department of Health, so as to improve the coherence of their alignment with the National Health Plan. During this period Papua New Guinea was pursuing the eight Millennium Development Goals, but with mixed results (Government of Papua New Guinea. Department of National Planning and Monitoring, 2015). Furthermore, significant governance challenges in the country affected the ability of agencies such as the UN to provide coherent support. As explained in a review of the UN's Development Assistance framework 2012–2015:

> At the time the UNDAF 2012-2015 was developed, alignment with government development strategies and policies was a challenge, due to the absence of an enabling environment. The Papua New Guinea context was considered fragile (ref para 13) and whilst the Vision 2050 was drafted along with the MTDP (2010-2015) and the DSP (2010-2030) there was limited continuity across these government strategies. As a result, the UN's positive engagement in strategic planning and the relevance of the UN's programme priorities to the country, different elements of the UNDAF were aligned to different strategic documents of Government and in some cases (as is the case of Gender) to the National Constitution. On this basis there is no shortage of government strategies with which to align programmes and the UNDAF as a whole provides clear pathways for strategic planning, encouraging positive alignment between government strategies and the UNDAF. The issue, however, is considered to be the lack of coordination between government departments, ministries and other stakeholders, compounded by varying levels of commitment, therefore implementation of their strategic priorities is affected and coordinated approaches to development overall is compromised, including implementation of the UNDAF (Lund & Lovai, 2016, pp. 14–15).

The effort to simplify UN engagement with the government of Papua New Guinea is expressed in Fig. 3.1:

By 2015, under the "delivering as one" structure, the UN's Papua New Guinea office and the Department of National Planning and Monitoring established a governance framework to ensure coordination of the work of the UN's 14 agencies working in the country:

1. Food and Agriculture Organization (FAO)
2. International Fund for Agricultural Development (IFAD)
3. International Labour Organization (ILO)
4. International Organization for Migration (IOM)

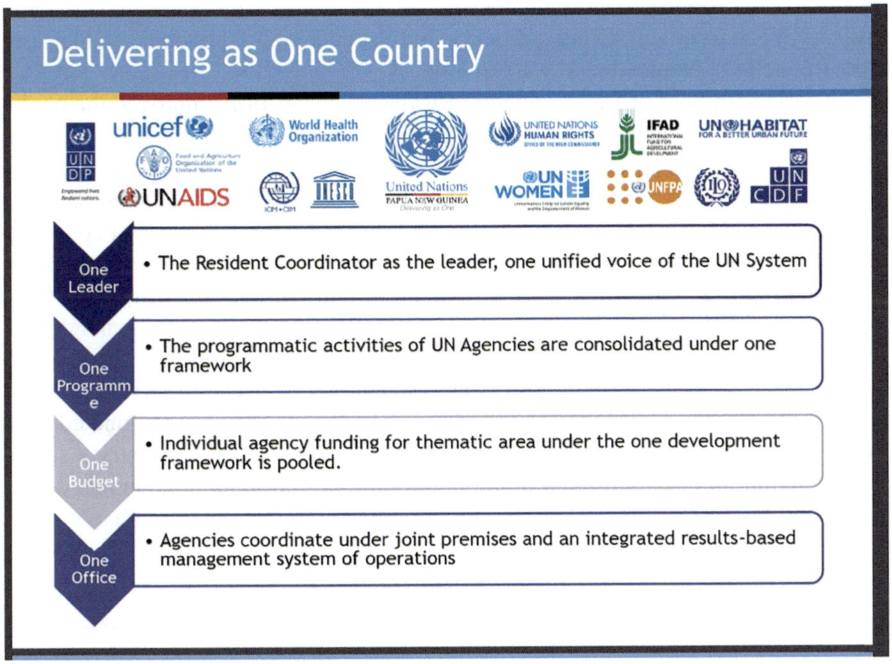

Fig. 3.1 "Delivering as One Country" in Papua New Guinea

5. Office of the High Commissioner for Human Rights (OHCHR)
6. Joint United Nations Programme on HIV and AIDS (UNAIDS)
7. United Nations Capital Development Fund (UNCDF)
8. United Nations Educational, Scientific and Cultural Organization (UNESCO)
9. United Nations Development Programme (UNDP)
10. United Nations Population Fund (UNFPA)
11. United Nations Human Settlements Programme (UNHABITAT)
12. United Nations Children's Fund (UNICEF)
13. United Nations Entity for Gender Equality and the Empowerment of Women (UN WOMEN)
14. World Health Organization (WHO)

These agencies were sorted into four "clusters" and addressed 10 priority areas:

Cluster 1: "Governance for Equitable Development", with projects in Governance, SDGs & Population, and Peacebuilding.

Cluster 2: "Social Justice, Protection & Gender Equality" with projects in Promotion & Protection of Human Rights, Gender Equality & Women's Empowerment, Child Protection, and HIV & AIDS.

Cluster 3: "Access to Basic Services", with projects in Health and Education; and.

Cluster 4: "Environment, Climate Change & Disaster Risk Management", with projects in Environment, Climate Change and Sustainable Livelihoods; and Disaster Risk Reduction, Preparedness & Response.

The establishment of *Development Assistance Frameworks* improved the quality of coordination in Papua New Guinea, the first covering 2013–2017 and a second covering the period 2018–2022 (United Nations, 2018). During these years the country was seeking greater alignment with the vision of its constitutional founders, as re-stated in its *Vision 2050* strategic plan of 2009 (Government of Papua New Guinea. National Strategic Plan Task Force, 2009), the *Medium-Term Development Plan 2011–2015* (Government of Papua New Guinea. Department of National Planning and Monitoring, 2010), and the Sustainable Development Goals globally agreed for the period 2015–2030. *Vision 2050* made no specific reference to UN collaboration apart from setting the aspirational target of reaching the top 50 countries on the Human Development Index (in 2020 Papua New Guinea ranked 155). While these initiatives have not yet had visible impact on the country's rate of development, they have resulted in improved relations between government departments and the many UN offices they are in partnership agreements with.

References

Aiafi, P. R. (2017). The nature of public policy processes in the pacific islands. *Asia & the Pacific Policy Studies, 4*(3), 451–466. https://doi.org/10.1002/app5.196

Government of Cook Islands. (2007). Te Kaveinga Nui (Pathway for sustainable development in the Cook Islands) living the Cook Islands vision - A 2020 challenge. In *National Sustainable Development Plan 2007-2010*. Government of Cook Islands.

Government of Kiribati. Ministry of Home Affairs and Rural Development. (n.d.). *Abaiang island development plan 1991–1995*. Government of Kiribati.

Government of Papua New Guinea, & UNDP. (1981). *Standard basic assistance agreement between the Government of Papua New Guinea and the United Nations Development Programme*. Government of Papua New Guinea and the United Nations Development Programme.

Government of Papua New Guinea. Department of National Planning and Monitoring. (2010). *Medium term development plan 2011-2015 building the foundations for prosperity*. Department of Natoinal Planning and Monitoring.

Government of Papua New Guinea. Department of National Planning and Monitoring. (2015). *Millennium development goals 2015. Summary report for PNG*. Government of Papua New Guinea.

Government of Papua New Guinea. National Strategic Plan Task Force. (2009). *Papua New Guinea vision 2050*. Papua New Guinea.

Government of the Solomon Islands, United Nations Development Progamme, & United Nations Capital Development Fund. (2008). *National Provincial Governance Strengthening Programme (PGSP) Joint programme document*. Ministry of Provincial Government and Institutional Strengthening.

Lund, A., & Lovai, B. (2016). *Independent evaluation UN development assistance framework (2012 – 2017) UN Country team in Papua New Guinea.*

Mara, R. S. K. (1997). *The Pacific way a memoir*. Center for Pacific Islands Studies, University of Hawai'i; Pacific Islands Development Program, East-West Center; University of Hawai'i Press.

Pacific Islands Association of Non-Governmental Organizations. (1994). *Pacific region Non-Government Organizations Conference on Population and Sustainable Development, Nadi, Fiji, 2–4 March. Conference Papers*.

Samoa. Ministry of Finance, E. a. P. D. (2008). *Strategy for the development of Samoa 2008–2012*. Ministry of Finance, Economic and Planning Division.

Tonga, Kingdom of, & UNESCAP. (2006). *Eua strategic development plan 2006–2006*.

United Nations. (2018). *Papua New Guinea. United Nations Development Assistance Framework 2018–2022*. The United Nations.

United Nations Development programme, & Government of Papua New Guinea. (2001). *Papua New Guinea Common country assessment*. United Nations Development Programme.

United Nations. Economic and Social Council. High-Level Political Forum on Sustainable Development. (2020). Iem 2 of the provisional agenda. Accelerated action and transformative patahways: Realizing the decade of action and delivery for sustainable development. Compilation of main messages for the 2020 voluntary national reviews. *E/HLPF/2020/5*.

Chapter 4
Regional Development and the UN

4.1 Regional Architecture of the Pacific

Chapter 3 reviewed the evolution of relations between the United Nations Organization and individual Pacific Island Countries. It demonstrated how initial project-based arrangements were eventually consolidated through *Standard Basic Agreements* and how these, in turn, provided the basis for *Common Country Assessments* and *Development Assistance Frameworks* – all put in place to secure shared understanding of the development challenges and the implementation strategies put in place to address them. The current chapter adds an additional layer of complexity, by examining the Pacific's "regional architecture" that has evolved through interactions between PICs, regional organizations, and their many intergovernmental arrangements for leadership, agenda setting, decision-making, funding, implementation, administration, and accountability. After setting out the Pacific regional institutions the chapter also reviews the UN's physical presence across the region.

UN engagement with the Pacific Islands region has progressed through at least three phases. The first, from 1947 to 1971, was the simplest because the SPC was the only regional organization in existence, and because its mandate was strictly limited to issues of health, and social and economic development. The creation of the South Pacific Forum in 1971 (later renamed the Pacific Islands Forum) opened a second phase, which prioritized political dialogue and the establishment of additional Pacific regional organizations, each engaged with UN agencies on a sectoral basis. This second period was characterised by *ad hoc* expansion of programs, projects, and platforms, leading to much activity yet little central coordination. A third stage emerged from approximately 1990, with initiatives taken both within the region as well as globally, to clarify development values, coordinate networks, improve data standardization and collection, and to establish more effective platforms for development dialogue. These initiatives included establishment of the Council of Regional Organizations of the Pacific, post-Forum dialogue partner meetings, and

G. Hassall, *The United Nations and the Pacific Islands*, United Nations University Series on Regionalism 24, https://doi.org/10.1007/978-3-031-34155-7_4

Table 4.1 Three phases of UN-Pacific regional engagement

Phase	Major themes
1. South Pacific Commission ascendency (1974–1971)	Social and Economic Development; technocrats ascendant; minimal civil society engagement.
2. South Pacific Forum and additional regional organizations (1971–1989)	Social, Economic, and Political Development; uncoordinated regional policy; sector-based development dialogue; technocrats engaged but national leaders and bureaucrats ascendant.
3. Council of Regional Organizations of the Pacific (1990–present)	Social, Economic, and Political Development; coordinated regional policy processes; technocrats and civil society engaged; national leaders and bureaucrats ascendant; increased participation in global policy dialogue.

PIC/Development partner annual consultations. At global level, they included the Human Development Report and Index, the "Paris Principles for aid effectiveness", and global development plans and related concerns – the MDGs, SDGs, global conferences since 1992, leading to the evolution of the SIDs framework, and issues of climate change and oceans. Some of the elements in these three phases are depicted in Table 4.1.

4.1.1 South Pacific Commission

The South Pacific Commission was established in 1947 by the metropolitan powers to promote the social and economic advance of their Pacific territories.[1] Because consultation on the proposed Commission commenced in 1942 when the prospective United Nations Organization was also the subject of high-level negotiations, and certainly because the Australian and New Zealand advocates of the idea were the internationalists H.V. Evatt and Peter Fraser, the 1947 agreement (Art. XV) mentioned the relationship the Commission was to have with the newly established international inter-governmental body:

> 55. The Commission and its auxiliary and subsidiary bodies, while having no organic connection with the United Nations, shall cooperate as fully as possible with the United Nations and with appropriate agencies on matters of mutual concern within the competence of the Commission.
>
> 56. The participating Governments undertake to consult with the United Nations and the appropriate specialized agencies at such times and in such manner as may be considered desirable, with a view to defining the relationship which may in future exist and to ensuring effective cooperation between the Commission, including its auxiliary and subsidiary bodies, and the appropriate organs of the United Nations and specialized agencies dealing

[1]These metropolitan powers were Australia, New Zealand, the United Kingdom, the United States of America, France, and the Netherlands. The story is told in (Smith, 1972) The Netherlands withdrew from membership in 1962 and the United Kingdom in 2004.

> with economic and social matters (Governments of Australia France Netherlands New Zealand United Kingdom and United States of America, 1947).

According to SPC's first historian, T.R. Smith, Australia and New Zealand were obliged to reduce their initial aspirations for the Commission's scope to gain support from the Great Britain, France, United States and the Netherlands. Consequently, the Commission was not empowered to report on administration of the territories within its jurisdiction (since this would infringe on the rights of the metropolitan powers), nor could it, consequently, engage with efforts toward self-government (Smith, 1972).[2] While the Commission's later accomplishments in social and economic development are widely acknowledged, this restricted scope, as well as its composition and power structure, became problematic over time. The Commission's tone and format, for instance, privileged the voices of donors rather than those of emerging Pacific leaders:

> The control of the SPC budget was in the hands of the metropolitan powers and since the constitution excluded discussion of political matters, Pacific leaders naturally grew impatient at the way their views were discounted. In 1966, for example, Fiji sought to have the budgetary debate opened up. By the late 1960s when the French programme of nuclear testing again led to attempts by Pacific leaders to raise political matters, the South Pacific Commission cracked under the strain. There were not yet enough independent voices of the new Pacific to change the SPC from within. It was clear that another "forum was needed" (SPEC, 1985, p. 4).

Emergence of the SPC coincided with emergence of the UN as a force in Pacific regional decision-making. A later report on aid effectiveness in the Pacific noted that one result of the UN discovering the Pacific region "in a big way" (SPEC, 1976, p. 22; Finau et al., 1976) was gradual diminution of the status and influence of the SPC:

> Dissatisfaction with the South Pacific Commission grew of a feeling that its potential as a regional political forum was being stifled. SPC's previous glamour as a 'multilateral' channel for regional activities was being eroded by increased UN and other multi-lateral activities (with much more money), and criticism of SPC's cost-effectiveness was rather freely thrown around, sometimes by people whose own costs seemed rather high in relation to their productivity" (SPEC, 1976, p. 22).

Notwithstanding these structural and political impediments, SPC's programmes grew and strengthened in the 1950s and 1960s. It focused on technical, social, and scientific knowledge and practices that were particularly useful in the immediate post-war years (Secretariat of the Pacific Community, 1999). The organization continually adapted its consultation processes as the number of actors increased and as expectations around dialogue and consultation increased. Formal decision-making centred on the "South Pacific Conference" held every three years (later changed to every two years) and attended by most heads of government. These conferences also provided opportunities for informal meetings with such partners as ESCAP, PIDP, FFA, ADB, CCOP/SOPAC, and SPEC. As will be noted in

[2] See Chap. 4: Agreement reached.

subsequent chapters, SPC managed to deflect early concerns expressed by the region's political leaders, and consolidated its position as an essential regional development organization. The Pacific Community has charted its direction through a series of 5-year strategic plans. In 2019 it employed a "strategic foresight" methodology to draft a ten-year Strategy, *Strategy 2021+*. The first phase included a useful mapping of international and regional policies and initiatives (The Pacific Community, 2021, p. 13), as set out in Table 4.2:

This *Strategy 21+* demonstrates the productive mapping of regional and national programmes within the framework of international plans and policies concerning the environment and sustainability. The plan, which is in place for the period 2022–2031, builds on four goals and six key focus areas interconnected by the themes of sustainable systems, climate action, and knowledge integration. The four long-term goals are:

1. Pacific people benefit from sustainable economic development
2. Pacific communities are empowered and resilient
3. Pacific people reach their potential and live long and healthy lives
4. One SPC delivering integrated programmes through streamlined services

SPC's six key focus areas - Natural Resources and biodiversity; Food systems; Blue Pacific livelihoods and economies; Transformative institutional effectiveness; Blue Pacific health (planetary health); and Equity, education and social development - place the work of the Pacific Community at the centre of efforts to achieve the SDGs (The Pacific Community, 2020).

4.1.2 South Pacific Forum

The second major regional organization, the South Pacific Forum, was founded in 1971 through a desire by the leaders of the new Pacific nations to discuss political issues that were beyond the SPC's mandate.[3] Speaking at its' inaugural meeting in 1971, New Zealand Prime Minister Sir Keith Holyoake described the Forum as "an opportunity for private and informal discussions rather than a new international organization" (South Pacific Forum, 1971, p. 1). This statement was literally true, for it was later necessary to reconstitute the organization to give it legal personality. Apart from an allocation of Crown Land in Suva by the Fijian government in 1973 for the establishment of a secretariat, and the signing of a formal agreement by participating countries at a meeting in Tonga in March 1973, the Forum only gained legal existence by treaty in 1991 (Pacific Islands Forum Secretariat, 1991a). For the

[3] Australia and New Zealand were also founding members of PIF, although as already pointed out, occupying delicate positions given that whilst geographically located in the southwest corner of the Pacific Islands region, they are not "small island developing states", nor are their populations predominantly of Pacific Island heritage.

Table 4.2 Pacific Community integration of International and Regional Development Agreements

International	Regional SPC lead	Regional SPC partner or contributor	Pacific centres
• Decade for Ocean Science for Sustainable Development • 2030 Agenda for Sustainable Development (2015–2030) • Sendai Framework for Disaster Risk Reduction (2015–2030) • Paris Agreement (2016) • SIDS Accelerated Modalities of Action (S.A.M.O.A.) Pathways (2014–2030) • Aichi Biodiversity Targets - Strategic Plan for Biodiversity (2011–2020) • Voluntary Guidelines for Securing Sustainable Small-Scale Fisheries (2015)	• 10 Year Pacific Strategy for Statistics (2010–2020) • Pacific NCDs Roadmap (2013) • Pacific Legislative Framework on NDCs (under development) • Regional Culture Strategy (2010–2020) • Pacific Platform on Action on Gender Equality and Women's Human Rights (2018–2030) • Regional Strategy for Pacific Women in Maritime (2020–2024) • Framework for Action on Energy Security in the Pacific (2010–2020) • Regional Framework for accelerating action on food security & nutrition in Pacific SIDS (in draft)	• Council of Regional Organizations of the Pacific (CROP) Charter (2018) • Pacific Leaders Gender Equality Declaration (2012) • Boe Declaration on regional security (2018) • Kainaki II Declaration for urgent climate action now (2019) • 2050 Blue Pacific Strategy (2020–2050) • Framework for Pacific Regionalism (2014) • Framework for Resilient Development Partnership (FRDP) (2017–2030) • Regional Roadmap for Sustainable Development • Pacific Disability Rights Framework (2016–2025) • Regional Roadmap for Sustainable Pacific Fisheries (2015) • Framework for a Pacific Oceanscape (2010) • Healthy Islands Framework (2005) • A New Song for Coastal Fisheries (2015) • Pacific Regional Education Framework (PacREF) (2018–2030)	• Pacific Data Hub (2019) • Pacific Centre for Renewable Energy & Energy Efficiency (2017) • Pacific Community Centre for Ocean Science (2017) • Centre for Pacific Crops and Trees (2007) • Maritime Technology Cooperation Centre Pacific (2017) • Pacific Climate Change Centre (2018)

first two decades the "Forum" (i.e., the annual meeting of PIC leaders) existed solely through ongoing goodwill and commitment amongst them until a desire to gain observer status at the UN forced the issue of establishing a treaty (Graham, 2008).

The Forum's first "...ad hoc gathering of Island Leaders" from Western Samoa, Tonga, Fiji, Cook Islands, Nauru, Australia and New Zealand, in Wellington in August 1971, engaged in "... private and informal discussion of a wide range of issues of common concern ...matters directly affecting the daily lives of the People of the Islands of the South Pacific, devoting particular attention to trade, shipping, tourism, and education" (South Pacific Forum Secretariat. Leader's Meeting, 1971). In fact, the first Forum discussed issues that were to define the region's engagement with UN agencies for the next five decades and more: trade, shipping, civil aviation, foreign investment, tourism, the Law of the Sea, environmental protection, ocean resources, education and human development, telecommunications, disaster preparedness, a nuclear-free Pacific, decolonization, establishment of the appropriate mechanisms for joint diplomatic representation, and regional cooperation.

At this time Fiji was the lone PIC member of the UN. On the topic of Law of the Sea the Leader's first communique welcomed an offer by Australia and New Zealand, as members of the UN Seabed Committee."...to keep Island Governments informed of progress of the Committee's deliberations and to draw the attention of the Seabed Committee to the special needs of the South Pacific Islands."[4] Although the Forum aspired to exhibiting 'The Pacific Way" (a subtle and conflict-avoiding process of determining a consensus view), it communicated its meeting outcomes in formal diplomatic language in its first years, in the hope of having "...greater impact in places like New York" (Piddington, 1986, p. 22). Several PIC leaders had commenced attending UN meetings, according to Fijian Prime Minister Ratu Mara, and United Nations practices had begun creeping into Forum meetings: "Long documents were produced by officials for leaders to read," he writes disapprovingly, "... worded like presentations to the General Assembly. Indeed, I think the Forum is now run by the bureaucrats, and the leaders come to talk on the subjects they are allowed to talk on" (Mara, 1997).

The desire for trade expansion was one of the first issues that brought the Forum into early engagement with the UN. At its second meeting, in Canberra in 1972, Forum Leaders decided to establish a Suva-based Bureau to study and advise on regional issues, commencing with trade, transport, communications, and economic development. Bureau costs were apportioned one third each to Australia, New Zealand. and the Islands as a group (South Pacific Forum, 1972). This South Pacific Bureau for Economic Cooperation (SPEC) was in 1988 renamed the South Pacific Forum and in 1999 the Pacific Islands Forum Secretariat.

The quest for economic growth in the PICs has focused on the processing of such raw materials as timber and minerals (gold, nickel, oil, copper); the harvesting of oceanic resources especially fish; agriculture; light industry; and service industries, notably tourism. Economic growth was a key pillar of the 2004–2013 Pacific Plan –

[4] https://www.forumsec.org/south-pacific-forum-wellington-5-7-august-1971/

although projections of the potential for growth of Pacific Island economies by the Asian Development Bank and the World Bank have always been moderately optimistic, at best. A second feature of this quest for economic growth has been pressure to accept the neo-liberal agenda according to which growth required the removal of barriers to trade, the creation of or entry into free trade agreements, reduction in size of the public sector, and divestment of state holdings in the domestic economy. PICs were expected to join the WTO, and to complete their own free trade agreements, which became PICTA and PACER.

This pro-growth agenda was moderated by other influences: (a) that growth be "sustainable"; (b) that PICs qualify for preferential treatment due to being former colonies (such as arrangements with the EU: Lomé I, Lomé II, and EPAs under the Cotonou agreement; access to Australian and new Zealand markets through SPARTECA; and the compact arrangements between the Micronesian states and the US economy), "less developed countries" (as calculated by the World Bank, OECD, and the UN's Committee for Development Policy), or small island states (as acknowledged within the UN system); and (c) that growth be inclusive (such as access to transport services and communications technologies be made available to all through the principal of "universal access" rather than by market forces). The competing needs from so many directions made the task of establishing a free trade agreement between the PICs extremely challenging (Scollay, 2001). The region's economic philosophy – to the extent that there can be said to be one – demonstrates a preference for promoting social welfare, environmental protection, and cultural sustainability, over the pursuit of ever-greater GDP. These preferences often are implied more than expressed, and co-exist with the many efforts by UN and other agencies to expand economic productivity.

In 1973 SPEC made a formal request for a UN preliminary mission to assist with a survey of regional trade. After four weeks in the region, a team of experts from ECAFE, UNCTAD, the United Nations Development Advisory Team (UNDAT) and UNDP (G. Marzouk, O. Knowles and R. Giry) published a 90-page report (UNCTAD, 1974) recommending a three-year plan of action toward regional industrial cooperation for processing or manufacture of coconuts, animal feed, fishing, milling, biscuits and sweets, sugar, paper, timber, textiles, fertilizers, cement, and ships; and for regional trade in such agricultural products as citrus fruits, bananas, pineapples, flowers, vegetables, nuts, spices, cereals, cocoa, and sugar. The report also envisaged regional trade in products from forest, sea, livestock, and dairy. The practise of basing plans on rapid assessments by external advisors who had little exposure to the Pacific region had begun! In reference to a report by a UNDP team SPEC had commissioned to examine prospects for the expansion of trade, SPEC's director alluded to tensions in the UN-Pacific relationship:

> The Director felt that any reports done for the region should consider as far as was possible the views of the people who had worked in the region for a long time. The services of experts for SPEC activities should be made available only at SPEC's request, and they should concentrate only on those topics which they were required to investigate. They should not impose their own ideas on the people of the region and should be practical in their approach.

> However, the Director emphasised that, on the whole, SPEC relations with UNDP were good (South Pacific Bureau for Economic Co-operation, 1973a, p. 16).

The SPEC director reported at one meeting that "...subsequent events provided SPEC with a salutary experience or the procedural and bureaucratic realities or the UN system...", an apparent reference to lengthy delays in report completion, protocols required before being permitted to deal directly with UNDP and other UN agencies, and clarification over author attribution of the reports (South Pacific Bureau for Economic Co-operation, 1973b, p. 5). Despite such challenges, SPEC reported making progress with a framework for the development of policies for regional trade expansion and "ultimately perhaps, economic integration" (South Pacific Bureau for Economic Co-operation, 1973b, p. 6). The leaders' 1974 communique noted the view of the SPEC Committee that priority be given to projects that assisted in filling "...the nutritional gap in the region" and directed SPEC to use the report as a basis for further study.[5]

By 1975 SPEC had articulated three categories of functioning as a secretariat: (1) servicing the South Pacific Forum and the SPEC Committee; (2) servicing regional ministerial bodies created by the Forum and such subordinate technical bodies as the South Pacific Regional Shipping Council and its Regional Shipping Advisory Board and The Pacific Regional Civil Aviation Council, and (3) regional technical meetings such as the UNDP/FAO Pests and Diseases survey, and the UNDP/ITU Regional Telecommunications Projects.

Studies aimed at improving the region's trading arrangements continued in the 1980s. Prospects for closer economic cooperation amongst SPEC members: a Trade Promotion Advisory Services and Training project funded by UNCTAD/ITC-UNDP and SPEC; and an Industrial survey of the South Pacific, ESCAP/ADB/SPEC (SPECn, 1981a). ITC/ESCAP helped PICs expand and establish trade information services, but this required infrastructure and service facilities beyond their resources, if a central information network was to be created (SPEC, 1980a). In 1989 The South Pacific Trade Organization gained access to the UN Information systems operated by international Trade Centre in Geneva, which allowed quick access to Australian financial media (Pre-Forum Officials Committee, 1989, p. 10 SPTC Work Programme).

The Forum gradually created a range of regional institutions in addition to SPEC, and instituted additional policy dialogue processes: a Regional Shipping Council (1974), a Regional Natural Disaster Relief Fund (1975), the Pacific Forum Line (1976), the South Pacific Regional Trade and Economic Cooperation Agreement (SPARTECA – 1980) (SPEC, 1980b), and the *Tarawa Declaration on Drift-Net fishing* (1989). From 1989 a "Post-Forum" Dialogue process began, with Canada, France, Japan, the United Kingdom, and the United States attending the first meeting. These annual dialogues proved an efficient process that allowed all partners to understand the transactional requirements, including their policy preferences at the UN.

---

[5] https://www.forumsec.org/fifth-south-pacific-forum-rarotonga-cook-islands-20-22-march-1974/

4.2 Additional Regional Organizations

The three additional regional organizations that emerged from the international collaboration efforts of the 1970s and 1980s with most direct UN engagement were the South Pacific Geoscience Organization (SOPAC), the South Pacific Regional Environment Programme (SPREP), and the Forum Fisheries Agency (FFA). Details of the origins and evolution of these organizations will now be reviewed in brief.

4.2.1 Committee for Co-Ordination of Joint Prospecting for Mineral Resources in South Pacific Offshore Areas/South Pacific Geoscience Organization (SOPAC)

The Committee for Co-ordination of Joint Prospecting for Mineral Resources in South Pacific Offshore Areas (CCOP/SOPAC) was established in 1972 as a UNDP Regional Project to coordinate and promote regional cooperation on research into the region's non-living marine resources. Up to 1991, SOPAC received approximately USD 10 million in funding support from UNDP, in addition to another USD 10 million from the US, Australia and New Zealand (SOPAC, 1987, p. 4). The original focus on assessment of deep-sea minerals and hydrocarbon potential later expanded to include assessment of the potential of ocean and onshore mineral resources, coastal protection and management, geohazard assessment, water, wastewater, sanitation, energy, and disaster risk management. In 1979 UNDP established a project to investigate the mineral potential of the South Pacific with USD 2.5 million in funding for three years (1979–1981), and with ESCAP as executing agency. UNDP continued to fund the project through the 1980s (SPEC, 1982, p. 1).

Until 1980, Pacific leaders were under the impression that the Committee for Co-ordination of Joint Prospecting for Mineral Resources in South Pacific Offshore Areas (CCOP/SOPAC) was a United Nations agency. However, the UN Office of Legal Affairs advised that it had not been constituted as an international body. Because this advice threatened to interrupt the programme's operations, SOPAC was quickly reconstituted and a first annual conference under its new status as an autonomous regional body was held in Apia in November 1984.[6] In 1999 SOPAC became a regional organization with the same legal standing as SPEC and FFA. Following a later review of the Pacific's regional institutional framework, it was integrated into SPC in 2011, and into SPC's Geoscience, Energy & Maritime Division in 2017.[7] Through this circuitous route this scientific organization which

[6]Fiji, Tonga, New Zealand, Papua New Guinea, Solomon Islands, Tuvalu, Kiribati, Cook Islands, Vanuatu, Western Samoa – with Australia assisting (South Pacific Forum Secretariat, 1985b, p. 1).

[7]https://gem.spc.int/about/gem-history

began as a UNDP technical programme barely known to the Pacific general public, has developed into an organization highly regarded for its contribution to scientific knowledge for the Pacific nations.

4.2.2 South Pacific Regional Environment Programme (SPREP)

The South Pacific Regional Environment Programme (SPREP) was established by decision of the 1977 Forum Leaders meeting in Port Moresby. It was to be developed jointly by SPEC, SPC, UNEP and ESCAP, with assistance from UNDP, based on national reports of environmental resources and problems and program priorities; a regional action plan and a regional conference on the Human Environment proposed by UNEP/ESCAP, which would recommend activities to be undertaken by SPREP. Additional support would be provided by the International Union for the Conservation of Nature (IUCN) and WHO (SPEC, 1978a, pp. 1–2).

The Programme commenced in 1979 with funding in the first year from UNEP, SPEC, SPC and ESCAP. IUCN, WHO, and UNDP were also requested to provide support (SPEC, 1979b). Forum Leaders decided "that SPEC should act as Chairman of the management body set up to oversee the programme, the SPREP Co-ordinating Group, which met twice yearly. SPC was allocated the role of Secretariat, utilizing the services of the SPC Regional Ecological Adviser. UNEP, and to a lesser degree ESCAP, are the main financial resources" (South Pacific Bureau for Economic Co-operation, 1981, p. 1). The Conference on the Human Environment, proposed in 1977, convened in Rarotonga in March 1982 (Dahl & Baumgart, 1982). By 1984 SPREP facilitated Pacific Participation in the UNEP's Global Environment Monitoring System (GEMS). In November 1986 the Convention for the Protection of the Natural Resources and Environment of the South Pacific (SPREP convention) was adopted by 16 states in the region.

In 1995 SPREP was re-established as an independent regional organization, with legal status allowing it to assist PICS in international environmental negotiations. By 2021 SPREP had 21 PIC members, five developed country members, and an annual budget of approx. USD 30 million. Its five priority areas (Climate Change Resilience, Environmental Governance, Island And Ocean Ecosystem Services, Waste Management And Pollution Control, and Organizational Goals) require close cooperation with UN agencies. SPREP has become a central actor in the region's engagement with UN climate change negotiations.

4.2.3 *Forum Fisheries Agency*

The third regional organization established in the 1970s with UN assistance was the Forum Fisheries Agency. Pacific leaders decided at their November 1977 meeting in Port Moresby to establish a regional fisheries organization based in Solomon Islands. UNDP/FAO were among the international organizations that pledged financial support.[8] Concerned at what they regarded as departures from the negotiated provisions in the UNCLOS text, Pacific Leaders (at the initiative of Fiji and Papua New Guinea) decided in 1977 in the interests of the region and for the protection of its fisheries resources, in particular the highly migratory stocks, to establish the FFA. FFA commenced operations in 1979, well ahead of the coming into force of the UN Convention on the Law of the Sea (UNLOS) in 1982, and notwithstanding the expressed opposition of foreign fishing nations (USA and others). FFA focused on frameworks for surveillance, and the conservation and management of migratory fish stocks, while SPC's interest in fisheries continued, with its Marine Resources division developing programmes in oceanic fisheries and coastal fisheries.

4.3 The Quest for a Single Regional Organization

All through this period in which regional organizations were being established to coordinate the region's interests on a sector-by-sector basis, attempts were being made to establish a single regional organization to provide coherence. The quest was sparked partially by institutional rivalry between SPC and PIF, but more fundamentally, by a desire for coherence in the region's governance and external representation (Hassall, 2021).

The Pacific Islands Forum, although established much later than SPC, sought to assert its primacy on the basis that it comprises the democratically elected leaders of the independent Pacific states. The SPC, on the other hand, had deeper pre-existing connections across the region and with the United Nations. There was merit to both options, since SPC's membership included dependent as well as independent members, and received ongoing funding from its metropolitan-country members. Concerned about the problem of duplication of effort between SPC and SPEC, an SPEC Committee meeting in Honiara in July 1979 recommended that a committee be established with resources provided by UNDP, to examine the SPEC Agreement and the Canberra Agreement (by which the SPC was established) "...with a view to ensuring that overlapping responsibilities is minimised", and to examine other ways in which the operations of both SPEC and SPC could be modified "to ensure more

[8] A South Pacific Islands Fisheries Development Agency had previously been established at SPC headquarters in Nouméa in 1969, with assistance from UNDP and FAO, but by 1972 differences had arisen over its usefulness. Although it was terminated in 1993, SPC continued its activities in fisheries development (Capiez et al., 2007, p. 84).

effective use of total human and financial resources available to the region" (SPEC, 1979a, pp. 2–3). Three decades on, the issue remains unresolved.

The 16th Forum, in Rarotonga in 1985, heard CMNI's view that an SPC-SPEC merger would be "inappropriate, uneconomical, premature and unfair" (South Pacific Forum Secretariat, 1985a, p. 2). The territories in the North Pacific were concerned about potential impact on their relations with the USA, about potential cutbacks on aid, or that decisions taken by an SRO with metropolitan members might have more weight than decisions taken by the Forum. There was, too, the prospect that "some countries might move to create another regional organization as had occurred prior to the formation of the Forum when only the SPC was in existence" (South Pacific Forum Secretariat, 1985a, p. 4). This came to pass, with formation of the Pacific Islands Development Forum (PIDF), initiated by Fiji in 2012.

Other countries pointed to flaws in SPC's membership and mandate: "The concept of having a single regional organization for the South Pacific is important to us," Papua New Guinea's Ambassador Vagi informed UNGA, "…we can no longer accept the colonial-type assumptions implicit in the constitution, the structure and the activities of the SPC. We do not want to prevent Governments or other interested parties in metropolitan countries from playing a constructive role in our region …but we are opposed to great-power rivalry in the South Pacific. we cannot accept foreign domination" (United Nations General Assembly, 1985, pp. 29–39). A regional meeting of officials held in May 1987 to plan for that year's Forum concluded:

> …there was an urgent need to strengthen the South Pacific Forum and SPEC in regard to the recognition accorded to both institutions by other Governments and organizations. After 17 years, the South Pacific Forum had developed as the paramount political institution in the South Pacific. However, this was insufficiently recognized by a number of other Governments and international Organizations, and steps should be taken to correct this. In addition, as the membership of the Forum grew and the relationship of the region with the rest of the world became more complex, it was necessary to look at changes in the Institutional arrangements in the South Pacific" (South Pacific Forum, 1987b, p. 1).

The meeting proposed the establishment of a committee on regional institutional arrangements to consider:

- means of increasing the recognition accorded the Forum by other Governments and Organizations outside the region, including in particular the governments of Japan and organizations such as UNDP, World Bank, EEC and the United Nations, including the question of Forum or SPEC observer status at the UN;
- the concept of a single regional organization and its implications for strengthening the Forum and SPEC;
- means of strengthening SPEC as the Forum Secretariat, including a possible change of name (South Pacific Forum, 1987b).

Papua New Guinea prepared a discussion paper for presentation to senior officials meeting in Suva in May 1987 that reiterated the purposes of an SRO which had been considered for the past decade:

- to ensure region-wide cooperative arrangements, including inter-governmental organizations and activities, are controlled by the Governments, and meet the needs of, the region (as is the case with major regional inter-Governmental organizations in other parts of the world);
- to coordinate inter-governmental cooperation so that priorities can be effectively sent between activities which are currently caried out by different regional organizations, and economies of scale can, where possible, be achieved; and
- to reduce inefficiency and waste – caused not only by duplication but by failure to integrate both the programmes and the support services provided by existing regional organizations (Government of Papua New Guinea, 1987).

The paper was sufficiently bold as to point out the division of PICs into those who championed the quest for an SRO and those who resisted it – at least in part due to influence from metropolitan countries which would lose their influence in the region if SPC's prominent role was diminished. The paper was one input amongst others to a Committee on Regional Institutional Arrangements which first met in September 1987. The Committee was also tasked with reporting on the procedure for gaining observer status at the UN. As it turned out, a conservative view prevailed, and the committee concluded that "for a variety of political and other reasons, the amalgamation of SPEC and the SPC was not achievable at this time" (Pacific Islands Forum Secretariat, 1988a, p. 2). The reader should also note that these consultations on regional organizations took place in the context of Fiji's first military coup, in May 1987.

Following its second meeting, in February 1988, the Committee noted: "...that the issue of Forum/SPEC observer status at the United Nations would be included in its report to the Forum on the basis that it was not a matter for priority action at this stage but would be placed on a check list for future action" (Pacific Islands Forum Secretariat, 1988b, p. 2). The most that this SRO initiative achieved was a change of SPEC's name to "Forum Secretariat."

A major review of the Pacific's "regional institutional framework" completed in 2006 (Tavola et al., 2006) appears to have given up the quest for a "single regional organization", recommending instead that regional organizations be organized into "three pillars"

- A political and general policy institution – the Pacific Islands Forum and its secretariat;
- A sector-focused technical institution – the Pacific Community and its secretariat;
- Academic and training organizations – the Fiji School of Medicine, the Pacific Islands Development Programme, and the University of the South Pacific.
- a very expensive institutional review exercise for not very much consequent institutional reform.[9]

[9] A six-person team conducted more than 100 interviews across the Pacific. Just three UN agencies (UNDP, UNESCO and UNFPA) were among the organizations consulted, but the report made no comment on the region's relations with the UN.

The region's leaders have periodically commissioned additional reviews of the "Pacific Regional Architecture." In 2004–2005 this resulted in initiation of the "Pacific Plan" and minor institutional revision (Tavola et al., 2006). The Pacific Plan commenced with a strong vision statement: "The Pacific region can, should and will be a region of peace, harmony, security and economic prosperity, so that all of its people can lead free and worthwhile lives. . ." The Plan's key implementation strategies were to promote partnerships (involving PICs, sub-regional groupings, development partners, international organizations, and non-state actors) in pursuit of its "four pillars" - good governance, economic growth, sustainable development, and security - and to promote national ownership and commitment, such as by mainstreaming regionalism at national level, and through oversight by a Pacific Plan Action Committee. Key to the Plan's success was upgrading frameworks for monitoring and evaluation, and review. Collaboration with UN agencies was integrated into the activities under each of the Plan's four pillars.

The Pacific Plan remained operational for nearly a decade, until a significant review in 2013 headed by former Papua New Guinea Prime Minister Sir Mekere Morauta. The resulting Morauta report had little to say about relations between the region and the United Nations or any other international intergovernmental organization, but expressed the view that progress with global targets as set in the MDGs and then in the SDGs (SDGs) were of more relevance to individual countries than to the region:

> The. . .Millennium Development Goals (e.g. improving material health, achieving universal primary education, reducing child mortality), while critically important for tracking global progress in a number of important areas of deprivation, are not that useful for tracking progress on regional integration and regional cooperation (Morauta, 2013, p. 125).

The Morauta Review was highly critical of the slow pace of regional integration and alluded to the need to consider more binding arrangements, including a free trade area, a customs union and eventually a federation. Although Sir Mekere personally communicated his findings to each of the region's national leaders, the report gained very little traction and is barely referred to in later PIFS reports on the state of Pacific regionalism (Pacific Islands Forum Secretariat, 2017; Pacific Islands Forum Secretariat, 2019).

4.4 Pacific Sub-Regionalism and National Interests

Pacific sub-regions assert their own identities at the same time as they continue to support a whole-Pacific regionalism. Sub-regional groups include the *Small Island States*, established in 1987,[10] the *Melanesian Spearhead Group*[11] in the Southwest,

[10]Cook Islands, Kiribati, Niue, Tuvalu, the Republic of Marshall Islands, Nauru and Palau.

[11]Papua New Guinea, Solomon Islands, Vanuatu, Fiji, and FLNKS (New Caledonia)– see http://www.msgsec.info/

the *Polynesian Group* to the east, the *Micronesian Chief Executives Summit* in the North Pacific.[12]

The Pacific's smaller states (Cook Islands, Federated States of Micronesia, Kiribati, Marshall Islands, Nauru, Niue, Palau and Tuvalu) have also insisted on receiving special attention. In 1981 SPEC proposed to the Forum's leaders the need for a process for considering their special needs. This idea resulted in the first instance in the convening of a meeting on this topic in Alofi, Niue, in February 1982, and subsequently to the formation of a distinct grouping within the Pacific Forum's development dialogue processes.[13] An inaugural summit was held at Rarotonga in January 1992, a Small Island States Programme has been based at PIFS since 2006, and an SIS strategic plan was endorsed in 2016.[14]

Sub-regional rivalries also surface on occasion within PIF. Choice of Forum Secretary General has periodically exposed inter-state rivalries; in 2021 the Micronesian countries threatened to leave the organization when Cook Islander Henry Puna was selected in preference to a Micronesian candidate.[15] Fortunately, an agreement was reached the following June that avoided this split in regional unity. Similar tensions have surrounded the appointment of the SPC Director.[16]

Commitment to Pacific regionalism does not put an end to competitiveness between nation-states. Papua New Guinea's prime minister raised at the 1981 Forum Leaders meeting the sensitive issue of regional solidarity versus individual national interest, by noting that PICs were slow to coordinate their distant-water fishing negotiations, and preferred to negotiate individually (SPEC, 1981b), a

[12] CMNI, Guam, Federated States of Micronesia and its states, Marshall Islands and Palau. In 2011 the Micronesian Chief Executives summit issued its 15th communiqué which announced, amongst other things, the establishment in Guam of the Office of the Micronesian Center for a Sustainable Future, a Regional Workforce Development Council and Pacific Workforce Investment Group.

[13] Initial topics were trade, and freight/passenger carriage, industry, fruit and vegetable processing, and prospects for a free trade agreement for goods wholly manufactured within the region.

[14] https://www.forumsec.org/smaller-island-states/

[15] SPEC Directors: Mahe Tupouniua (Tonga, 1972–1980); Gabriel Gris (PNG 1980–1982); John Sheppard (acting – Australia 1982); Mahe Tupouniua (Tonga 1983–1986) and Henry Naisali (Tuvalu 1986–1988); PIFS Secretaries-General: Henry Naisali (Tuvalu) 1986–1992; Ieremia Tabai (Kiribati) 1992–1998; Noel Levi (Papua New Guinea) 1998–2004; Greg Urwin (Australia) 2004–2008; Feleti Teo (acting – Tuvalu 2008); Tuiloma Neroni Slade (Samoa) 2008–2014; Meg Taylor (Papua New Guinea 2014–2021); Henry Puna (Cook Islands) 2021–2022.

[16] SPC Directors: William D. Forsyth (Australia 1948–1951); Leslie Freeston (United Kingdom 1951–1954); Ralph Bedell (USA 1955–1958); Thomas R. Smith (New Zealand 1958–1963); William D. Forsyth (Australia 1963–1966); Gawain Bell (United Kingdom 1967–1969); Afioga Afoafouvale Misimoa (Western Samoa 1970–1971); John E de Young (acting – USA 1971); Gustav F.D. Bentham (Western Samoa 1971–1975); E Macu Salato (Fiji – 1975–1979); M. Young Vivian (Niue 1979–1982); Francis Bugotu (Solomon Islands 1982–1986); Palauni M. Tuiasosopo (American Samoa 1986–1988); Jon Jonassen (Acting - Cook Islands 1989); Atanraoi Baiteke (Kiribati 1989–1993); Ati George Sokomanu (Vanuatu 1993–1996); Robert B. Dun (Australia 1996–2000); Lourdes Pangelinan (Guam 2000–2006), Jimmie Rodgers (Solomon Islands 2006–2014); Colin Tukuitonga, (New Zealand 2014–2020); Stuart Minchin (Australia 2020–).

tendency that continued through to their individual and collective negotiations toward Economic Partnership Agreements with the European Union.

Fiji has also adeptly promoted regionalism as well as its national interest. It facilitated the establishment of such regional institutions as the University of the South Pacific, and later the South Pacific Bureau for Economic Cooperation, which successive Fijian governments have treated as national institutions. Their internet addresses, for instance are ".fj" rather than ".int" or ".org", and their employees require Fijian work permits. Fiji has used selective and focused engagement with UN bodies as an element of its foreign policy. In the 1970s, at the 6th Session of the Law of the Sea conference, it proposed itself for the seat of the International Seabed Authority – an organization eventually set up in Kingston, Jamaica.

Pacific states have jostled to secure international partners and projects. After agreeing to host a Regional Telecommunications Training School, for example, Fiji proposed that it be renamed the "Telecommunications Training School" on the argument that Fiji would end up paying most of its costs (SPEC, 1977). At the same time, Papua New Guinea advocated establishing a Telecommunications training college at Lae in collaboration with ITU/UNDP "...because the volume of need for Papua New Guinea could not be met through the Fiji centre without limiting its availability to other regional countries" (South Pacific Bureau for Economic Co-operation, 1975, p. 15).

The epitome of Fijian-Samoan rivalry was reached in the race to host the SIDS III conference in 2014 (the conference is the subject of Chap. 11). Samoa's then prime minister Tuila'epa Sa'ilele Malielegaoi recalled in his memoir that the topic of hosting country was raised at the Forum Leader's meeting in 2011 when Fiji was suspended from the group due to the 2006 military coup. On hearing of concerns that Samoa was not a suitable venue due to having too little accommodation, Malielegaoi reminded the leaders that the ACP agreement now known as the Lomé Convention was originally scheduled to be signed in Fiji and was only moved to Lomé in Togo due to Fiji's previous coup, in 2000. If not for this change of venue, he sought to point out, that famous agreement would have put the Pacific on the map as the "Nadi Convention" rather than the Africa-themed "Lomé Convention." The leaders took his point and agreed that SIDS III should be held in Apia. The Government of New Zealand supplied a cruise ship to provide the necessary additional accommodation (Malielegaoi, 2017, p. 215). In order to restore Pacific unity, Fiji and Samoa came to a bilateral agreement on the matter in the side-lines of UNGA in 2012.

4.4.1 The Position of the Metropolitan Countries: Australia and New Zealand

Australia and New Zealand occupy ambivalent positions in Pacific Islands regionalism. Physically, Australia and New Zealand are the two largest island groups in the Oceanic region. Politically, they have exercised ascendent influence over the PICs – through colonial rule, or through their hegemonic power in diplomatic, trade, and cultural relations. On the one hand Australia and New Zealand are founding members of the Pacific Islands Forum and the Pacific Community, in addition to a host of other regional organizations, and in all these contexts insist on their status as fellow Pacific Island Countries. At the same time, they belong to the Western European and Other Group (WEOG) within the UN while all other PICs are grouped with the Asian member countries (in 2011 the PICs succeeded in changing the formal name of this group to "Group of Asia and the Pacific Small Island Developing States"). Australia and New Zealand have exercised, nonetheless, a nurturing role. These contrasting roles, sometimes as regional hegemon and sometimes as partner in development assistance, mean that Australia and New Zealand are accepted as "Pacific Island Countries" in some instances yet decidedly not so in others.

Some PICs would prefer that these two countries "step back" from their positions of dominance in the region (Fry, 2015). Yet the PICs tolerate this ambivalent status in return for the benefits they bring to the region, especially development assistance, humanitarian support in times of crises, and diplomatic alliance on most issues. At the time of the creation of the South Pacific Forum in 1971, for instance, Island leaders welcomed Australia and New Zealand's offer to keep them informed of progress of the United Nations Seabed Committee (South Pacific Forum Secretariat. Leader's Meeting, 1971). This ambivalence toward the Australian and New Zealand presence in Pacific Island affairs pervades the welcome Cook Islands' Premier Albert Henry offered participants to a regional telecommunications meeting in Rarotonga in 1978:

> Since I have been Premier of this country, I have had arguments with New Zealand but I cannot get away from the fact that New Zealand and Australia about forty years ago realised the need to share what they have with those that have not. They were the first two countries to begin this new concept. It was Australia and New Zealand in 1947 in Canberra that signed an agreement that they must look to the scattered islands of the Pacific to see what they could do to help them. The spirit of colonialism was not there anymore. What was needed was the belief in the spirit that 'we can help'. There were, at that time, countries that had small colonies in the Pacific and so they consulted each other; metropolitan powers that had colonies in the Pacific such as the United States, the United Kingdom, France and the Netherlands. The funny thing, they all had that same spirit in them. Australia and New Zealand put the idea to the other metropolitan governments. The other metropolitan governments accepted the idea, and hence was formed in 1947 the South Pacific Commission.
>
> With the orbit of that ideal came the desire to help, and slowly the spirit of ruling, telling, began to cool off. I remember that two years after the Commission had started, these people who were trying to do good were wondering how it was that when they gave money for the social development of these islands, these people did not seem to care for it. Why? We give them something and they do not want it! Why? The spirit of colonialism existing that time indicated that what was good for Fiji should also be good for the Cook Islands. Yet when the same thing was offered to the Cook Islands, they say they don't want it. So, some people said, may be we are giving them the wrong thing.
>
> And so they said, well let's ask them to send the people of their own country to tell us what they want. That meeting was held in Suva in 1950 where the people of the countries were asked what they wanted and not the people in the administration. In the past, people in the administrations were sent to such meetings and the administrations were controlled by the colonial governments. The administration officers or the 'magicians' as you have called them, were people trained by the metropolitan people. They told the metropolitan governments what they thought they wanted to know, not what the people wanted.

> ...The South Pacific Commission of course, we all know, have done a great job ...However, we were not allowed to speak our own minds in the South Pacific Commission. We were not supposed to talk 'politics' in the South Pacific Commission. So we decided to form a group so that we could talk 'politics' and hence was born the South Pacific Forum. We started off with four and then five – Fiji, Tonga, Western Samoa and the Cook Islands – those were the first four. Then we asked Zealand to host the first meeting and then my friend Hammer DeRobert of Nauru asked why we left them out. We weren't really leaving them out – he was so far away and hard to get in touch with. Hence was born the desire to upgrade regional telecommunications... (SPEC, 1978b, pp. 64–65).

Australia and New Zealand provide two-thirds of the Pacific Islands Forum's budget, and a considerable proportion of the Pacific region's total development assistance. They also provide a majority of development assistance to other regional organizations and to many PICs bilaterally. Up to at least 1991, for instance, Australia and New Zealand were contributing 74% of all member contributions to FFA (Pacific Islands Forum Secretariat, 1991b, p. 2 Agenda item 4. FFFA Financial Policy Review). In 2001 90% of SPC's budget was provided by the donor members (Australia, France, New Zealand, USA and United Kingdom), with the remaining 9.4% from its 22 PIC members (Secretariat of the Pacific Community, 2000). Such reliance on core funding from Australia and New Zealand has at times created tension within the Forum, notably when seeking consensus positions on climate change not accepted by conservative Australian and New Zealand governments.

4.5 UN Recognition of Pacific Regional Organizations

4.5.1 Pacific Islands Forum

Because membership in the UN is reserved for nation-states, the UN's recognition of intergovernmental regional organizations came much later, and through initiatives from the regions more than from the UN itself. The UN Charter's Chap. 8 addresses "regional arrangements", but only in the context of peace and security issues, and only in the context of the Security Council. In 1987 the Pacific Islands Forum decided to investigate how it could best present its views internationally. A working group had identified nine strategies then in use, which commenced with the despatch of missions at ministerial and officials' level, through to attendance at and presentations to regional and international meetings, accrediting foreign country missions in the capitals of Forum countries, establishing forum groups to pursue regional positions, such as the Oceania Group on the Law of the Sea, and Forum Communiqués and press releases (SPF, 1987a).

The seven of the Forum's 13 members who were also members of the UN (Australia, New Zealand, Fiji, Papua New Guinea, Solomon Islands, Vanuatu, and Western Samoa) designated themselves as the "Forum UN Group" and undertook to keep the remaining Forum members updated on UN affairs three times annually. Key concerns were ensuring that New Caledonia remained on a path toward independence, that the Trusteeships in the North Pacific be terminated, and that the Pacific member states work together to maximise UNDP's financial allocation to PICs.

"Political emancipation and the assertion of a say in their own destiny have led the South Pacific countries to early recognition of the desirability of cooperation with each other," New Zealand PM Geoffrey Palmer informed UNGA in 1985, before adding "...Political independence in the Pacific, as elsewhere, is not automatically accompanied by economic self-determination, but acting together has strengthened the voice of the region in international forums" (United Nations General Assembly, 1985, pp. 69–70). In the years prior to formal recognition of the Pacific Islands Forum as a regional organization, the Forum UN members were already speaking with one voice. At the 46th General Assembly in 1991, for instance, a representative from the Federated States of Micronesia spoke on behalf of the nine Pacific UN member states to the Third Committee, on the topic of Narcotic Drugs (Pacific Islands Forum Secretariat, 1992). Statement by James T. Stovall III, Deputy Permanent Representative of the Federated States of Micronesia on behalf of the nine members of the South Pacific Forum which are also member states of the United Nations, in the Third Committee, 6 November)

Addressing the question of gaining observer status at the UN, the SPEC Committee on Regional Institutional Arrangements noted that under UNGA rule 14, a request that an organization be granted observer status could be made by any member state. The Commonwealth Secretariat had been granted observer status in 1976 following a motion put by Papua New Guinea, and the European Community had also gained observer status in this way (SPEC, 1987, p. 1). SPEC and its successor, PIFS, only gained legal existence by treaty in 1991. The "Forum" (i.e., the annual meeting of PIC leaders) continued to exist solely through ongoing goodwill and commitment amongst them, until a desire to gain observer status at the UN in 1994 forced the issue of establishing a treaty. In 2005 this treaty was reviewed and updated to state the Forum's purpose as being "...to strengthen regional cooperation and integration, including through the pooling of regional resources of governance and the alignment of policies, in order to further Forum members' shared goals of economic growth, sustainable development, good governance, and security" (PIF, 2005). The treaty clarified the PIF's status as an intergovernmental organization, but made no mention of the United Nations, apart from stating that it would be registered with the UN's Secretary-General (Alison, 2007).

Having obtained observer status at the UN in 1994, PIFS established communications with the New-York based Department of Political Affairs, and in Suva was in regular contact with the UN Resident Coordinator and UNDP Resident representative. Also in 1994, the region's leaders issued the *Suva Declaration on Sustainable Development,* and the NGO community issued the *Port Vila Declaration on Population and Sustainable Development*, as the Pacific region's contributions to the International Conference on Population and Development held in Cairo that year.

Observer status was only the first step toward closer engagement with the UN. In 2001 Kiribati President Teberuroro Tito, on behalf of the Pacific nations, asked the UN General Assembly to add to its agenda the subject of "cooperation between the

United Nations and the Pacific Islands Forum", stating "The Forum Island countries are committed to establishing this partnership in the interests of global peace and the prosperity of all peoples." This, together with other overtures, achieved their purpose. At UNGA's 57th Session on 15 October 2002, the UN Secretary-General tabled a 12-page report on "Cooperation between the United Nations and the Pacific Islands Forum", which outlined PIF collaborations with 16 UN agencies (United Nations. General Assembly, 2002). Recognition of the relationship between the UN and the Pacific Islands Forum was then consolidated in UNGA resolution 57/37 of 21 November 2002, which reads in part:

> The General Assembly
>
> ... *Recalling* that one of the purposes of the United Nations is to achieve international cooperation in addressing international problems of an economic, social, cultural or humanitarian character,
>
> *Bearing in mind* that one of the purposes of the Pacific Islands Forum, established in 1971, is to promote regional cooperation among its members through trade, investment, economic development and political and international affairs,
>
> *Welcoming* the ongoing efforts towards closer cooperation between the United Nations on the existence of regional arrangements or agencies for dealing with such matters relating to the maintenance of international peace and security as are appropriate for regional action and other activities consistent with the purposes and principles of the united Nations,
>
> ... *Affirming* the need to strengthen the cooperation that already exists between entities of the United Nations system and the Pacific Islands Forum in the areas of economic and social development, as well as in political and humanitarian affairs,
>
> Mindful of the need for the coordinated utilization of available resources to promote the common objectives of the two organizations.
>
> 1. *Takes note* of the report of the Secretary-General on cooperation between the United Nations and the Pacific Islands Forum, as well as his efforts to strengthen that cooperation;
> 2. *Invites* the Secretary-General of the United Nations to take the necessary measures, in consultation with the Secretary-General of the Pacific Islands Forum, to promote and expand cooperation and coordination between the secretariats of the two organizations in order to increase the capacity of these organizations to attain their common objectives;
> 3. *Recommends* that further consultations be undertaken by the secretariats of the Pacific Islands Forum and the United Nations to develop contacts and promote cooperative activities in a more structured, regular and transparent manner, including an examination of the option of formalizing cooperation and coordination between the two organizations in the future;
> 4. *Calls upon* the relevant United Nations bodies, in consultation with the Pacific Islands Forum, to implement the framework for cooperation in peace-building adopted at the fourth high-level meeting between the United Nations and regional organizations in 2001 by developing long-term peace-building programmes to address the security threats in the Pacific Islands Forum region;
> 5. *Urges* specialized agencies and other organizations and programmes of the United Nations system to cooperate with the Secretary-General of the United Nations in order to initiate, maintain and increase consultations and programmes with the Pacific Islands Forum and its associated institutions in the attainment of their objectives... (United Nations. General Assembly, 2003, pp. 2–3).

With this recognition of the UN-Forum relationship in place, the Forum introduced resolutions to UNGA in subsequent years to set out what PICs invited, welcomed, requested, called upon, even urged, the UN and the international community to do for the development of the Pacific small developing nations and region (United Nations General Assembly, 2006b). Content for these resolutions was prepared by the Pacific Islands Group in New York and coordinated by the Forum Chair. The Pacific Islands Forum also commenced attending sessions of ESCAP in Bangkok, Thailand.

UN departments and Pacific regional institutions settled into a steady relationship: regional support assisted UN agencies implement their global programs, and the UN agencies in turn were essential facilitators of global funding, expertise, and knowledge. The interplay between the Pacific region's support for the UN's global initiatives and its request for specific assistance is exemplified by Agenda item 108 (r) submitted to UNGA's sixty-first (2006) session (United Nations General Assembly, 2006a). (See appendix A) 2006 is also the year in which UN commenced observer status with PIF. It first officially participated in a Pacific Islands Forum meeting when ECOSOC's Executive Secretary attended the Forum Leader's meeting in Alofi, Niue, in August 2008.

Just when these clear lines of engagement were being established, relations between the United Nations, the Pacific Islands Forum, and individual Pacific member states of the UN became intertwined with political developments in Fiji. In December 2006 Fiji's Military, led by Commodore Bainimarama, removed from office the democratically elected government of Laisenia Qarase, and commenced rule by decree. When protracted political dialogue involving the Pacific Islands Forum, the Commonwealth Secretariat, and numerous other interested parties failed to restore democracy, the Forum suspended Fiji's membership and this action spurred Fiji to establish new relationships on the international stage, and to engineer changes in regional organizations. At the UN, Fiji and the other PICs distilled the "Pacific Small Islands Developing States" group from the existing "Forum UN member countries" in order to side-line Australia and New Zealand. Within the Pacific region, Fiji created the Pacific Islands Development Forum in an effort to side-line the Pacific Islands Forum (Wallis, 2016, p. 35).

Since 2007 the UN's relations with Pacific regional organizations and initiatives, particularly those based in *Fiji*, have thus incorporated the additional complexity of being mindful of Fijian sentiments. Two Secretaries-General have now visited the Pacific, but not Fiji. In 2011 UNSG Ban Ki-moon visited Solomon Islands and Kiribati and addressed the 42nd Pacific Islands Forum in Auckland. The Pacific leaders marked this first meeting between a UN Secretary General and PIF, four decades into the regional organization's existence, by issuing the *Waiheke Declaration on Sustainable Economic Development*, as well as a first joint statement with the UNSG reaffirming the importance of this relationship and noting the areas in which cooperation between the UN and Pacific states would support the national priorities identified by PIF Leaders and the BPOA, Mauritius Strategy, Istanbul Programme of Action, and the Pacific Plan:

> ...enhancing the productive capacity of Pacific SIDS; food security; renewable energy; disaster preparedness and risk reduction, and emergency relief and recovery; non-proliferation and disarmament, including controlling the illicit trade in small arms and light weapons and negotiation of an Arms Trade Treaty; addressing the ongoing impacts of nuclear testing in the Pacific; preventing and addressing the impacts of non-communicable diseases in the region; and promoting good governance, democracy, the rule of law and human rights, and addressing violence against women (Pacific Islands Forum Secretariat, 2011).

The statement provided, New Zealand representative to the UN Jim McLay suggested when thanking the Secretary-General for his first visit to the Pacific, "...a firm foundation for achieving a new level of cooperation and engagement between the United Nations and the Forum in terms of regular, high-level dialogue and practical cooperation and assistance tailored to the region's priority needs" (United Nations General Assembly, 2010, p. 10). Subsequent dialogue between the UN Secretary General and the Forum Leaders has centred on development issues as well as on administrative issues. A joint statement issued after a meeting on the margins of the 67th Session of UNGA in September 2012, for instance, acknowledged the need for "greater accountability and reporting of the support provided by the United Nations system to Pacific small island developing states" (United Nations. Secretary General, 2012).

Links between the Forum and the UN were given additional strength through the establishment of a Permanent Observer Office at the United Nations Office at Geneva (UNOG) in 2018.

4.5.2 *Pacific Community*

The Pacific Community appears to have taken a different approach to its engagement with the United Nations. Rather than attempting to establish a position as premier regional organization, it has focused on engagement with specific UN agencies. Thomas Smith, SPC's fourth Secretary-General (1954–1958), was enthusiastic about strengthening cooperation with the UN, particularly training programs. Although his successor visited WHO's regional headquarters in Bangkok in 1959 to strengthen the partnership between the two organizations, no basic agreement or formal framework was established, and cooperation continued on specific projects: health education courses, and surveys such as maternal and child health. Thereafter agreements appear to have been made on a program and project basis.

This failure by the Pacific Community to gain official status – or failure of the UN to establish recognition of this significant regional intergovernmental body - had consequences as spelt out in the UN's Independent Evaluation Office 2015 investigation into the contribution of Human Development Reports to public policy processes:

> An issue specific to the Pacific region (especially the small island countries) was inadequate coverage in the global HDRs. Due to lack of national statistical capacities and human resources in individual countries, some countries did not report data in prescribed formats

> for inclusion in the UN Statistics. Exacerbating this problem, UN agencies do not recognize data coming from the Secretariat of the Pacific Community, a regional agency tasked with compiling and supporting statistics in the Pacific. The exclusion is based on the grounds that the Secretariat is not a national statistical agency, and hence not the official national counterpart agency. The Secretariat was not ratified by the UN, even though it is maintaining a large statistical database for all countries in the region, supporting national agencies in data collection and is implementing a ten-year regional statistics programme in partnership with UN agencies. As a result, the rankings of Pacific countries were either missing or based on non-Secretariat approximations and projections. At the small island developing states conference in August 2014, the Pacific Forum resolved to ratify the Secretariat as their official data provider of the UN. It is hoped that future global HDRs will have more complete data on the Pacific countries (UNDP Programme. Independent Evaluation Office, 2015, p. 59).

In 2014 UNGA invited the Pacific Community to participate in its sessions as an observer, but SPC has for the most continued to focus on technical, social, and economic themes.

4.5.3 Pacific Islands Development Forum

The Pacific Islands Development Forum is the third Pacific inter-governmental organization to gain accreditation with the United Nations. Fiji had been under military rule since December 2006, and after political pressure was put on Vanuatu to cancel the annual MSG meeting in 2010 to prevent its leader from assuming the position of chair, Fiji convened in that year a regional meeting "Engaging with the Pacific" to explore "new modes of regionalism and new international partnerships" (Tarte, 2015, p. 80). The 2010 gathering laid the foundation of a 2012 meeting which endorsed the idea of establishing the PIDF, and an inaugural meeting in August 2013 was planned in collaboration with PIPSO, UNESCAP, IUCN, and the Solomon Islands. No summits were held in 2017 or 2018 and some commentators are skeptical about the organization's future viability: it has struggled to attract full political support from PICs, struggled to secure funding, and struggled to clarify the value it adds to the work of existing regional organizations (Aqorau & Batley, 2019). A summit and forum held in Fiji in July 2019 on the theme 'South – South Cooperation for a Resilient Pacific' was nonetheless attended by a number of UN agencies, including the UN Office for South-South Cooperation.

4.6 Discussion

This chapter has set out the origins of the Pacific's main regional organizations and explored the basis of their informal and formal relationships with the United Nations and its agencies and programmes. The Secretariat of the Pacific Community and the Pacific Islands Forum took different approaches their relationship with the UN –

with the former restricting itself to the social and economic development fields and the latter engaging in political dialogue as well as overall policy coordination. By entering into formal agreement with the Pacific Islands Forum the United Nations has given implicit recognition to the primacy of its position in overall policy coordination for the Pacific Island Countries, and as subsequent chapters will demonstrate, this partnership has become increasingly important to both sides in the pursuit of such global policy projects as addressing climate change, and pursuit of the sustainable development goals.

References

Alison, D. (2007). Moving Towards or Turning Away from Institutions? The Future of International organizations in Asia and the Pacific. *The Singapore year book of international law, 11*, 177.

Aqorau, T., & Batley, J. (2019). The Pacific Islands development forum: A shaky future? *Department of Pacific Affairs: In Brief, 8*.

Capiez, A., Templeton, A., Purvis, K., & Hoffmann, R. (2007). *Meeting House of the Pacific: The story of SPC 1947–2007*. Secretariat of the Pacific Community.

Dahl, A. L., & Baumgart, I. L. (1982). *Conference on the human environment in the South Pacific, Rarotonga, Cook Islands, 8–11 March 1982: Report*. UN.

Finau, P. P., Craw, C., Morgan, L. R., Crocombe, R. C., Hughes, A. V., Qionibaravi, M., & Spratt, R. G. (1976). *More effective aid: A report to the South Pacific Forum*. South Pacific Bureau for Economic Co-operation.

Fry, G. (2015). Recapturing the spirit of 1971: Towards a New Regional Political Settlement in the Pacific. *SSGM DISCUSSION PAPER, 3*.

Government of Papua New Guinea. (1987). Towards a single regional organization for the South Pacific: An overview of purposes, progress, and possible proceses for implementation of the proposal: Senior Officials Meeting on a Single Regional Organization, SPEC Headquarters, Suva 5 May 1987. Meeting papers. *(87)MP.12*.

Governments of Australia France Netherlands New Zealand United Kingdom and United States of America. (1947). *Agreement establishing the South Pacific Commission*. Canberra

Graham, K. (Ed.). (2008). *Models of regional governance for the Pacific: Sovereignty and the future architecture of regionalism*. University of Canterbury Press.

Hassall, G. (2021). *Government and Public Policy in the Pacific Islands*. Emerald Publishing Limited.

Malielegaoi, T.'e. S.'i. (2017). *Pālemia: Prime Minister Tuila'epa Sa'ilele Malielegaoi of Samoa, A Memoir*. Victoria University of Wellington Press.

Mara, K. R. S. (1997). *The Pacific way a memoir*. Center for Pacific Islands Studies, University of Hawai'i; Pacific Islands Development Program, East-West CenterUniversity of Hawai'i Press.

Morauta, M. (2013). *Pacific Plan Review 2013. Report to leaders. Volume 1*. Pacific Islands Forum Secretariat.

Pacific Islands Forum. (2005). *Agreement establishing the Pacific Islands Forum*. Pacific Islands Forum.

Pacific Islands Forum Secretariat. (1988a). Pre-Forum SPEC Committee Meeting, Nukualofa, Tonga 14–16 September 1988. Meeting papers. Folder 2. *(88) MP. 1 Folder 2*.

Pacific Islands Forum Secretariat. (1988b). Summary record. In Committee on Regional Institutional Arrangements Second Meeting, SPEC Headquarters, Suva 22–26 February 1988. Meeting papers. *(88) MP. 2*.

Pacific Islands Forum Secretariat. (1991a). Agreement establishing the South Pacific Forum Secretariat. *(91)NN.4.*

Pacific Islands Forum Secretariat. (1991b). South Pacific Organizations Coordinating Committee (SPOCC) Third Annual Meeting, Honiara 14–16 April 1991. Meeting papers. *(91) MP. 5.*

Pacific Islands Forum Secretariat. (1992). Forum Regional Security Committee Meeting, Forum Secretariat Headquarters, Suva 18–21 February 1992. Meeting papers. Folder 1. *(92) MP. 4 Folder 1.*

Pacific Islands Forum Secretariat. (2011). *Joint Statement of Pacific Islands Forum Leaders and UN Secretary General.* Pacific Islands Forum Secretariat.

Pacific Islands Forum Secretariat. (2017). *State of Pacific regionalism report 2017.* Pacific Islands Forum Secretariat.

Pacific Islands Forum Secretariat. (2019). *State of Pacific regionalism report 2019.* Pacific Islands Forum Secretariat.

Piddington, K. (1986). *South Pacific Forum: The first 15 years.* SPEC.

Pre-Forum Officials Committee, M. (1989). Pre-Forum Officials Committee Meeting, Tarawa, Kiribaiti 3–5 July 1989: [meeting papers]. *(89) MP. 1.*

Scollay, R. (2001). *Regional trade agreements and developing countries: The case of the Pacific Islands' proposed Free Trade Agreement* (Vol. 10). UN.

Secretariat of the Pacific Community. (1999). *The Canberra Agreement and Related Resolutions Adopted By Conference.* Secretariat of the Pacific Community.

Secretariat of the Pacific Community. (2000). Summary of Decisions from the Thirtieth Meeting of the Committee of Representatives of Governments and Administrations. Annex 2. *SPC Records*, 20–24 November 2000, 6p.

Smith, T. R. (1972). *South Pacific Commission. An analysis after twenty-five years.* Price Milburn for the New Zealand Institute of International Affairs.

SOPAC. (1987). Letter to SPF, 14 April 1987. Eighteenth South Pacific Forum, Apia, Western Samoa 29–30 July 1987 Meeting papers. Folder 2. *(87) MP. 2 Folder 2.*

South Pacific Bureau for Economic Cooperation. (1976). More effective aid: A report to the South Pacific Forum. *SPEC(76) 11(SPEC(76)11).*

South Pacific Bureau for Economic Cooperation. (1977). Report of the SPEC Committee Meeting, Port Moresby, Papua New Guinea, 24–26 August. *SPEC(77)41.*

South Pacific Bureau for Economic Cooperation. (1978a). Draft Agendia Item 5: South Pacific Environment Programme. SPEC Committee Meeting, Alofi, Niue 11–13 September 1978: [meeting papers]. *(78)MP.7.*

South Pacific Bureau for Economic Cooperation. (1978b). Opening Statement by Albert Henry. Summary record of the Fifth South Pacific Regional Telecommunications Meeting 4–10 July 1978 Rarotonga, Cook Islands. *SPEC(78)16*, 134p.

South Pacific Bureau for Economic Cooperation. (1979a). Regional Institutional Arrangements. SPEC Committee Meeting, Honiara, Solomon Islands 5–6 July 1979: [meeting papers]. *SPC (79) SCH.1 (Draft).*

South Pacific Bureau for Economic Cooperation. (1979b). SPEC Committee Meeting, Honiara, Solomon Islands 5–6 July 1979: [meeting papers]. *SPC(79) SCH.1 (Draft).*

South Pacific Bureau for Economic Cooperation. (1980a). South Pacific Provisional Regional Proposal for EEC Assitance for Trade Promotion Activities. ACP/EEC Pacific Group Ministerial Meeting on Regional Projects to be financed under the LOME II Convention, Apia, Western Samoa 28–29 August 1980: [meeting papers]. *(80) MP. 1.*

South Pacific Bureau for Economic Cooperation. (1980b). South Pacific Regional Trade and Economic Co-operation Agreement (SPARTECA). *(ND)NN.2 SPEC(81).*

South Pacific Bureau for Economic Cooperation. (1981a). Regional Committee on Trade First Meeting, Rarotonga, Cook Islands 9–10 July 1981: [meeting papers]. *(81)MP.3.*

South Pacific Bureau for Economic Cooperation. (1981b). Twelfth South Pacific Forum. Port Vila, Vanuatu, 10–11 August 1981. Summary Record. *SPF(81)REP.*

South Pacific Bureau for Economic Cooperation. (1982). Statement by Cruz A. Matos, Technical Secretariat, CCOP/SOPAC Regional Conference of Development Planners. SPEC Headquarters, Suva, 7–11 June. *(82) MP. 8.*

South Pacific Bureau for Economic Cooperation. (1985). *South Pacific Forum: The first 15 years.* South Pacific Bureau for Economic Cooperation.

South Pacific Bureau for Economic Cooperation. (1987). Committee on Regional Institutional Arrangements Inaugural Meeting, Suva 21–23 September 1987. Meeting papers. Agenda Item 7: Observer Status at the United Nations *(87) MP. 14.*

South Pacific Bureau for Economic Co-operation. (1973a). *Minutes of SPEC Committee Meeting, Suva 21–23 November.* South Pacific Bureau for Economic Co-operation.

South Pacific Bureau for Economic Co-operation. (1973b). *SPEC Committee Meeting, Suva 21-–23 November Draft Agenda.* South Pacific Bureau for Economic Co-operation.

South Pacific Bureau for Economic Co-operation. (1975). *Report of the South Pacific Regional Telecommunications Meeting 4–9 September 1975, Nuku'alofa Tonga.*

South Pacific Bureau for Economic Co-operation. (1981). *Twelfth South Pacific Forum, Port Vila, Vanuatu, 10–11 August 1981: South Pacific Regional Environment Progamme.* South Pacific Bureau for Economic Co-operation.

South Pacific Forum. (1971). South Pacific Forum. Wellington, 5–7 August, Summary record of proceedings. *(71) NN. 1.*

South Pacific Forum. (1972). *South Pacific Forum, Canberra, 23–25 February. Communique.* Pacific Islands Forum Secretariat.

South Pacific Forum. (1987a). Agenda Item 9: Presentation of Forum Views internationallally: Eighteenth South Pacific Forum, Apia, Western Samoa 29–30 July 1987. Annotated Agenda. *(87) MP. 2.*

South Pacific Forum. (1987b). Pre-Forum SPEC Committee Meeting, Apia 25–27 May 1987. Agenda item 5: Agreed conclusions. Eighteenth South Pacific Forum, Apia, Western Samoa 29–30 July 1987. Annotated Agenda. *(87) MP. 2.*

South Pacific Forum Secretariat. (1985a). Foreign Ministers' Committee on the Single Regional Organization Proposal - second Interim Report. Sixteenth South Pacific Forum, Rarotonga, Cook Islands 5–6 August 1985. Meeting papers. Folder 2. *(85) MP. 1(85) MP. 1.*

South Pacific Forum Secretariat. (1985b). Relationship between the South Pacific Forum and the Comittee for Co-ordination of Joint Prospecting for Milerals Resources in South Pacific Areas. Sixteenth South Pacific Forum, Rarotonga, Cook Islands 5–6 August 1985. Meeting papers. Folder 2. *(85) MP. 1((85) MP. 1).*

South Pacific Forum Secretariat. Leader's Meeting. (1971). *Communique.* South Pacific Forum Secretariat.

Tarte, S. (2015). A new pacific regional voice? The Pacific Islands Development Forum. In G. Fry & S. Tarte (Eds.), *The new pacific diplomacy* (pp. 79–88). ANU Press.

Tavola, K., Baaro, M., Bogari, L., Mangelinan, L., Simcock, A., & Tuioti, E. (2006). *Reforming the pacific regional institutional framework.* Pacific Islands Forum Secretariat.

The Pacific Community. (2020). *Fiftieth meeting of the committee of representatives of governments and administraions. Working Paper. Agenda item no. 4: Pacific Community Strategic Plan 2021+ and results reporting.* The Pacific Community.

The Pacific Community. (2021). *Pacific community strategic plan 2021+ planning for our future our SPC our blue pacific.*

UNCTAD. (1974). *Report of the United Nations preliminary mission to the South Pacific Bureau for Economic Co-operation.* United Nations Conference on Trade and Development.

United Nations Development Programme. Independent Evaluation Office. (2015). *Evaluation of the contribution of the global and regional human development reports to public policy processes.* United Nations Development Programme.

United Nations General Assembly. (1985). *Provisional Verbatim record of the 7th meeting: Held on Tuesday 24 September 1985, New York, General Assembly, 40th session.* UN.

United Nations General Assembly. (2006a). *Cooperation between the United Nations and regional and other organizations: Report of the Secretary-General.* UN.

United Nations General Assembly. (2006b). Cooperation between the United Nations and the Pacific Islands Forum: Draft resolution. In *A/61/L.20*. United Nations.

United Nations General Assembly. (2010). *General Assembly official records, 65th session: 63rd plenary meeting, Monday, 13 December 2010*. UN.

United Nations. General Assembly. (2002). *Cooperation between the United Nations and the Pacific Islands Forum: Report of the Secretary-General*. UN.

United Nations. General Assembly. (2003). *Cooperation between the United Nations and the Pacific Islands Forum: Resolution*. UN.

United Nations. Secretary General. (2012). *Joint Statement by Leaders of Pacific Islands Forum*. UN Secretary-General.

Wallis, J. (2016). Is oceania a regional security complex? In A. Holtz, M. Kowasch, & O. Hasenkamp (Eds.), *A region in transition politics and power in the Pacific Island Countries*. Saarland University Press.

Chapter 5
UN Offices in the Pacific

The need for representation in such a large diverse number of small island countries as exist across the Pacific region has been an ongoing challenge for the UN. We have suggested that prior to the 1980s, UN agencies and programs engaged with Pacific Island governments and regional organizations on a sectoral basis and with less concern for how their activities related to those of other UN and non-UN development partners. Their goal during this period was to maximize the impact of their respective programs in hopes of ensuring their continuance or expansion. Prior to 1984, a Pacific Liaison Office was based in Nauru and a United Nations Development Advisory Team based in Fiji. These were combined as the ESCAP Pacific Operations Centre (EPOC) and relocated to Port Vila between 1984 and 2005, before being relocated to Suva and renamed the ESCAP Subregional Office. Country offices were established in three locations: (1) Samoa (soon after Samoan independence in 1962, and covering also Cook Islands, Niue, and Tokelau); (2) Fiji (covering ten countries: Fiji, the Federated States of Micronesia, Kiribati, Marshall Islands, Nauru, Palau, Solomon Islands, Tonga, Tuvalu and Vanuatu); and (3) Papua New Guinea.

By the 1990s some 26 UN Funds and Programmes had established a presence in the region – 16 in Fiji and 10 in Samoa. Rationalization through the establishment of multi-country offices, particularly in Suva, improved efficiency, coherence, and access to the Pacific's regional organizations and other international organizations, but at the cost of in-country representation elsewhere. UN Bodies located in Suva include:

1. United Nations Resident Coordination (UNRC).
2. United Nations Development Programme (UNDP).
3. United Nations Children's Fund (UNICEF).
4. International Strategy for Disaster Reduction (UNISDR).
5. United Nations World Health Organization (WHO).
6. International Labour Organization (ILO).
7. United Nations Population Fund (UNFPA).

G. Hassall, *The United Nations and the Pacific Islands*, United Nations University Series on Regionalism 24, https://doi.org/10.1007/978-3-031-34155-7_5

8. United Nations Women/Entity for Gender Equality and the Empowerment of Women (UN Women).
9. Joint United Nations Programme on HIV/AIDS (UNAIDS).
10. Office of the UN High Commissioner for Human Rights (UNOHCHR).
11. UN-EPOC (United Nations ESCAP Pacific Operations Centre).
12. UN Office for the Coordination of Humanitarian Affairs (UNOCHA).
13. United Nations Department of Safety and Security (UNDSS).
14. United Nations Volunteers (UNV).
15. World Meteorological Organization (WMO).
16. International Fund for Agricultural Development (IFAD).
17. United Nations Agency for Human Settlements Program (UNHABITAT).
18. International Monetary Fund and Pacific Financial Technical Assistance Centre (IMF/PFTAC),
19. United Nations Environment Programme (UNEP).
20. UN Refugee Agency (UNHCR).
21. International Agency for Migration (OIM).
22. United Nations Educational, Scientific and Cultural Organization (UNESCO).
23. United Nations Capital Development Fund.

Many of these UN bodies also maintain offices in Port Moresby and Apia, and a few are also present in such smaller capitals as Honiara and Port Vila. Non-resident agencies (IAEA in Geneva),, UN Environment, UN-Habitat (Fukuoka), UNIDO, UNCTAD, and UNODC (Bangkok)) implement projects and programs in the region in partnership with resident agencies. Through the 2000s the General Assembly supported the expansion of the UN's field presence in the Pacific Islands. A 2006 resolution included at its article 16:

> *Welcomes*, in this regard, the endorsement by the Secretary-General of the establishment of an expanded and joint in-country presence of the United Nations in Kiribati, the Marshall Islands, Micronesia (Federated States of), Nauru, Palau, Solomon Islands, Tuvalu and Vanuatu (United Nations. General Assembly, 2007).

Pacific leaders have made clear their desire for a stronger physical UN presence. The Federated States of Micronesia's President Urusemal, for instance, when addressing a High-Level Plenary Meeting of the 60th Session of the United Nations General Assembly in 2005, stated:

> While grateful for the assistance of the United Nations in our national development efforts, we recognize the need for further strengthening of cooperation and closer linkages between the United Nations and many of its member states in the Pacific, like my country. I refer here to the need to establish a physical presence of the United Nations system within our national borders. The United Nations system must be seen as directly involved in our development process and in our national efforts at achieving the Millennium Development Goals, on a sustained and regular basis. Such presence can only contribute to better coordination, strengthening our capacity, a more effective use of resources and an effective delivery of assistance. It would demonstrate the UN system's commitment to its most vulnerable members" (Federated States of Micronesia, 2005).

A 2019 review of multi-country offices recommended opening an additional office, in the North Pacific, and this was established in Pohnpei, Federates States of Micronesia, in 2021, to service Palau, Marshall Islands, Nauru and Kiribati. This constitutes a fourth UN centre in the Pacific region, alongside those in Fiji, Samoa, and Papua New Guinea.

5.1 Early Engagement

The following section introduces ten UN agencies engaged with the Pacific region by the 1970s: UNDAT, UNDP, ITU, ESCAP, UNEP, WHO and FAO. Each had unique governance systems, and individual work programs across diverse sectors. A Host of additional UN entities that commenced their activities slightly later are also described in brief.

5.1.1 United Nations Development Advisory Team

A United Nations Development Advisory Team (UNDAT) was established in 1950 but only commenced South Pacific operations in 1971, as part of ESCAP's Suva office. UNDAT's three initial reports addressed Intra-regional Trade in Cement; a Preliminary Review or the Prospects or Intra-regional Trade in Fruits and Vegetables; and Trade Development of South Pacific Handicrafts. In 1976 the work of UNDAT was deemed so vital to regional development planning that SPEC wrote to the UNSG requesting that funding be continued during 1976 (this was facilitated by the UN drawing UNDAT funds from UNDP's 1977 budget) (South Pacific Bureau for Economic Co-operation, 1976b, p. 3). SPEC was pleased with UNDAT's work but concerned about the high cost of maintaining UNDP experts (USD 200,000 in 1977), a cost which reduced the funds available for existing regional projects (SPEC, 1976, p. 3). By 1978 UNDAT had assisted most PICs, as well as SPEC, SPC, and USP. SPEC's relations with UNDAT had been particularly fruitful and three reports had been jointly prepared. UNDATs continued to function for specific purposes in later periods, although with less prominence, due to the arrival of more formalized UN programmes and agencies.

5.1.2 United Nations Development Programme (UNDP)

The United Nations Development Program, established by UNGA in 1966, rapidly became the largest and most pervasive presence of the UN across the Pacific Islands. Its broad "development" mandate has resulted in its engagement in every significant sector of government activity. In 1971 the UNDP Governing Council shifted support from individual projects to integrated 'Country Programmes' of between 3 and 7 years (see Chap. 3). This shift to medium-term programming required close collaboration with planning and finance departments in individual countries, and also required improvements in forward budgeting for personnel, training, equipment, and government capital inputs. UNDP funds for technical assistance development activities globally were sourced from an annual Pledging Conference for UN member-states. These were then allocated to recipient countries and regions by UNDP's Governing Council, whilst allocations within countries were determined through consultation between that country and the UN representative, who was also responsible to decision-makers in New York. UNDP provides funds for its own programs, for programs in collaboration with member-countries, with other UN agencies, and for programs of organizations outside the UN framework, such as the International Civil Aviation Organization (ICAO) and the European Economic Community (the EU).

In 1973 there were just six UNDP experts in the region (Pacific Islands Forum Secretariat, 1973, p. 23). By 1975 Fiji agreed to a request from UNDP that it be permitted to establish a secretariat in Suva, but indicated that it would not provide facilities free of cost and agreed, in turn, to raise its annual contribution to UNDP from USD 3000 to USD 15,000 (South Pacific Forum, 1975, p. 19; South Pacific Bureau for Economic Co-operation, 1975). By 1978 there were 110 UNDP sponsored projects in the Pacific, including 104 country projects and 6 regional projects (South Pacific Bureau for Economic Co-operation, 1978a, p. 65). Funding extended to infrastructure projects, such as Western Samoa's Faleolo Airport, and telecommunications equipment for Papua New Guinea and Fiji. By 1988 UNDP was one of the Pacific region's key dialogue partners, together with Japan, the US, Canada, France, the UK, and the Asian Development Bank. It was funding regional research and policy in petroleum, civil aviation, and shipping. Activities under UNDP auspices continued to expand in subsequent decades.[1]

5.1.3 World Health Organization (WHO)

The World Health Organization epitomizes a global agency whose knowledge of a particular field (in this case, obviously – health) is essential to improving the quality of life of all Pacific Islanders. The organization was established by UNGA in 1948 through amalgamating a number of existing international health organizations, and a Regional Office for the Western Pacific established in 1951. Within that structure an Office for the South Pacific was established in Sydney in 1956 and transferred to Suva in 1965. WHO's membership is not limited to sovereign nations (health has no regard for political boundaries!), and by 2019 fourteen dependent and independent

[1] https://pacific.un.org/; https://papuanewguinea.un.org/; https://samoa.un.org/

Table 5.1 Pacific Island Country date of entry into WHO membership

Country	WHO membership (by date of entry)
Samoa	16 May 1962
Fiji	1 January 1972
Tonga	14 August 1975
Papua New Guinea	29 April 1976
Vanuatu	7 March 1983
Solomon Islands	4 April 1983
Cook Islands	9 May 1984
Kiribati	26 July 1984
Marshall Islands	5 June 1991
Federated States of Micronesia	14 August 1991
Tuvalu	9 May 1993
Niue	4 May 1994
Nauru	9 May 1994
Palau	9 March 1995

countries had become members, (World Health Organization, 2020, pp. 227–231) as depicted in Table 5.1.

At global level, the organization's decisions are taken at an annual World Health Assembly. Through experience, WHO has identified four core governance functions that provide a sense-making framework for any health system: financing, resource generation, service delivery, and stewardship. Each of these core functions has numerous additional elements: the six sub-functions of Stewardship, for instance are (1) generating intelligence; (2) formulating strategic policy direction; (3) ensuring tools for implementation: powers, incentives and sanctions; (4) building coalitions and partnerships; (5) ensuring a fit between policy objectives and organizational structure and culture; and (6) ensuring accountability. Such guiding principles would no doubt be useful for any organization – particularly if they inform its practice!

Given the extent of WHO's work in PICs, it is simply not possible to attempt a comprehensive survey of WHO's interaction with PICs and Pacific regional organizations, apart from noting some key themes. The significant challenge confronting WHO as the capable "agent", and PICs as "principals" seeking its assistance, has been how to best convert knowledge into effective policies and services. WHO's global goals include building "Universal Health Coverage" (i.e., ensuring that everyone has equal access to quality health services), building healthy environments (which includes quality of water, sanitation, and health promotion), eliminating or controlling disease (such as through immunization programs), and responding to health emergencies (such as disasters or pandemics). All these conditions apply to the Pacific region, where WHO engages with PICs at a political level with (1) Ministers for Health, and administratively with (2) Health departments, regarding national health policies and health governance. WHO also deals with (3) regional organizations; (4) the health professions, particularly concerning training standards and staff procurement and retention (Buchan, Connell, & Rumsey, 2011); and (5) the public, particularly through awareness campaigns about non-communicable diseases – combatting high levels of diabetes, hypertension, obesity, smoking and alcohol consumption (in 2011 the Pacific Health ministers issued a "Honiara communique on the Pacific Non-Communicable Disease Crisis").

A few specific examples will illustrate WHO's engagement at regional level. In July 1979 a SPEC committee meeting in Honiara noted a WHO//SPEC Task Force Report's recommendations covering such matters as the bulk purchasing of drugs, central warehousing, quality assurance, laboratory testing, and a drug information service. However, the committee also noted the "significant practical and geographic implications" of the proposals, the technical details of which were yet to be considered and resolved (SPEC, 1979e, p. 3). These remain critical issues four decades on.

Other projects have proven equally challenging. In 1996 WHO partnered with SPC to establish the voluntary 22-member Pacific Public Health Surveillance Network.[2] WHO also introduced the global "National Health Accounts" project to the Pacific in 2007. Pilot projects in Fiji, Federated States of Micronesia and Vanuatu revealed the complexity of this exercise (World Health Organization. Regional Office for the Western Pacific, 2010), with few subsequent completions to date. The lack of established information sharing mechanisms even within Ministries reflects a great need for strengthening this domain.

Since 1995 WHO has convened biennial meetings of PIC Ministers of Health. In 2017 the Twelfth such meeting endorsed a *Health Islands Monitoring Framework* and noted the need for a "Pacific Health Information Network (PHIN) to strengthen their Health Information Systems. This was established in 2006 to strengthen handling of digital health data (Pacific Health Information Network, World Health Organization, & Pacific Community, 2018). WHO quickly capitalized on these decisions by convening experts on e-health to finalize a *Regional Action Agenda on Harnessing E-Health for Improved Service Delivery* and to consult on progress with the monitoring framework (World Health Organization. Regional Office for the Western Pacific, 2018). The 13th meeting, held in French Polynesia in 2019, consulted on strengthening primary health care systems, universal health care, the Pacific Islands Action Plan on Climate Change and Health, health security, non-communicable diseases, immunization, the WASH program, human resources for health, and e-health and health information systems.

5.1.4 International Telecommunications Union

The International Telecommunications Union (ITU) has played a central role in establishing technical standards and providing both policy and technical support, but its extensive contribution to the Pacific Region's telecommunications infrastructure is not known to the general public, and is quite possibly also unfamiliar to the

[2] https://www.pphsn.net/

contemporary technical community (Purcell & Hassall, 2017). In the 1970s ITU proposed establishing a regional telecommunications training centre (South Pacific Bureau for Economic Co-operation, 1973b, p. 10). In 1974 UNDP approved assistance for a preparatory phase that focused on collecting traffic data. During this era telecommunications covered telephone, telegraph, and telex services. Funds were also approved in 1973 for a maritime telecommunications coordinator (although this project was terminated by 1976) (South Pacific Bureau for Economic Co-operation, 1976a, p. 7).

In 1978 the ITU noted South Pacific telecommunications assistance was "very well organized under the auspices of SPEC." It was considering appointing a Regional Adviser, but feeling at the 6th regional telecommunications meeting in Wellington in November 1979 was that the provision of an advisor, plus facilities, was not cost effective and that a strategy of engaging specialists from time to time as the need arose was preferable (SPEC, 1979c, p. 10).

ITU also assisted with assessing tariff structures. Each country has the sovereign right to determine its own rating policies, but there is also benefit in being aware of tariffs being set by one's neighbours. Too much imbalance (i.e., making charges too high) can lead to a loss of revenue due to callers in high tariff countries requesting those they want to speak with to "call back" through cheaper services (SPEC, 1979c, p. 13).

An ITU report in 1983 identified growth in Pacific regionalism as a consequence of working together to meet telecommunications needs:

> Because of the diverse nature of the region and the realities of the political situation, strong regional commonality of approach is not often very discernible, even though an underlying current exists. On the other hand, the past few years have witnessed an increasing sub-regional cooperation on a broad front. So far some well defined country groupings have emerged namely ASEAN, Pacific Island developing countries and the LDC group, whose influence was strongly felt in the prolonged consultative of UNDP with the Governments, regional commissions, intergovernmental and sub-regional groupings and the agencies…" (South Pacific Regional Telecommunications, 1982, p. 2).

The work of ITU in the Pacific region gained impetus through projects associated with the World Summit for the Information Society (its conferences in 2003 and 2005, and its post-conference monitoring as explained in Chap. 9), then with the MDGs and the SDGs (Purcell & Hassall, 2017).

5.1.5 United Nations Economic and Social Commission for Asia and the Pacific (ESCAP)

The Economic Commission for Asia and the Far East (ECAFE) was established in 1947 to assist the social and economic development of, at that time, 32 members and nine associate member countries in a region that extended from Iran in the west to the Cook Islands in the east. At its 1974 annual session the Commission changed its name from ECAFE to ESCAP (Economic and Social Commission for Asia and

Pacific) to incorporate the full scope of its work in social and economic development and to more accurately depict its geographic coverage. In 1978 Cook Islands Premier Sir Albert Henry claimed some credit for the more inclusive name, telling a regional telecommunications meeting:

> Up to now the regional system has gone a long way because of the work of the Bureau [SPEC] and the assistance of Australia and New Zealand which we are very thankful for, and the assistance of the other agencies, such as United Nations agencies. It would not be fair for me to name one agency. The UNDP of course has been the most prominent in the Pacific. But the United Nations has other agencies that have also assisted us such as the once upon a time ECAFE, which meant Economic Commission for Asia and the Far East. I think the Cook Islands delegation suggested at Sri Lanka in 1974 that the name of that organization might be changed. Why go Far East? Why not go Pacific? So the name was changed to the Economic Commission for Asia and the Pacific to become ESCAP. We are very thankful for that agency in the Cook Islands and for the assistance that it has given us (SPEC, 1978b, p. 67).

ESCAP's decision to increase its work in the Pacific was welcomed at the Forum Leader's meeting in Niue in September 1978. Initial areas were the newly established regional environment program, shipping, and industrial development. A first joint UNDP/ESCAP/SPC/SPEC Senior Planning Officials' meeting was held at SPEC Headquarters in February 1979 to review inter-country planning for the period 1979–1981, including review of nine existing and 25 projected projects around the region (SPEC, 1979a, p. 1; SPEC, 1979d, p. 28). USD 12 million was allocated for a five-year programming cycle 1977–1981, and a meeting was to be held in 1981 to consider a further cycle for the period 1982–1986 (SPEC, 1979b, p. 1). ESCAP sought to cultivate close relationships with the Pacific regional agencies. It assisted SPEC investigate the establishment of a regional Shipping line. It increased collaboration with SPEC in regional environmental, shipping, and industrial development projects, and the Director of SPEC was appointed to ESCAP's Board.

However, ESCAP struggled to secure its position with the PICs, and possibly also with other UN entities in the region. On several occasions in the 1970s SPEC refused ESCAP's request to host an office at its newly built headquarters in Suva (SPEC acceded to a request from the Commonwealth Secretariat to host an office and felt that it could not in future continue to decline a request from ESCAP (South Pacific Bureau for Economic Co-operation, 1978b, pp. 10–11). SPEC's view was that whereas it had excellent working relations with ESCAP, it doubted that this relationship could be improved through the establishment of such an office (SPEC, 1975, p. 6). Perhaps because of this resistance, ESCAP's Pacific Operations Centre operated from Port Vila, Vanuatu, from 1984 until a 2002 review that considered the location not cost effective resulted in a shift to Suva (2003, p. 4). In 2000 the Vanuatu Office oversaw 53 advisory missions to 10 PICs on corporate planning, strategic planning, social development, development planning, economic reform, regional planning, and youth policy, banking, horticulture, privatization, governance, and food technology (United Nations. Economic and Social Council, 2001, p. 19). In 2002 ESCAP adopted a work program with three major themes: poverty-reduction, managing globalization, and emerging social issues. It also established a "Special Body on Pacific Island Developing Countries", which met in Bangkok simultaneously with the Commission's sessions. In March 2008 the Forum and ESCAP co-organized a meeting of sub-regional organizations "to strengthen modalities for cooperation to address inter-subregional disparities." (United Nations. Economic and Social Commission for Asia and the Pacific, 2008, p. 5) and through these and other initiatives ensured that the small states of the Pacific gained ongoing value from their engagement with ESCAP.

By 2022 ESCAP's Pacific programme was spread across eight sectors, including continued work in trade, transport, energy, and national planning, but also now including ICTs, environment and development, statistics, and social development.[3] Although the prospect of duplication between the activities of ESCAP and other development agencies remains – including replication of projects by other UN agencies - ESCAP's programmes bring the PICs into contact with the practices of Asian countries, and its Pacific regional reports make important contributions to global assessment processes, such as those associated with Further Implementation of the Barbados Program of Action for Sustainable Development of SIDs (United Nations. Economic and Social Commission for Asia and the Pacific, 2010).

5.1.6 Food and Agriculture Organization (FAO)

The Food and Agriculture Organization (FAO) was established as an autonomous specialised agency of the United Nations in October 1945, focused mostly on agriculture, forestry, fisheries, nutrition, commodities, economic and social activities, and investments, and aligned with other agencies through the Economic and Social Council. In 1951 FAO's headquarters were transferred from Washington, D.C., to Rome, and an Asia Pacific Office was established in Bangkok. By the 1970s FAO had assisted the South Pacific in agriculture, forestry, fisheries, and food and nutrition. It completed the UNDP/SPC Control of the Rhinoceros Beetle Project technical assistance to the College of Tropical Agriculture in Western Samoa. A UNDP/SPEC survey of Agricultural Pests and Diseases commenced which in July 1974 and shifted from Apia to Suva in June 1975 presented an interim report in March 1976 (United Nations. FAO & UNDP, 1976). FAO provided a Regional Fisheries coordinator, and from 1977 a project on Root crop Development. In 1980 FAO surveyed the coconut industry in Fiji and reported serious concerns about the poor quality of the stock and its management, leading to doubts about the viability of the industry (SPEC, 1982). These are just some of FAO's initial projects in the Pacific, which laid the foundations of institutional relationships that resulted in more far-reaching programmes in later decades.

---

[3] https://www.unescap.org/subregional-office/pacific#

The World Food Programme, established by FAO in 1962 to provide food aid around the globe as required, has assisted PICs since the 1970s, including the Seaqaqa sugar cane production and settlement scheme which commenced in Fiji in 1976 (SPEC, 1978a, p. 55). In 2019 the Food Programme launched its first Pacific regional programme to address food security vulnerabilities resulting from natural disasters, and in response to SDGs concerning environmental protection, gender equality, sustainable management of natural resources, biodiversity conservation, and strengthened resilience to climate change impacts and disasters (World Food Programme, 2019). In 2017 FAO issued a multi-country framework for the Pacific Islands for the period 2018–2022, focused in improving the production and marketing of domestic agriculture products and the healthy consumption of safe and nutritious food (Food and Agriculture Organization of the United Nations, 2017).

5.1.7 United Nations Population Fund (UNFPA)

The United Nations Population Fund was established in 1969 to assist in promoting awareness of social and economic implications of population problems and had a coordinator based in Suva by 1971. The fund's projects expanded to include census/demography/statistics, maternal and child health and family planning, population education, information and communication, research on relationships between population and development, research on human reproduction, and evaluation of population projects (SPEC, 1978a, p. 71) During 1981–1982 SPC worked with UNFPA to assist PICs and other organizations develop policies strategies and programs to strengthen women's' roles in national development. Measures included improving research and data collection, program evaluation, strengthening the SPC community education training centre in Suva (UNDP, 1980, p. 2). All such projects were planned in five-year cycles, the third being for 2003–2007, the fourth for 2008–2012.

UNFPA has also focused on educating leaders about population issues. It has, for instance, collaborated with SPC in convening a regional network and Conference for Pacific Parliamentarians to strengthen their understanding of population and development, reproductive health and gender issues. For a number of years the network, the Pacific Parliamentary Assembly on Population and Development (PPAPD), was joined to a second network, the Forum Presiding Officers and Clerks Conference (FPOCC), in an effort to build ownership of these issues by Pacific parliamentarians. Ministers for Health were a particularly important target audience for UNFPA which sought to educate them about HIV/STI and "reproductive Health Commodity Security", and the importance of ensuring that their countries had adequate supplies of contraceptives and socially appropriate means for their distribution. For the period 2018–2022 UNFPA's sub-regional programme for the Pacific projects USD 30 million on its three focus areas: sexual and reproductive health, gender equality and women's empowerment, and population dynamics (United Nations. Executive

Board of the United Nations Development Program, United Nations. Population Fund, & Services, 2017).

5.1.8 United Nations Environment Programme (UNEP)

The United Nations Environment Programme (UNEP) was established following the United Nations Conference on the Human Environment in Stockholm in June 1972, and through the initiative of Dr Arthur Dahl, who had attended the Stockholm Conference and who was in 1974 appointed to the position of Regional Ecological Advisor to SPC. Dahl visited UNEP in Nairobi to establish relations on behalf of SPC and UNEP's director Maurice Strong subsequently visited Noumea. Strong referred to this new Pacific Islands collaboration in his address to the Pacific Science Congress in Vancouver in August 1975 and in October SPC made a formal request to UNEP for assistance with developing a comprehensive programme for environmental management in the region (Johnson 2012).

A preliminary proposal submitted to UNEP in March 1976 received a favourable response. With the endorsement of the South Pacific Forum and the South Pacific Conference, SPC and SPEC jointly planned a five-year Comprehensive Environment Management Programme (CEMP) (SPEC, 1978a, p. 68). UNEP indicated willingness to help finance part of the programme. It funded, for instance, a 1976 Regional Ecosystems Survey conducted by SPC's Regional Ecological Advisor, and made available two experts to the region to evaluate the SPC's environment programme.

UNEP had also been approached by Papua New Guinea to carry out an environmental study of the Purari project in Papua New Guinea and to fund an environmental audit of the Bougainville copper project. It had also provided US$14,500 to the Papua New Guinea Government for the 1975 Waigani Seminar and the Wood Preservation Seminar (US$23,737). By 1988 UNEP was funding 41 of SPREP's 111 activities (Pacific Islands Forum Secretariat, 1988). Its presence in the region continues through an Asia Pacific office in Bangkok, and through collaboration with SPREP.

5.1.9 United Nations Conference on Trade and Development

In the important trade sector the Pacific region has engaged with the UN Conference on Trade and Development, a platform for countries to promote their integration into the world economy. Although no PICs attended UNCTAD's first meeting in Geneva in 1964, at which seventy-seven countries signed a joint declaration and established the network known at G77, a majority joined in later years (Fiji, Federated States of Micronesia, Marshall Islands, Nauru, Papua New Guinea, Samoa, Solomon Islands,

Tonga and Vanuatu). Palau joined in 2002 but exited in 2004, claiming that UNCTAD had not provided any demonstrable incentives to membership.

UNCTAD convenes a quadrennial meeting of members to renew the mandate of the organization through an outcome document that is negotiated by member-states' delegates. Soon after the establishment of SPEC its director was being invited into UNCTAD programs, such as a 1973 panel to "study particular problems of developing Island countries, to make recommendations and to prepare a report to the Trade and Development Board" (South Pacific Bureau for Economic Co-operation, 1973a, p. 10). Another regional study undertaken at about the same time, a UN/UNCTAD South Pacific Regional Transport Survey, was of such scale as to require coordinated response by SPEC and SPC (the former having economists and the latter statisticians). In the 1970s UNCTAD assisted SPEC with studies in trade expansion, a mobile training programme for professional development in trade promotion for governments and business community personnel, and an integration study for Micronesia (Trust Territories of Pacific Islands). A UNDP/FAO/SPEC regional plant pests and diseases survey required cooperation of the efforts of the SPC plant protection officer under SPEC coordination. A meeting between the two agencies in April 1975 agreed "that SPC was involved in economic and statistical services, agriculture, including livestock, fisheries and forestry. SPEC had been particularly interested in telecommunications....transportation, particularly shipping..." (*Special Meeting Between SPEC Representatives and SPC Staff 20–21 May 1974*, 1974)

In 1985 the Forum had raised with the UN the issue of attending to the special circumstances of small island countries and received a reply that summarized an UNCTAD report summarizing the various piecemeal initiatives underway to that point in time. Australia and New Zealand, the region's major bilateral contributors, were assisting the Pacific Forum Line and the matter of trans-shipment problems. UNCTAD had been requested to continue its program of in-depth studies of common problems of island economies (Pacific Islands Forum. Standing Committee of Officials on the Decolonisation of New Caledonia, 1985 Item 14). In 1990 UNCTAD financed a meeting of the Regional Committee on Trade to facilitate consultation the Uruguay Round of the General Agreement on Tariffs and Trade negotiations.

In 2012 PIFS became an observer organization and participated in UNCTAD XIII in Doha, Qatar, the theme for which was 'Development–Centred Globalisation: Towards Inclusive and Sustainable Growth and Development'. Pacific participants included Samoa; Solomon Islands; Tuvalu, Vanuatu Papua New Guinea, and the Pacific Islands Forum Secretariat (as an observer organization). Associated events included a World Investment Forum, a Women in Development Forum, and a Civil Society Forum. In his statement at the General Debate, the Deputy Prime Minister of Samoa expressed his appreciation for the work and support offered by UNCTAD to date and encouraged further future engagement:

> Samoa has also been vigilant in advancing critical mandates identified and agreed upon within UNCTAD and deemed critical for integration of economies into the global economy.

> Our recent accession to the WTO in December 2011 demonstrates that, Samoa has delivered on UNCTAD's overarching mandate.

At a high-level forum held as part of UNCTAD XIII, Vanuatu's Ambassador to the European Union outlined Vanuatu's experience with respect to the issue of graduation from Least Developed Country (LDC) status. UNCTAD XIII's outcome document, "The Doha Mandate", recognised the trade and development challenges of SIDS on issues of climate change, attracting investment and transport and trade logistics and reaffirmed UNCTAD's role in the implementation of the Barbados Programme of Action and the Mauritius Strategy for the Sustainable Development of SIDS. Papua New Guinea has been a member of UNCTAD's Trade and Development Board.

5.1.10 United Nations Educational, Scientific and Cultural Organization (UNESCO)

The United Nations Educational, Scientific and Cultural Organization was established in 1946 (succeeding the International Institute of Intellectual Cooperation – IIIC – established in 1922) to strengthen the "intellectual and moral solidarity of humankind" for the achievement of peace. UNESCO established a Pacific Office in Apia in 1984 with six PICs as members. By 1995 all PICs were members of the Organization, and each has established a National Commission recognized in law and comprising government and non-government members. UNESCO has a surprisingly wide mandate. It has, for instance, promoted the concept of "cultures of peace" across the region. In Education, it has promoted literacy and numeracy from pre-school to tertiary levels and supports the "Education for All" and "Information for All" global campaigns. In 1977 it convened the first meeting of the Pacific Heads of Education System (PHES) network, and in 2013 established the Pacific Islands Universities Research Network (PIURN).[4] In the field of ICTs UNESCO has promoted the preservation of cultural heritage and indigenous knowledge. In 2016 UNESCO launched a *Plan of Action and Implementation Strategy for the Small Island Developing States (SIDS*), and in 2018 a *Pacific Strategy 2018–2022* – both of which are closely aligned with the SDGs.

5.2 Later Engagement

At least ten additional UN agencies have subsequently established a presence in the Pacific region in addition to the ten profiled in the previous section. Each has its own format for membership, governance structure and mode of engagement with

[4] https://piurn.org/

government and civil society, and most have expanded over time to include all PICs in the region.

5.2.1 International Labour Organization

In 1975 the International Labour Organization established its office in Suva at the invitation of the Fijian government. The ILO has evolved a unique system of governance that includes government, workers and employers in a tripartite consultative system to find agreement on the implementation of labour standards in each member country. It has established eight fundamental conventions that member countries are expected to implement (covering freedom of association, the right to organize, minimum age, child labour, equal remuneration, and discrimination) as well as approximately 200 additional conventions and 200 non-binding recommendations.

ILO's conventions collectively known as the "international Labour code" provide the foundation for programs in its Pacific Member countries. By the 1970s ILO advisors were working in Fiji, Papua New Guinea, Solomon Islands, Western Samoa, and the Trust Territory of the Pacific Islands (SPEC, 1978a, p. 51). A first meeting of Pacific labour ministers convened in Sydney in 1973, and subsequent regional conferences, have focused on labour administration, labour laws and labour relations, and safety codes (International Labour Organization, 1991, p. 27). The 7th Conference of South Pacific Labour Ministers held in Kiribati in July 1981 resolved to seek the help of SPEC and ILO to establish for itself a permanent secretariat. The SPEC pre-forum committee in August 1981 accepted this proposal and put it to the Forum Leaders. In February 2016 Tonga became the ILO'*s* 187th member, and the 8th Pacific Island nation to join in the previous 15 years. The date of membership of PICs in ILO is noted in Table 5.2.

The gradual acceptance of ILO standards and procedures has not been without challenges. In the Cook Islands, the Cook Islands Workers' Association lobbied government for a decade to join the ILO so that core workers' rights could be

Table 5.2 Date of entry to International Labour Organization

Fiji	1974
Papua New Guinea	1976
Solomon Islands	1984
Kiribati	2000
Vanuatu	2003
Marshall Islands	2007
Samoa	2008
Tuvalu	2008
Palau	2012
Cook Islands	2015
Tonga	2016

recognised, before an application for membership was lodged in 2015 ("Cooks Set to Join International Labour Organization: Union: ILO bid 'significant leap' for government, employers, workers," 2015).

In 2015–2016 tense negotiations took place in Fiji concerning delays to a joint implementation plan involving the Government, the Fiji Commerce and Employers Federation (FCEF) and the Fiji Trades Union Congress (FTUC). Fiji had refused to agree to a plan after unions argued that the government was infringing workers' rights. The ILO issued a report in January 2016 listing 14 concerns about amendments to Fiji's *Employment Relations Promulgation (ERP)* legislation, such other issues as non-compliance with the Tripartite Agreement, an assault on Fiji Trades Union Congress national secretary Felix Anthony, pending cases against Daniel Urai, the use of bargaining units to undermine trade unions, the denial of prison officers' right to form and join unions, nomination of representatives in the Employment Relations Advisory Board (ERAB), and the reinstatement of disputes terminated under the Essential National Industries Decree (Radio New Zealand International, 2016). An ILO Mission visited Fiji in January 2016 and on the basis of revisions to legislation passed in Fiji's parliament the following month, the full board of ILO decided that it was not necessary to proceed with a commission of inquiry ("ILO Decision On Fiji Of 'Huge Significance': Minister: International body ends commission of inquiry into labor issues," 2016).

In the case of Samoa, an ILO inquiry pointed to difficult issues concerning cultural acceptance of child labour. A Rapid Assessment on Child Street Vendors in Apia, Samoa in September 2014 found children involved in child labour, particularly as street vendors, but also as sex workers. The following May and December, the ILO brought together participants from government ministries, Police, Immigration & Customs, Samoa Workers Congress, Samoa Chamber of Commerce, employer representatives from the private sector; civil society groups, community groups, and journalists, to validate the Assessment's findings, and to also extend the research to include child labour activities on the island of Savai'i ("Samoa, ILO Tackling Child Labor Problem Head On. Forum to validate issues raised in child labor study," 2015). Existing legislation outlawed children working in "a formal place of work" during school hours but not after hours, and did not say anything about the informal sector, such as street vendors, and thus required reform ("ILO Helps Samoa Develop Anti-Child Labor Legislation: Laws needed to address growing street vendor problem," 2015).

5.2.2 United Nations Children's Fund - UNICEF

The United Nations Children's Fund, more commonly referred to as UNICEF (*United Nations* International Children's Emergency Fund), was established in 1946 to provide relief for suffering of children in war-devastated Europe and expanded into a permanent UN program in 1953. Initial work in the Pacific was managed by WHO, and then in 1970s a school curriculum development project was implemented in collaboration with UNDP. Thereafter direct relations were established with PIC governments and UNICEF's multi-sectoral, multi-dimensional programme, evolved. By 1985 UNICEF had 72 projects throughout the region under programmes for child health, water supply and sanitation, nutrition, and early childhood education. This multi-country programme for the Pacific was managed from Manila until a sub-office was established in Suva in 1993. In 1997 this sub-office was upgraded to a Pacific multi-country office. UNICEF now works with partners in all PICs to improve the conditions of the region's vulnerable children.[5]

5.2.3 UN Women

UN Women, the United Nations entity dedicated to gender equality and the empowerment of women, was created in 2010 through amalgamation of four previously distinct UN bodies: Division for the Advancement of Women (DAW); International Research and Training Institute for the Advancement of Women (INSTRAW); Office of the Special Adviser on Gender Issues and Advancement of Women (OSAGI); and United Nations Development Fund for Women (UNIFEM). Prior to this amalgamation, UNIFEM was the lead UN agency promoting the advancement of women in the Pacific region. It conducted programmes on Peace and Security, and vigorously promoted the adoption of "special temporary measures" to include women in Pacific Parliaments (Palmieri, 2016).

5.2.4 UN-Habitat

Amongst the most recent of the major UN programs to establish itself in the Pacific region is the United Nations Human Settlements Programme (UN-Habitat), which was established by stages after a global conference on housing in Vancouver in 1976, ("Habitat I"). UN-Habitat's engagement with governments and other social actors bears some similarities to WHO's: to achieve its vision of "a better quality of life for all in an urbanizing world". UN-Habitat cultivates networks at regional, national and global levels, that bring together all state and non-state actors engaged in aspects of urbanization, to systematize and share knowledge of common issues and solutions.[6]

At global level, impetus toward addressing the issue of urbanization grew following Habitat II which convened in 1996, and then through General Assembly Resolution A/56/206 of 2002, which converted UN-Habitat into a fully-fledged

[5] https://www.unicef.org/pacificislands/about-us

[6] https://unhabitat.org/

programme in the UN system. Following a third global conference, Habitat III, in Quito in 2016, UN Habitat initiated the World Urban Forum, the sixth of which was held in Italy in 2012 and the eleventh in Poland in 2022.

These global conferences influenced developments at regional level. The Pacific Urban Agenda (PUA) adopted at the first Pacific Conference on Urban Management in Nadi in 2003 and endorsed by ESCAP at its sixtieth session in Shanghai in 2004 (resolution 60/7) was assigned a high priority in the Pacific Plan by the Pacific Islands Forum (Initiative13.5). The 2005 Forum Economic Ministers Meeting requested that "member countries develop policies and plans for urbanization" – a call which had not previously been issued by any of the PIC regional bodies. In reality, the adoption of resolutions had little impact on action within individual PICs, and much stronger advocacy and enticement was required before there was much movement.

In April 2007 the second Pacific Conference on Urban Management prepared a more targeted PUA (PUA2), and mobilized increased development partner support from PIFS, SPC, CLGF and bilateral agencies such as NZAID and AusAID. Another 2007 regional workshop, of Pacific planners, convened by the Planning Institute of Australia, AusAID and UN-Habitat, refined PUA2 into a Regional Action Framework (RAF) which was endorsed by the Forum's Pacific Plan Committee as an agreed "regional urban management strategy" for implementation by PICs over five years with support from a coalition of international development partners: UN-Habitat, ESCAP's Pacific Operations Centre (EPOC), Pacific Islands Forum Secretariat (PIFS), the Secretariat of the Pacific Community (SPC) and the Commonwealth Local Government Forum (CLGF). Subsequent progress has been monitored by a fifth Pacific Urban Forum, convened in Nadi in 2019, and a 2021 Forum (held virtually due to the COVID-19 pandemic) (Pacific Partnership for Implementing the New Urban Agenda, 2021).

UN-Habitat has also responded to PIC requests from individual PICs for urban sector policy advice, research, and technical cooperation in support of the full spectrum of human settlements issues - from housing to urban development, and from post-disaster shelter to slum prevention. It also produced "rapid urban profiles" for cities and towns, including Goroka (UN-Habitat, 2010a) Honiara (United Nations Human Settlements Programme, 2012), and Port Moresby (UN-Habitat, 2010b).

5.3 Additional UN Entities

Additional UN entities with a presence in the Pacific Islands include the United Nations Office for the Coordination of Humanitarian Affairs (OCHA), which is concerned with human rights, humanitarian assistance, disarmament, and dialogue amongst civilizations; the Office of the High Commissioner for Human Rights (OHCHR), which established a regional office in Suva in 2005 to provide advice to 16 countries (including Australia and New Zealand) on human rights treaties, particularly CAT, CCPR, ESCR and CMW; the UN Capital Development Fund (UNCDF), which commenced in the Pacific with a Local Climate Adaptive Living Facility project in Tuvalu.

Another UN office closely aligned with environmental conditions, the United Nations Disaster Relief Office, was established by UNGA to assist with mobilisation of international emergency relief assistance in disaster-stricken countries, and the improvement of disaster preparedness and disaster prevention. The Office has provided financial and substantive support since an initial South Pacific Disaster Preparedness and Relief Seminar in Suva in September 1976, co- sponsored by the League of Red Cross and the Commonwealth Fund for Technical Co-operation (CFTC).

Other highly specialized agencies working in the region include the International Maritime Organization (IMO), which has regional seas programs monitoring the South West Pacific and South East Pacific, and the UN Industrial Development Organization (UNIDO), established by UNGA in 1966 to promote and accelerate the industrialisation o£ developing countries, with particular emphasis on the manufacturing sector. Since the 1980s UNIDO has assisted the Pacific region with surveys of industrial development possibilities, formulation of industrial development plans and programs, pre-investment and feasibility studies, identification and development of export-oriented industries and training assistance through individual fellowships, technical workshops and in-plant training (PIFS, 1982).

This review of UN agencies and programmes in the Pacific Islands suggests some general observations about their presence and impact. First, it demonstrates the challenges associated with monitoring the activities of UN agencies operating at Pacific regional level, let alone seeking to coordinate them to ensure maximum efficiency and best use of resources. These two challenges are related to a third, which is assessment of their impact on development progress. An additional observation is the pervasive presence of UN agencies, which have come to operate in virtually every sector of government activity and across all phases of policy process.

Notwithstanding the severity of these coordination challenges, these UN programmes and projects produce multiple benefits. There is, for instance, the enhancement of policy literature, some of which has found its way into libraries but a considerable proportion of which remains in government offices and occasionally on agency websites. This corpus of feasibility studies, project plans, progress reports, completion reports, and evaluations, has not been fully enumerated or collated, and constitutes a valuable reservoir of institutional memory concerning regional public policy and project implementation.

Yet another beneficial outcome has been the employment and professional development of PIC nationals. Some of these have stayed employed within the UN system in permanent positions or on fixed term contracts, while others have taken their skills into government positions, or have leveraged their UN experience to obtain positions overseas. Some view such pathways as nothing more than "brain drain" of Pacific Islanders out of essential positions with national governments and regional organizations, but the counterfactual alterative - not having Pacific Islanders obtain deep understanding of regional and international organizations – is surely less desirable, whether for the individuals concerned or for their countries and the region. A final benefit that can be noted is the capacity development obtained by Pacific Islanders – whether from the public sector, provide sector, or civil society – through their engagement with UN processes, projects, programmes, networks, platforms, and conferences.

5.4 Regional Development Frameworks

Having introduced no less than sixteen distinct UN agencies and programmes active in the Pacific region, we now consider the efforts made to improve their coordination and coherence. Collectively, Forum Leaders had minimal engagement with the UN throughout the 1970s. The Forum's second meeting, in Canberra in February 1972, was presented with a list of UN agencies as part of a "select list of organizations of actual or potential assistance in the development of the South Pacific" (FAO, UNDP, UNIDO, UNICEF, WHO, WMO, GATT, ICAO, ILO, ITU, UPU, and ECAFE). At this time the only PICs that were members of any of these bodies were Fiji (FAO and ECAFE), Nauru (UPU), Tonga (UPU), and Western Samoa (WHO).

By the 1970s, as has been noted in previous chapters, UNDP was at the forefront of convening regional dialogue with PICs on policy issues, but the challenges of development assistance governance were becoming increasingly complex. In March 1990 UNDP brought together 30 high-level government officials and the South Pacific Organizations Coordinating Committee (SPOCC) at the Pacific Islands Development Programme (PIDP) in Hawaii, to identify priorities for UN regional Assistance. Those deliberations, plus a comprehensive and influential paper by ESCAP head Savenaca Siwatibau, (Siwatibau, 1991) together with other inputs, such as a large meeting (76 representatives from 12 PICs, six regional organizations, ten donor countries, and five international aid organizations) at the Forum Secretariat headquarters in Suva in February 1991, shaped the UN's Regional Program for the Pacific for the period 1992–1996.

The importance of a regional consultation meeting involving UNDP/ESCAP/SPC/SPEC held in February 1979 was demonstrated by attendance by UNDP Administrator Bradford Morse from New York, UNDP Regional Bureau for Asia and the Pacific Director Andrew Joseph from Bangkok, and ESCAP Executive Secretary J.B.P. Maramis, also from Bangkok. UNDP had allocated USD 12 million to Pacific operations for the five-year programming cycle 1977–1981, (UNDP 1979, p. 3) and this was the first opportunity that Forum Officials had had, "...to exchange views with senior executives from UNDP and other organizations on the development objectives and priorities of the region" (UNDP, 1979, p. 24).

Another influential gathering - a UN Mini-MAC meeting at SPC in *Nouméa* in June 1991 that brought 35 officials from 19 PIC governments, 6 SPOCC members, 6 bilateral and three multilateral donor organizations, 5 UN organizations and agencies and 4 funding UN organizations (UNDP, WHO, UNICEF, UNFPA) - identified four themes that were to drive policy in the region for the next four decades: human development and poverty alleviation; environment and natural resource management; economic and financial management and reform; and intra-regional cooperation (Pacific Islands Forum Secretariat, 1991).

Participants at these large regional meeting made significant progress in understanding not only the social and economic development changes facing the region, but in understanding the need for improved coordination of development assistance (Pacific Islands Forum Secretariat, 1991, p. 1 Agenda Item 5(1): Development Priorities for the Region - Annex One). Additional significant developments in the coordination of regional development dialogue resulted from a 1993 Pacific Island Countries/Development partners meeting. First was the decision that PICs would take lead responsibility for coordination of consultations with their donor partners. Second was that UNDP, ADB and The World Bank would assist with the documentation of these national and sector-level consultations. Third - and most significant to this study - was the decision that regional development assistance roundtables would be held in conjunction with other regional meetings as often as possible, to reduce costs, with national development partner consultations to be held in intervening years (PIFS, 1994, p. 5).

5.4.1 UNDAF 2008–2012

The UN's *Development Assistance Framework for the Pacific 2008–2012* (United Nations, 2007) was designed to align with the four major themes of the Forum's Pacific Plan: Economic growth, sustainable development, good governance, and security. In 2009–2010 alone, the UN Secretary General reported that UN agencies collaborated with Pacific regional organizations on several major consultations to advance one or other key objective: Regional Conference on Security Sector Governance (Tonga, with Asia Pacific Security Centre and others); workshops with ADB in Fiji, Papua New Guinea and Tonga on aid effectiveness and UNCAC; Pacific Conference on the Human Face of the Global Economic Crisis (Vanuatu, with UNDP, PIF and others); Development Partner's Round Tables on Climate Change (Fiji, with UNDP, PIF and others); a high-level Pacific regional preparatory meeting for the Five-Year review of the Mauritius Strategy for the Further implementation of the Barbados Programme of Action for the Sustainable Development of Small Island States and a Pacific Millennium Development Workshop; DPA and UNDP provided electoral administration assistance; UNFPA was working on MDG goal reporting, Cairns compact development partner reporting; Regional Statistics Benchmarking Study and Implementation Plan; WHO, UNICEF and FAO partnered PIF in delivering a Food Summit in Vanuatu, ILO helped deliver a Tri-partite High-level meeting on Decent work for sustainable Development in the Pacific in Vanuatu, and the Office for Disarmament Affairs assisted PIF with a workshop on implementation of the Programme of Action on Small Arms and Light Weapons (United Nations General Assembly, 2010, pp. 21–22). Similar levels of engagement continued through the remaining years of the MDGs, and into those of the SDGs (2015 and after) (United Nations General Assembly, 2012, pp. 21–22).

5.4.2 UNDAF 2013–2017

The UN's second generation of regional coordination, UNDAF 2013–2017, included eleven PICs (Cook Islands, Fiji, Federated States of Micronesia, Kiribati, Palau, Marshall Islands, Samoa, Solomon Islands, Tonga, Tuvalu, and Vanuatu). It did not include Papua New Guinea, which had a separate UNDAF, nor Nauru. UNDAF 2013–2017 was the first programme to be based on extensive in-country consultation between UN teams and member countries (United Nations, 2013). It had five "Outcome Areas": (1) Environmental management, climate change, and disaster risk management; (2): Gender equality; (Poverty reduction and inclusive economic growth; (4) Basic services (Health and Education); and (5) Governance and Human Rights.

In their introduction to the Framework, the Resident coordinators of the Apia and Suva multi-country offices stated: "Pacific Island countries have reminded us to continue to work to avoid fragmentation, reduce the transaction and reporting burden on small island administrations and increase the efficiency. Ultimately, this is the essence of UN Reform. Therefore, the UNDAF 2013–2017 will scale up the use of joint programmes as the premiere UNDAF delivery tool. It will also enhance novel operational arrangements for efficient delivery at country level, such the UN joint presences and the Kiribati One Fund" (United Nations, 2013, p. 5). Morauta's review of the Pacific's regional architecture had commented on the desirability of reducing the burden of reporting on the MDGs/SDGs, and had suggested that just one of the SDG indicators could be chosen as a suitable proxy (Morauta, 2013, p. 124). UNDAF included PIFs "Principles of Good Leadership and Accountability", which were subsequently built into such UNDP projects as the Strengthening of the National Parliament of Solomon Islands (UNDP, 2012).

5.4.3 UN Pacific Strategies 2018–2022 and 2023–2027

In 2018 the UN commenced a third five-year strategy, *United Nations Pacific Strategy (UNPS) 2018–2022* (The United Nations, 2018), involving 14 PICs and 26 UN entities. Again, the strategy excluded Papua New Guinea, which had a separate UN office and Development Assistance Framework. But it did include non-UN-member PICs Cook Islands, Niue, and Tokelau. The framework resulted from consultation with 14 PICs and the CROP Sustainable Development Working Group (SDWG) and was intended to support localised responses to the global 2030 Agenda for Sustainable Development. It also sought to align the UNPS priorities and regional accountability and reporting against, the *Framework for Pacific*

Regionalism, the SDGs and SAMOA Pathway (see Chap. 11). The strategy had six outcome areas: (1) Climate Change, Disaster Resilience and Environmental Protection; (2) Gender Equality; (3) Sustainable and Inclusive Economic Empowerment; (4) Equitable Basic Services (health, education, social protection); (5) Governance and Community Engagement; and (6) Human Rights.

In 2023 the region adopted the *UN Sustainable Development Cooperation Framework 2023–2027* with emphasis on pursuit of the SDGs under a mnemonic of four development priorities: planet, people, prosperity, and peace.

5.5 Cumulative Impact of UN Offices in the Pacific Islands

Some twenty UN agencies and programmes have contributed to the development of Pacific Island Countries in several useful and interrelated ways. Key benefits have included not only direct impact on social and economic development, but the enhancement of policy literature and policy processes, the employment and professional development of Pacific Islanders, human resource development more broadly, and the facilitation of connections with global networks. The development of human resources is an often unstated but always implicit objective in UN engagement with PICs. It is often referred to as building capability and capacity, and opportunities for training exist in virtually every sector, and for both state and non-state actors. Tuvalu's report to the Pacific Islands Law Officer's Meeting for 2009 provides an example of the opportunities available in that year for that country's legal officers: twelve opportunities were accepted, most of which were UN affiliated: UNDP, Universal Periodic Review Training, Auckland, January to February; UNHCR, Adoption of Tuvalu National Periodic Report, Geneva, March; UNODC, Counter-Terrorism Workshop, Suva, April/May; UNODA, Small Arms and Light Weapons Seminar, Sydney, June; FFA, Sub-regional Workshop Western and Central Pacific Fisheries Commission, Nadi, July; FFA, Workshop for Legal Options giving effect to WCPFC Obligations, Honiara, August; and UNITAR, Regional Training Programme to Enhance the Conflict Prevention and Peacebuilding Capacities of Indigenous Peoples' Representatives of the Pacific, Cairns, August (Government of Tuvalu. Office of the Attorney General, 2009). Similar scenarios are reported in countless reports from diverse sectors from PIC departments and agencies across the Pacific region. Indeed, the extent of "duty travel" by government officials, whether for meetings or training purposes, can be intense in small states where an individual's skills are highly valued and in short supply.

References

Buchan, J., Connell, J., & Rumsey, M. (2011). *Recruiting and retaining health workers in remote areas: Pacific Island case studies* (Vol. 1). World Health Organization.

Cooks Set to Join International Labour Organization: Union: ILO bid 'significant leap' for government, employers, workers. (2015, June 9). *Cook Islands News.*

Food and Agriculture Organization of the United Nations. (2017). *Multi-country programming framework for the Pacific Islands 2018–2022.* Food and Agriculture Organization of the United Nations.

Government of the Federated States of Micronesia. (2005). *Address by H.E. Joseph J. Urusemal, President of the Federated States of Micronesia before the High-Level Plenary Meeting of the 60th Session of the United Nations General Assembly, Friday 16 September.*

Government of Tuvalu. Office of the Attorney General. (2009). *PILON Report of the Office of the Attorney-General Tuvalu.*

ILO Decision On Fiji Of 'Huge Significance': Minister: International body ends commission of inquiry into labor issues. (2016, March 29). *Fiji Times.*

ILO Helps Samoa Develop Anti-Child Labor Legislation: Laws needed to address growing street vendor probleme. (2015, December 3). *Samoa Observer.*

International Labour Organization. (1991). *ILO in the Pacific. Review of ILO activities 1975–1990.* International Labour Organization Office for the South Pacific.

Johnson, S. (2012). *UNEP. The first 40 years. A narrative.* United Nations Environment Programme.

Morauta, M. (2013). *Pacific plan review 2013. Report to leaders. Volume 1.* Pacific Islands Forum Secretariat.

Pacific Health Information Network, World Health Organization, & Pacific Community. (2018). *Evaluation and renewed vision and strategy (2019–2021) for the Pacific Health Information Network (PHIN).* World Health Organization.

Pacific Islands Forum Secretariat. (1973). Fourth South Pacific Forum, Apia, Western Samoa, 17–18 April. Proceedings. *(73)NN.1.*

Pacific Islands Forum Secretariat. (1982). Regional Committee on Trade. Second Meeting Canberra, Australia, 5–7 July. *(82)MP.7.*

Pacific Islands Forum Secretariat. (1988). SPREP Co-ordinating Group Meeting/Intergovernmental Meeting on the SPREP Action Plan, Nouméa, New Caledonia 25 June/27 June–1 July 1988. Meeting papers. *(88) MP. 23((88) MP. 23).*

Pacific Islands Forum Secretariat. (1991). Forum Officials Committee Pre-Forum Session. Pohnpei, Federated states of Micronesia 22–24 July. Meeting Papers. *MP(91) MP.14.*

Pacific Islands Forum. Standing Committee of Officials on the Decolonisation of New Caledonia. (1985). Standing Committee of Officials on the Decolonisation of New Caledonia Inaugural Meeting, SPEC Headquarters, Suva 6–8 November 1985. Meeting papers. *(85) MP. 12 ((85) MP. 12).*

Pacific Partnership for Implementing the New Urban Agenda. (2021). *2021 virtual Pacific Urban Forum: Accelerating the Pacific New Urban Agenda. Regional progress and priorities in 2021. Executive summary.* UN-Habitat.

Palmieri, S. (2016). *Temporary special measures to increase women's political participation in the Pacific: Case studies of implementation in the region.* United Nations Development Programme.

Purcell, G. F., & Hassall, G. H. (2017). Global policies and initiatives for e-government in Pacific SIDS. In R. Cullen & G. Hassall (Eds.), *Achieving sustainable e-government in Pacific Island States* (pp. 33–54). Springer International Publishing.

Radio New Zealand International. (2016, January 11). *Fiji Unions call on government to accept ILO mission: Failing to take advantage of opportunity will hurt Fiji: Anthony.*

Samoa, ILO Tackling Child Labor Problem Head On. Forum to validate issues raised in child labor study. (2015, 1 December). *Talamua.*

Siwatibau, S. (1991). *Report on consultations with Pacific governments: Identification of priorities for U.N. Regional Assistance.* United Nations Development Program. Regional Bureau for Asia and the Pacific.

South Pacific Bureau for Economic Cooperation. (1975). SPEC Committee meeting, Nukualofa, Tonga, 26–27 June. Agreed record.

South Pacific Bureau for Economic Cooperation. (1976). SPEC Committee meeting, Nauru 22–23 July 1976: UNDP Cutbacks on Regional projects. *SPEC(76)12(SPEC(76)12).*

South Pacific Bureau for Economic Cooperation. (1978a). Directory of aid agencies. *(78)FSCOL4.*

South Pacific Bureau for Economic Cooperation. (1978b). Opening statement by Albert Henry. Summary record of the Fifth South Pacific regional telecommunications meeting 4–10 July 1978 Rarotonga, Cook Islands. *SPEC(78)16*, 134p.

South Pacific Bureau for Economic Cooperation. (1979a). Relations with other international organizations - United Nations Economic and Social Commission for Asia and the Pacific (ESCAP). SPEC Committee Meeting, Honiara, Solomon Islands 5–6 July 1979: [meeting papers]. *SPC(79) SCH.1 (Draft).*

South Pacific Bureau for Economic Cooperation. (1979b). Relations with other International Organizations: United Nations Development Programme. SPEC Committee Meeting, Honiara, Solomon Islands 5–6 July 1979: [meeting papers]. *SPC(79) SCH.1 (Draft).*

South Pacific Bureau for Economic Cooperation. (1979c). Sixth south pacific regional telecommunications meeting, Wellington, New Zealand 1–7 November, 1979: Summary record. *SPEC(79) 24.*

South Pacific Bureau for Economic Cooperation. (1979d). SPEC Committee Meeting, Honiara, Solomon Islands 5–6 July 1979. Agreed report. *SPEC(79)19, 60p.*

South Pacific Bureau for Economic Cooperation. (1979e). Work Programme 1979/80: Trade and Marketing. SPEC Committee Meeting, Honiara, Solomon Islands 5–6 July 1979: [meeting papers]. *SPC(79) SCH.1 (Draft).*

South Pacific Bureau for Economic Cooperation (1982). Regional meeting of coconut interests, SPEC Headquarters, Suva, Fiji, 15–19 March 1982: [meeting papers]. *(82) MP. 2.*

South Pacific Bureau for Economic Co-operation. (1973a). *Director's preliminary report 1972/73.* South Pacific Bureau for Economic Co-operation.

South Pacific Bureau for Economic Co-operation. (1973b). *Report of the South Pacific regional telecommunications meeting 28 November–5 December 1973.* South Pacific Bureau for Economic Co-operation.

South Pacific Bureau for Economic Co-operation. (1975). South Pacific Foum Tonga 1–3 July 1975. .

South Pacific Bureau for Economic Co-operation. (1976a). *SPEC Committee meeting Naurau 23–24 July, 1976. Agreed record.* South Pacific Bureau for Economic Co-operation.

South Pacific Bureau for Economic Co-operation. (1976b). *SPEC Committee meeting Nauru 22–23 July, 1976. UNDP financial cutbacks on regional projects.* South Pacific Bureau for Economic Co-operation.

South Pacific Bureau for Economic Co-operation. (1978a). *Directory of aid agencies.* South Pacific bureau for Economic Co-operation.

South Pacific Bureau for Economic Co-operation. (1978b). *Special SPEC Committee meeting on membership of the South Pacific Forum and SPEC, 18–19 April 1978.* South Pacific Bureau for Economic Co-operation.

South Pacific Forum. (1975). *Sixth South Pacific Forum, Nuku'alofa, Tonga: Meeting papers.*

South Pacific Forum Secretariat. (1994). *Review of international support to ECDC: Experiences of developed and developing countries, their problems and prospects.* UNCTAD.

South Pacific Regional Telecommunications, M. (1982). ITU - Meeting of NDP aid coordinators of Asia and the Pacific. Inter-country telecommunication program and projects 1983–1986. Ninth South Pacific regional telecommunications meeting, Alofi, Niue: 28 October–3 November 1982: [meeting papers]. *(82) MP. 9.*

Special Meeting Between SPEC Representatives and SPC Staff 20–21 May 1974. (1974). Suva.

The United Nations. (2018). *United Nations in the Pacific. United Nations pacific strategy 2018–2022. A multi-country sustainable development framework in the Pacific Region.* Suva and Apia: Office of the UN Resident Coordinator -Federated States of Micronesia, Fiji, Kiribati,

Marshall Islands, Nauru, Palau, Solomon Islands, Tonga, Tuvalu, Vanuatu, and Office of the UN Resident Coordinator - Cook Islands, Niue, Samoa Tokelau.

UN-Habitat. (2010a). *Papua New Guinea: Goroka city profile*. United Nations Human Settlements Programme.

UN-Habitat. (2010b). *Papua New Guinea: Port Moresby urban profile*. United Nations Human Settlements Programme.

United Nations. (2007). *United Nations Development Assistance Framework (UNDAF) for the Pacific SubRegion 2008–2012*. United Nations.

United Nations. (2013). *United Nations Development Assistance Framework (UNDAF) for the Pacific Region 2013–2017*. United Nations.

United Nations Development Programme. (1979). *UNDP/ESCAP/SPC/SPEC Meeting of Senior Planning Officials to Discuss UNDP Inter-Country Programme 1979–81, SPEC Headquarters, Suva 2–6 February 1979: Report*. United Nations Development Programme.

United Nations Development Programme. (1980). *Subregional Follow-up meeting for Pacific Women on the World Conference of the United Nations for Women, Suva, Fiji, 30 October–3 November 1980. Statement by Mr Arthur N. Holcombe, Resident Representative of the United Nations Development Programme*. United Nations Development Progamme.

United Nations Development Programme. (2012). *Project document: Solomon Islands parliament project phase*, 3.

United Nations General Assembly. (2010). *Cooperation between the United Nations and regional and other organizations: Report of the Secretary-General. (A_65_382&S_2010_490-EN)*. UN.

United Nations General Assembly. (2012). *Cooperation between the United Nations and regional and other organizations: Report of the Secretary-General. (A_67_280--S_2012_614-EN)*. UN.

United Nations Human Settlements Programme. (2012). *Solomon Islands: Honiara urban profile*. United Nations Human Settlements Programme.

United Nations. Economic and Social Commission for Asia and the Pacific. (2003). *Outcome of the evaluation of the ESCAP Pacific Operations Centre*. UN.

United Nations. Economic and Social Commission for Asia and the Pacific. (2008). Annual report 24 May 2007–30 April 2008. *E/ESCAP/64/39*.

United Nations. Economic and Social Commission for Asia and the Pacific. (2010). *Sustainable development in the Pacific: Progress and challenges. Pacific regional report for the 5-year review of the mauritius strategy for further implementation of the barbados programme of action for sustainable development of SIDS (MSI+5)*. ESCAP Subregional Office for the Pacific.

United Nations. Economic and Social Council. (2001). *Regional cooperation in the economic, social and related fields: report of the Secretary-General: Addendum*. UN.

United Nations. Executive Board of the United Nations Development Program, United Nations. Population Fund, & Services, (2017). United Nations population fund. Subregional programme document for Pacific Island countries and territories. *DP/FPA/CPD/PIC/6*.

United Nations. Food and Agriculture Organization, & United Nations Development Programme. (1976). UNDP/FAO Survey of agricultural pests and diseases: interim report, March 1976. *SPEC(76) 4*, 2 v.

United Nations. General Assembly. (2007). Cooperation between the United Nations and the Pacific Islands Forum: Resolution. In *A/Res/1/48*. UN.

World Food Programme. (2019). *Pacific interim multi-country strategic plan (2019–2022)*. World Food Programme.

World Health Organization. (2020). *Basic documents* (49 ed.). World Health Organization.

World Health Organization. Regional Office for the Western Pacific. (2010). *Meeting on the implementation of the project on national health accounts in the Pacific, Nadi, Fiji, 25–27 May 2010: Report*. World Health Organization. Regional Office for the Western Pacific.

World Health Organization. Regional Office for the Western Pacific. (2018). *Meeting on health information system strengthening in the Pacific Island Countries, 27–29 June, Suva, Fiji*. World Health Organization.

Chapter 6
Governance and Pacific Islands Policy Networks

Chapters 1–5 set out the ubiquitous involvement of UN programmes in the Pacific Islands region and demonstrated how the UN has participated for such an extended period of time, in such a broad range of sectors, that its presence is now taken as a "given", sometimes even taken for granted. Much of the Pacific region's initial transport and communications infrastructure planning, for instance, as well as prospects for agricultural production and trade, emerged from UN-sponsored research programs, co-financing, and implementation. Furthermore, UN engagement greatly influenced patterns of development planning and administration at both national and regional levels. This engagement in the field of development administration changed over time through reflection on experience in the Pacific and globally. However, limited access to information, combined with constant change of personnel, have resulted in limited contemporary awareness of these regional and inter-governmental public administration undertakings.

This chapter focuses on the governance arrangements that evolved in response to conditions in the Pacific Islands region, as well as to the public policy processes that were emerging across public sector organizations in general. Governance is a challenging concept to define. In one useful formulation, Kooiman emphasizes the "quality of interactions" between actors:

> . . .the totality of interactions, in which public as well as private actors participate, aimed at solving societal problems or creating societal opportunities; attending to the institutions as contexts for these governing interactions; and establishing a normative foundation for all those activities. (Kooiman, 2003, p. 4)

The working definition that has evolved within the United Nations system emphasizes UN values. The UNDP definition put to the 2000 Pacific Regional conference of Parliamentarians stated:

> ...the exercise of economic, political and administrative authority to manage a country's affairs at all levels. It comprises the mechanisms, processes, and institutions through which citizens and groups articulate their interests, exercise their legal rights, meet their obligations and mediate their differences. (ADB, PIFS, UNDP, UNESCAP - Pacific Operations Centre, & United Nations Office for Project Services, 2000, p. 50)

G. Hassall, *The United Nations and the Pacific Islands*, United Nations University Series on Regionalism 24, https://doi.org/10.1007/978-3-031-34155-7_6

There have been two aspects to the development of governance capacity, one relating specifically to the capability of each state's institutions, such as parliament, public sector, and accountability officers, and the other relating to the capability of states to partner non-state actors in co-governance arrangements.

By the 1980s innovative governance arrangements involving Pacific regional organizations and international development partners sought to avoid duplication, minimize costs, and create synergies. New arrangements included the establishment of multi-party agreements that combined both state and non-state actors in networked planning and implementation of projects that would formerly have been the responsibility of government alone. This might be an energy survey mission by UNDP/EWC/EEC (South Pacific Bureau for Economic Cooperation, 1982a), or a road improvement and maintenance study in Tonga financed by UNDP and executed by ADB (South Pacific Bureau for Economic Cooperation,, and Forum Secretariat, 1975, p. 60). The South Pacific Regional Environment Programme was jointly administered by SPEC, SPC, ESCAP and UNEP, with each agency taking a specific role: SPEC was chair; UNEP as part of its Regional Seas programme, and SPC as secretariat (South Pacific Forum Secretariat, 1982, p. 1).

Concern for the size and capability of the public sector in Pacific Island countries is not new. A workshop on "Making government more effective: public sector reform in the South Pacific" held in Fiji in 1993 concluded that reform required a clear purpose, effective communication, planning and management, adequate training and staff development, successful implementation, and effective evaluation (World Bank & Institute of Social and Administrative Studies, 1993, pp. viii–ix). A 1994 PIFS report to an UNCTAD conference reviewing international support to developing countries was frank about the state of public sector cohesion in its member-countries:

> At the national level, FIC public services are currently fragmented among a great number of ministries, departments, agencies, authorities and state-owned enterprises with as many as thirty central government departments in some countries. The size and composition of FIC public services and their ad hoc expansion over the years reflect a number of factors many of which bear little functional relation to the actual needs of the communities they are intended to serve. Instead of improving performance, this has led to compartmentalization, excessive red-tape and costly duplication of management and administrative services. Rationalization of the public service will help to correct this situation. (South Pacific Forum Secretariat, 1994, p. 3)

In 2000, and again in 2003, the United Nations and partners hosted regional conferences on governance in the Pacific. The 2000 conference in Fiji held over 4 days and sponsored by the ADB, ESCAP and UNDP, brought together 80 senior officials and parliamentarians from thirteen PICs, plus 49 others from UN agencies, regional organizations. Papers were presented on parliaments and governance in the twenty-first century, parliament's role in upholding good governance, transformative leadership, the responsibilities of statutory office holders, the role of committee systems, and the role of civil society in the democratic process. Parliamentary speakers convened their own retreat, at which one decision was to continue such forums, perhaps in a "slimmer and less expensive fashion." (Asian Development Bank et al., 2000, p. 42).

A close reading of the conference report suggests the existence amongst the national leaders of considerable reticence about accepting the "good governance" program offered by the UN and the other agencies. Fijian Senator Vunibobo said PICs had adopted modes of government based on western systems which often were "...not the most appropriate in their individual contexts." Tongan parliamentary speaker Veikone pointed out that "...political agendas needed to be defined more by local political cultures and less by international agencies. Local political systems had to be developed from local political traditions and cultures rather than from outside." Participants also suggested that shifts in paradigms of international development assistance were determined by events, issues and policies in the larger donor nations and organizations but had major implications for small island countries with limited resources and should be examined carefully; the issues of good governance and accountability raised by the international community required resources that small land nations often did not have. NGOs were often only accountable to themselves and liable to manipulation, and global agendas and initiatives such as those on the sustainable management of forests placed further difficulties on small island nations which needed to exploit their resources for development and survival:

> In advocating the principles of good governance, one had to consider what was applicable, affordable and practical in the local contexts. ...International trade agreements such as the WTO complicated matters even further. Issues such as the demand for gender equality and for a greater role for NGOs were driven by communities and cultures of donor countries to meet their agendas; these too needed to be examined carefully in the context of small island nations. (Asian Development Bank et al., 2000, p. 5)

A final observation was that, in undertaking reforms, "...donors and governments needed to proceed at a pace that suited each country. And both should keep in mind the sequencing of changes and reforms; all "eight principles" [Forum Principles of Accountability], for example, could not be introduced at once." (Asian Development Bank et al., 2000, p. 5). All such "reservations" about the governance systems were nonetheless were voiced along-side presentations on the merits of public accounts and other parliamentary committees, leadership codes of conduct, the independence of statutory office holders and the roles of ombudsman and auditor-general, and the beneficial potential of an active civil society.

Such conferences designed to improve awareness of legislative practice were complemented by such other forums and assessments, such as the periodic meetings of the Pacific Public Services Commissioners Conference, the Pacific Islands Centre of Public Administration (PICPA) at the University of the South Pacific, the Forum Economic Ministers Meetings, and the Asian Development Bank. (ADB, 2009, 2010) All have generated valuable discussion and debate about issues and challenges of governance and reform, but often with only limited follow-up.

The second area of innovation has been in co-governance arrangements. Such new levels of collaboration between national governments, development agencies, professional bodies, and independent experts and consultants, meant that responsibility for decision-making as well as implementation and outcomes, had to be shared. Any inability of a network to agree on objectives, timetables, and work programs, imperilled outcomes vital to all, and while such co-dependencies could not guarantee success, they did create atmospheres of mutual accountability that were distinct from such traditional lines of accountability as between member of parliament and his or her constituent.

By the 2000s, the Pacific Islands Forum was keen to expand the use of networks and multi-level governance. A discussion paper at the time the Pacific Plan was being formulated suggested:

> ...governance used to be more or less defined as a concept of the nation-state. However, with the increasing responsibility of regional and global bodies in managing resources, it is now accepted that governance should include various forms and levels of local and national government, and regional and global organizations. This is more than the transparent and accountable management of regional organizations and processes. A regional focus on governance should also include an examination of the type of functions and processes that are best managed at, national, sub-regional, regional or global levels. (Pacific Islands Forum Secretariat, 2004, p. 5)

Successful implementation is, of course, always more challenging than is the mere formulation of abstract models. It requires adequate funding, leadership, management, and evaluation. Co-governed projects require clear articulation of responsibilities and are constantly destabilized by the tendency for organizations to strategically withdraw when prospects of failure appear to be outweighing prospects of success.

6.1 The Policy Cycle

The ideas of "policy discourse" and "policy cycles" have become increasingly important. They imply the ability to keep track of past experiences with its mistakes and accomplishments, make decisions on the basis of lessons learned, and share this learning with other actors into the future in such a manner that they understand those past contexts and decisions and do not spend the resources of the present generation repeating past errors and having to learn the same lessons. These are characteristics of knowledge societies, of learning societies. However, they are extremely difficult to embed in an organization let alone a bureaucracy or society in general. A counter argument could be that each generation has its unique challenges which solutions from the past won't solve, and faces new complexities that require innovative thought and action independent of decisions made under the conditions of previous historical periods.

Policy cycles assist organizations and their networks track results in relation to intentions, costs, and time. This desire to increase efficiency, effectiveness, and productivity, has combined with ever-expanding computing capabilities to produce time and data-driven models of policy processes. Although policy cycles can become complex when applied to the features of specific sectors, they generally have in common the fundamental stages of issue identification, formulation of policy

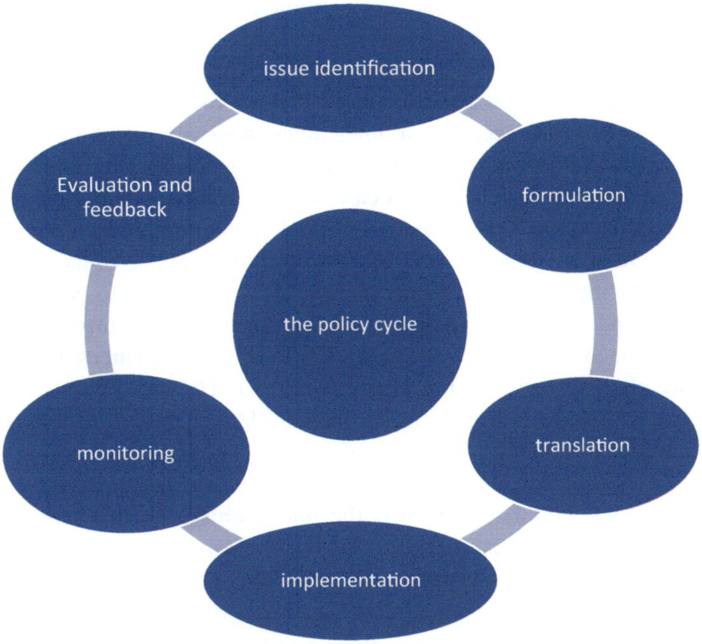

Fig. 6.1 The basic public policy cycle

options, "translation" of policy options into funded programmes, an implementation phase, accompanied by monitoring and evaluation processes.[1] Goals state what an organization wishes to do, with what resources, and in what time frame. They can be aspirational, but preferably, should also be achievable, and measurable.

Such process models, as depicted in Fig. 6.1, can bring conceptual clarity to work programs that otherwise appear chaotic and subject to all manner of disruptive influences. They are often prepared within broader analytical frames that appraise leadership and strategic thought, assessments of risk, the availability of resources, and historical, cultural, social, geographic, and political context. The advantage of using a "policy cycle" approach is that it is systematic, phased and sequential. It can assist financial planning and forecasting, can facilitate transparency, coordinate stake-holder participation, and built trust between actors in the network/platform. Disadvantages of adhering too closely to such a model, on the other hand, are that it may not reflect the "messy reality on the ground", may be used more with intent to attract funding than to implement, and may attempt to enforce a "rationality" on circumstances which are subject to change, and which can produce unanticipated alternative opportunities or unintended consequences. There can be a tendency for small states to view global public policy challenges (security, money laundering,

[1] Policy practices in Pacific Islands context are examined in detail in *Government and Public Policy in the Pacific Islands* (Hassall, 2021)

human trafficking, etc.) as someone else's problem, and rely on international agencies and experts to meet compliance requirements. There can be a tendency for vision-statements to be more visible and more easily available than the completion reports and evaluations that should follow them. Project failure can have many causes: poor design on account of inadequate data, over-ambitious objectives, inadequate funding, unsuitable governance arrangements, poor leadership and political will, or inability to cope with "shocks" such as staff turnover.

There are, too, policy issues that are "wicked", which require not merely "multi-level" and "whole of government" responses, but responses that reach beyond government into the private and voluntary sectors—in efforts to combat problems that don't respond to easy or quick fixes. Climate change, for instance, is a global problem with incontrovertible national and local consequences (sea-level rise, changing weather patterns, cyclones/typhoons, etc). The steady emergence of climate change as a major policy issue has increased the demand for effective policy settings, implementation, governance, and accountability (Karlsson-Vinkhuyzen et al., 2017). It has increased the need for both greater self-reliance as well as greater regional and global collaboration. It is the policy challenge that raised the island states of the Pacific to the status of full partners in global policy processes.

Perhaps more than for any previous global governance/policy challenge, adequate response to climate change requires (a) competent leadership; (b) both vertical and horizontal "whole of government" collaboration; (c) multi-stake-holder support (stake holders in the public, private, and voluntary sectors)—global to national and local levels; (d) evidence-based policy; and (e) accurate and timely monitoring and evaluation. Furthermore, climate change mitigation and adaption require many forms of development assistance, including financial transfers, all of which require sound financial management. Institutions handling climate change policy and implementation are as subject to accountability requirements as any other public institution (Cumming et al., 2012). They handle large sums of money, and unless there are clear indicators of what successful climate change policy looks like, there is the danger that considerable resources (and time) could be consumed with minimal positive result.

6.2 Planning and Consultation

This section illustrates several interactions between regional UN agencies and organizations on matters of planning and consultation. We have noted that this new institutional relationship commenced in the early 1970s. In 1974 a meeting between SPC and SPEC staff discussed relations with UN/UNCTAD over the South Pacific regional transport survey. The regional nature of the proposal necessitated increased cooperation between the agencies, such as in sharing data on trade and shipping that SPC's statistician could use—and they also saw the benefits of standardising their trade statistics (South Pacific Commission, 1974, p. 1). UNDP defined its role as developing national capability and collective self-reliance, as being a catalytic agent. Long term benefits of UNDP's interventions depended on active support and "in depth participation of the countries concerned." (South Pacific Regional Telecommunications, 1982a, p. 19). From participating country perspectives, on the other hand, time-bound or under-financed projects were evidence of agency reluctance to support essential projects. The regional telecommunications development project which was a "heavy programme of proposed development advice and assistance" sought by PICs, for example, met with "severe financial stringencies from the UNDP financed Inter-Country Programme" (Pacific Islands Forum Secretariat, 1983b, p. 1).

In a different context, "Regrettable delays" had occurred, an SPEC report on Civil aviation tabled at the 14th South Pacific Forum held in Canberra in 1983 noted, "while UNDP ascertained whether ADB wished to participate either as co-financer or as executing agent" for a regional civil aviation study, only to be in formed after six months that the ADB declined either role (South Pacific Forum Secretariat, 1983, p. 1). Cooperation was similarly required for the SPEC/UNDP Plant Pests and Diseases survey, with SPC's plant protection officer collaborating with FAO/UNDP experts, and SPEC activities oriented toward progressive relaxation of quarantine regulations to promote intra-regional trade in island products (South Pacific Commission, 1974, p. 3).

By 1974 SPEC was being assisted by experts provided and financed by ITU, CFTC, ESCAP and UNDP (South Pacific Bureau for Economic Cooperation, 1974, p. 2). In 1974 there was a UNDP/ESCAP Inter-country programming mission - SPEC had identified the areas of trade, transport and energy as a focus (South Pacific Commission, 1974, p. 6). A UNDP programming mission visited to identify projects for funding during 1979-81, prior to UNDP regional program being confirmed by officials meeting in January 1979 (South Pacific Bureau for Economic Cooperation, 1980). A UNDP/ESCAP mission in September 1981 to ascertain government requests for 1982-86 worth USD 16 million. There were occasional tensions over potential duplication of effort: SPEC informed the fifth Pacific leaders' meeting, for instance, of misunderstandings with the ITU which sought to conduct seminars, courses, and institutions in the Pacific region, which SPEC felt duplicated existing activities (South Pacific Bureau for Economic Co-operation, 1974a, p. 83).

Once development assistance started flowing to the Pacific states certain trends became apparent. The 1976 study *More Effective Aid* (South Pacific Bureau for Economic Cooperation, 1976a) observed that smaller, flexible and personalised forms of aid were preferred to programs of the big multilateral agencies, which were:

> ...too centralized in administration, heedless of small country facts of life, arrogant in assessing priorities, and unreliable in funding major projects (because the Pacific has a low global priority ...it is important to provide continuity of competent and respected staff assigned to the region and delegated to make decisions; so far only WHO has really achieved this. big programmes induce big administrations, and tend to become centralised, depersonalised and less effective at project and field level. Major donors should examine the characteristics of the more popular programmes to see what it is that makes them so. Inter-donor consultation may improve the effectiveness of everyone's programmes. (South Pacific Bureau for Economic Cooperation, 1976a, p. 10)

This advice remains pertinent to development assistance activities in the twenty-first century, and yet so little observed, that such reports deserve reviving and sharing for consideration by current development agencies and their staff. By 1979 Fiji was expressing the view to the other PICs that the Pacific leaders and not UN officials should be determining the region's development priorities. It cited the experience of the ACP countries in the negotiation of the Lomé convention with the European Union, which resulted in a request being put to ESCAP and other relevant UN Agencies to '...recognize the special problems facing island developing countries when preparing and implementing their regional projects." (South Pacific Bureau for Economic Co-operation, 1979, p. 30).

The difficulties faced in aligning national and regional planning preferences were addressed by J.P. Sheppard, the Director of SPEC, when opening a first regional conference of senior planners at SPEC Headquarters in Suva in June 1982:

> Under our Charter, the SPEC agreement, SPEC is required to "prepare studies for the development plans and policies of member governments in an effort to promote cooperation in the region; and investigate the scope for regional development planning . . ." it was in accordance with this mandate that the Bureau considered the time to be right to convene a meeting with the objective of examining the scope and priorities for regional development planning.
>
> We are aware that this is not an easy area to deal with. Whenever the issue of regional cooperation is raised, individual countries question the effect of regional moves upon their own national initiatives. Moreover, development planning on a national scale is a difficult task in itself, and to integrate a regional outlook into national plans would create further problems. (South Pacific Bureau for Economic Cooperation, 1982c, p. 2)

Following this June 1982 meeting, a *draft report on regional projects: criteria and priorities*, was consulted on at a regional meeting in Suva in February 1983. At issue was whether the structural issue (i.e. the existence of two regional bodies—SPEC and SPC) had been overcome, and whether it was possible to find agreement on regional development priorities. The meeting agreed that projects defined as "regional" and eligible for regional funding should "address common needs in areas of importance and relevance" to PICs, be developmental, and promote regional cooperation. Their priority would be determined by the number of countries that would benefit from them, and that would be of most benefit to the region (Pacific Islands Forum Secretariat, 1983c).

The 1982 *draft report* suggested that donors were more engaged in planning and appraisal of regional projects than the main regional organizations (SPC and SPEC). UNDP had an established tripartite procedure for project review that included government officials and donors. It held an annual Planning and Evaluation meeting with SPC which was more comprehensive than its annual budget meeting with SPEC (Pacific Islands Forum Secretariat, 1983d, p. 6). Whereas some PICs offered no views, Vanuatu was forthright in stating that the draft failed to address the "real problems and constraints", notably the continued competition and duplication of effort by SPEC and SPC:

> A number of areas are still the subject of competition or uncertainty. Amongst them are energy (both SPC and SPEC employ energy specialists); finance (under what umbrella will

> the Pacific Islands funds or the regional stabilisation scheme be set up – if ever?) maybe to some extent fisheries; environment (are the present arrangements cost-effective?) etc…Vanuatu continues to welcome steps that can be taken to clarify the regional institutional arrangements.…let us assume for example that we wish to set up a regional project to coordinate coconut timber production and marketing in the region. Which regional organization should we approach to this end? SPC, which has a programme item to "promote economic utilisation of coconut by-products", or SPEC, which has traditionally been concerned with trade and industry?
>
> More generally a delimitation of the activities between the two organizations, but sectors (i.e. health, agriculture, fisheries)—on which is based the statement that "cooperation has never been better" may not always be really beneficial to regional development. To some extent it hampers the integration of development activities, which can be seen as a major condition for the effectiveness of the development effort. To take an example, the SPC project "self sufficiency on atolls" is an integrated project and its outputs would be severely reduced if it were decided to withdraw the energy component and transfer it to SPEC. (In this particular case it is acknowledge that one arrangement could be to entrust the execution of the project to both organizations, but the anticipated effectiveness of such an arrangement would be questionable).
>
> The bottlenecks created by the present situation should not be underestimated, when one knows the tremendous physical difficulties inherent in regional planning in the Pacific, one can only say that the existing competition for sectors, followed by a competition for funds is a luxury which does nothing but add another handicap to regional development. Technically speaking it would be desirable to have a single regional organization. This however neglects the political aspects of the issue (Pacific Islands Forum Secretariat, 1983a, p. 1).

Responding to these concerns, New Zealand was in favour of reducing a tendency toward "horse-trading", and of outside donors who didn't have representation in the region, making greater use of SPC and SPEC in aid coordination (Pacific Islands Forum Secretariat, 1983c). United Nations Resident Representative Arthur N. Holcombe suggested that regional development projects should be matched with one of the three regional organizations—SPEC, SPC, or USP. It did not make sense to propose that WHO coordinate projects in primary health care and national administration (a proposal about which the UN consented and allowed direct coordination by WHO) (Pacific Islands Forum Secretariat, 1983d). It was "not by accident" that much informal collaboration was being achieved: "The 'horse-trading' referred to is not a function of donor programming procedures so much as the very nature of collective national self-interest within the Pacific region."—there was more 'horse-trading associated with the Lomé II program coordinated by SPEC than in the UNDP's ICP program of the third IPF Cycle coordinated directly by governments at an October 1980 meeting in Apia (Pacific Islands Forum Secretariat, 1983d, p. 3).

This desire to improve coordination has at times seen regional organizations resist efforts by international agencies to place additional staff in the region, such as when EEC sought in 1981 to place a trade promotion consultant in the Pacific when a UNDP/ITC project on trade promotion was already in place (South Pacific Bureau for Economic Cooperation, 1982b). The disagreement continued into 1982 (South Pacific Bureau for Economic Cooperation, 1982d).

> It is generally found convenient, and effective, to plan and implement development on the basis of sectoral programmes, with varying degrees of decentralisation of responsibility for

> planning and management. Thus all these phases involve, to varying degrees, interaction between the central planning office and sectoral ministries, and between national and district/ provincial levels of government where the latter exist. By the same token, there is a need for the central planning office to assist its partners in the departments or at the decentralised levels of government, in using the techniques for planning and management of development i.e. in project formulation, appraisal, programming, monitoring and control.
>
> The responsibilities and activities outlined above are largely confined to Government. More recently it has been recognized that since development is for people, the communities should be deliberately involved, to the extent feasible, in the development process, starting with identification/formulation of proposals. Unfortunately, while it may be easy to agree that this should be done, it is somewhat more difficult to determine in particular cases how it can best be done, and to actually do it. Experience suggests however that when the effort is made with due care, it is more than justified by the results (South Pacific Bureau for Economic Cooperation, 1982f, p. 1).

In 1982 challenges were faced in implementing projects under EEC's Lomé II convention: coordinating the preferences of 8 participating PICS, delays in getting EEC approval for project commencement, EEC's rejection of certain components of the Regional Communications and Regional Telecommunications projects, and EEC's insistence on appointing a consultant to the UNDP/ITC Project on Trade Promotion, Information and Advisory Service (South Pacific Bureau for Economic Cooperation, 1982d, pp. 1–2).

Passage of Lomé II, a global trade and development agreement between the European countries and the countries in Asia the Caribbean and the Pacific (hence ACP), introduced additional levels of complexity into Pacific regional policy processes, including challenges of coordination and prospects of duplication and institutional rivalry alongside the prospects of higher levels of development assistance. Priority areas for development included agricultural and integrated rural development (with emphasis on food self-sufficiency); commodity marketing and processing; development of mineral resources, energy resources; fisheries and marine and other seabed resources; industry; tourism; transport and communications; and education and training—all being sectors that UN programmes were similarly interested in. The Pacific states felt they held a minor position in the ACP group given the size of the African states and adopted a strategy that sought to increase and improve on the benefits obtained from Lomé I, or at least sustain existing benefits (Pacific Islands Forum Secretariat, 1984c, pp. 1–2).

Another early example of the coordination challenge was in the maritime sector. The April 1986 first meeting of the South Pacific Working Group on Maritime Affairs deliberated on how to maximise use of twelve projects on offer from IMO, ESCAP, ICAO, and UNCTAD: participation in a regional network of maritime training centres, training through the World Maritime University, Port Pricing and Investment Policy, a regional project on computerization in port, maritime safety training, a regional seminar on transport, handling and stowage of dangerous goods, courses on specimen examination, and ship-repair assistance (Pacific Islands Forum Secretariat, 1986a). Several of these projects were associated with the whole Asia-Pacific region, and ESCAP was commencing a "Decade of Transportation" inclusive of Pacific regional and bi-lateral projects. Whereas each of these development projects had their own merit, SPEC suggested to the Working Group that not all were relevant for the Pacific region at that time, and that those projects which were relevant would only be successful if conditions for successful implementation were carefully noted and adhered to. Collectively these projects were valued at several millions of dollars: the temptation was to accept Pacific participation so as to maximise the region's share, but this increased the potential for scarce human and material resources to become unavailable for more timely and appropriate use. Similar judgements were required across all sectors, whether social, economic, environmental, or administrative.

In 1988 SPEC and UNDP held monthly meetings in Suva at Deputy Director level. UNDP and ESCAP were at that time funding a Short-Term Advisory Services scheme (Approx. USD 75,000 per year), and the Pacific Energy Development Program (PEDP) together with SPEC's Energy Division. SPEC was insistent that it, rather than ESCAP, be the executing agent for South Pacific Maritime Development Projects (SPEC, 1988, p. 1).

The problem of inter-agency coordination also applied to coordination challenges within the UN system, such as between UNDP and other agencies having closely aligned mandates or projects. In the 1970s this applied to telecommunications projects in which both UNDP and ITU were heavily invested. These emerging issues concerning development administration, were later addressed under MDG goal 8: "financing for development", as well as the "Paris principles."

6.3 Pacific Policy Networks

The forgoing discussion builds a picture of the small Pacific states striving for solutions about how to handle expanding institutional engagements and opportunities. This "architecture of Pacific regionalism" emerged in response to the needs of the PICs and though interaction with metropolitan powers and inter-governmental organizations—principally the United Nations. In the absence of "deep integration" amongst the PICs, in such forms as economic, customs, or political union, the CROP agencies, with PIF as lead agency, exemplify "network governance" (Rhodes, 2001) that relies on "soft" instruments such as "declarations", "statements", and "frameworks", in preference to legally-binding treaties, to identify common purposes and interests and to resolve problems of coordination.

An OECD discussion paper on aid coordination presented to a 1988 discussion group on regional economic issues noted the emerging use of "consultative groups" by the World Bank and "round tables" under UNDP sponsorship. There were "donor's meetings", which could nonetheless include recipient country representatives, and "country aid group" meetings which more specifically included both donors and recipients (OECD, 1988). A PIFS presentation to UNCTAD in 1994 suggested that Papua New Guinea's planning processes were more reliant on World Bank consultations but "most other Forum Island countries" had focused on UNDP round tables (South Pacific Forum Secretariat, 1994, p. 4).

This pattern, in which a few formal institutions interweave with a larger number of networks, platforms, and partnerships that use soft mechanisms of governance, describes Pacific practice. Within the region there are eighteen Forum countries, five Forum Observer countries and territories, nine intergovernmental regional organizations within the Council for Regional Organizations in the Pacific (CROP); regional and national civil society organizations; and regional and national private sector entities and organizations. By 2020 the Forum's international engagement included 18 Forum Dialogue Partners; seven Forum Development Partners; international organizations and mechanisms, including the WTO, World Bank, IMF, ADB, and the European Community in addition to the United Nations. This network complexity is replicated for the Pacific Community, and each of the other CROP agencies, not least the Secretariat of the Pacific Regional Environment Programme.

From the outset the Pacific's regional organizations—spearheaded by SPC and SPEC—faced the challenge of weaving together PIC national development plans with opportunities for sector-specific development that were offered in overlapping ways by donor agencies and international inter-governmental organizations. PIFS has assumed the main responsibility for policy coordination, while SPC focused on technical projects and assisting PICs with implementation. The PIFS Secretary-General presides over annual Leaders Meetings and chairs the Council of Regional Organizations of the Pacific. The Annual leaders' meeting includes formal sessions that consider policy options generated by regional organizations and member countries, as selected by a Forum Officials Committee meeting; plus an informal day-long leaders retreat at which matters are considered in "Pacific style" and off the record.

Messaging and decision-making between so many national, regional, and international actors, is coordinated through formal meetings with Forum Leaders and Officials – the Associate Member/Observer Dialogue, Forum Dialogue Partners Dialogue, CROP Heads Dialogue, NSA-Leaders' Dialogue, Private Sector Dialogue; Regional policy consultations and submission processes across Forum Member countries as part of the *Framework for Pacific Regionalism*; Participation in global events such as the United Nations General Assembly and Conference of the Parties (COP) climate change conferences; workshops, seminars and events across the region on priority issues such as trade, fisheries and climate change; and digital surveys, updates and briefings.

6.3.1 Council of Regional Organizations of the Pacific (CROP)

By the 1980s the need to improve regional coordination of the multiple agencies, networks and processes described above had become obvious to all. A South Pacific Organisations Coordinating Committee was established as a point of contact between the growing number of regional inter-governmental organizations. In 1988 this was re-constituted as the "Council of Regional Organizations of the

Pacific" (CROP). CROP first met in 1989 and eventually comprised the Fiji School of Medicine (FSMed), Pacific Islands Forum Fisheries Agency (FFA), Pacific Islands Development Programme (PIDP), the Secretariat for the Pacific Community (SPC), PIFS , the Secretariat of the Pacific Regional Environment Programme (SPREP), South Pacific Tourism Organization (SPTO), The University of the South Pacific (USP), the Pacific Power Association (PPA) and the Pacific Aviation Safety Office (PASO).[2] The Council is chaired by the Pacific Islands Forum Secretariat and oversees sectoral working groups, which are introduced further below.

The select number of CROP agencies belies the extensive array of additional regional organizations. The *Pacific Association of Supreme Audit Institutions* (PASAI), the Oceanic Customs Organization (OCO), and the South Pacific Chiefs of Police Conference (PICP) are additional examples of organizations that could be included in the premier group of regional organizations.

6.3.2 Crop Working Groups

Whereas the Pacific's CROP agencies each have legal form and clear membership rules, participation by member countries in specific programmes is voluntary, and very few decisions taken at regional level are legally or administratively enforceable. "Regional consultations have their limitations", according to a 1994 PIFS report, "...in large part because there is no firm commitment on many of the issues. While recommendations are made and are normally endorsed by government leaders at the Forum, they are not binding on either island countries or the development partners..." (South Pacific Forum Secretariat, 1994, p. 8). This means that "successful" implementation of regional initiatives must rely on other drivers. The more that member countries identify their needs and ensure that these needs are addressed in regional programmes, the more they can be expected to participate in planning, implementation, and evaluation of such programs (Hassall, 2021, p. 139).

At regional level as much as national, governments tussle constantly with "siloization" versus "whole of government" approaches better suited to the cross-sector realities, and to the pursuit of "sustainable development." The Forum Secretariat recorded its perspective on this issue in a 1994 report to a conference convened by UNCTAD:

> ...by and large, regional organizations are organized along traditional sectoral lines reflecting the approach to aid and aid management that prevailed at the time of their establishment. There is also some misunderstanding of the term "sector" which is often loosely used to describe areas of development which are not single sector issues or even sectors at all. These include population, the environment and women. Separating them out as if they were sectors has sometimes created the impression that they are optional extras (which they are not). If governments and regional organizations are to ensure sustainable

[2] https://www.forumsec.org/council-of-regional-organisations-of-the-pacific/

> development, they must seek to fully integrate these critical, and multisectoral, issues into mainstream development planning." (South Pacific Forum Secretariat, 1994, pp. 3–4)

In response to such circumstances, CROP inter-agency and cross-sectoral working groups have been established to maximise synergies in their work programmes, and to avoid duplication of effort. Twelve working groups existed in 2020, although the number and purpose of groups changes as the region's need for collaboration changes. These groups, and the organizations they bring together, are set out in Table 6.1.

The informality of these networks delivers both the strength and the weakness of Pacific regionalism. On the one hand their flexibility provides a coping mechanism for diversity, contention, and dispute; on the other, lack of clear lines of authority and accountability can slow the pace of decision-making and implementation. This is a price that regional countries are willing to pay to preserve their political unity in the longer term, in the face of extremely polarizing political and economic pressures brought into the region by external actors.

The 2014 *Framework for Pacific Regionalism* defines Pacific regionalism as "The expression of a common sense of identity and purpose, leading progressively to the sharing of institutions, resources, and markets, with the purpose of complementing national efforts, overcoming common constraints, and enhancing sustainable and inclusive development within Pacific countries and territories and for the Pacific region as a whole", and which seeks to set out."...a robust process through which regional priorities will be identified and implemented" (Pacific Islands Forum Secretariat, 2014). However, the expansive policy agenda and growth in number of regional actors calls into question the ability of soft regionalism to deal with complex matters. Leaders' decisions are little more than non-binding declarations which are only implemented when leaders drive the necessary political, legal and administrative processes in their respective countries. There is no regional parliament empowered to enact regional law, nor executive to implement, nor any a formal dispute mechanism to adjudicate disputes amongst member states. In 2020, for instance, Pacific Island political leaders promised to tackle corruption by signing the Teieniwa Vision, which was in turn endorsed by PIF in 2021.

6.4 Financing for Development

Continuity of funding has always been a concern. So many projects are either proposed but not funded (due to lack of funds) or commenced but not completed (due to lack of funds). No doubt it was often the case that the relevant UN agency was short of funds; there was always the possibility that agencies without funding approval from projects they were less partial too, and that funds flowed for projects that fit neatly with current UN / agency priorities. It is one thing to have UNDP experts provide technical advice, the Premier of the Cook Islands reminded leaders at their second meeting in 1972, and another to fund the implementation of that

Table 6.1 Council of Regional Organizations of the Pacific Working Groups

CROP Taskforce/Working Group[a]	Purpose	Chair & Taskforce/Working Group Members
Nuclear Legacy Issues in the Pacific	To deliver on the Forum Leaders' decision to address the ongoing impacts and legacy issues of nuclear testing in the blue Pacific, with a primary focus on Republic of Marshall Islands and Kiribati, through effective CROP collaboration and utilisation of technical expertise and supportive resources.	***Chair:*** Republic of Marshall Islands ***Members:*** Fiji, Kiribati, Republic of Marshall Islands, PIFS, SPC, SPREP and USP
2050 Strategy for the Blue Pacific Continent	To engage CROP agencies in the development of the 2050 Strategy, particularly for the purpose of providing technical and specialist advice to assist in the discrete steps of the development of the 2050 Strategy.	*Members:* FFA, PASO, PIFS, PPA, SPC, SPREP, SPTO and USP *Members of the FOC Sub-Committee on 2050 Strategy for the Blue Pacific Continent may participate in the meetings of the Taskforce.*
International Engagement and Advocacy for Ocean Events in 2020–2021	To provide technical advice that will inform strategic engagement and advocacy messaging to Forum Members in preparation for, and participation at, the second UN Ocean conference and other ocean-related events in 2020 and 2021.	***Chair:*** OPOC and FFA (co-chairing) ***Members:*** PIFS, SPREP, SPC, FFA, USP, SPTO and OPOC
Socio-economic Impact Assessment Taskforce – COVID- 19	To oversee the design and delivery of a region-wide socio-economic impact assessment of COVID-19 pandemic, through a coordinated and collaborative approach, undertaken first and foremost through the CROP mechanism, to inform regional policy mechanisms, and underpinned by the outcomes and decisions of the Forum Economic Ministers Meeting of 11-12 August 2020.	***Chair:*** PIFS and SPC (co-chairing) ***Members:*** Relevant CROP agencies and Forum Members
Pacific Sustainable Development Working Group (PSDWG)	To provide technical advice and support to the Pacific Sustainable Development Steering Committee (PSDSC) in its role to promote the effective implementation of the Pacific roadmap for Sustainable Development (PRSD). It is responsible for supporting coordination,	*Members:* 4 key representative groups totalling 11 members: (a) CROP Agencies (4–1 rep each for social, environment, economic and governance) (b) UN agencies (3–1 rep each pillar) (c) Non-State Actors (2– Regional NGO, regional private

(continued)

Table 6.1 (continued)

CROP Taskforce/Working Group[a]	Purpose	Chair & Taskforce/Working Group Members
	implementation, monitoring and reporting on global sustainable development commitments including the 2030 Agenda for Sustainable Development and the SDGs and the S.A.M.O.A Pathway.	sector) (d) Development partner representatives (2 – Bilateral, multilateral)
Gender Working Group (GWG)	To encourage greater coordination and collaboration amongst CROP in addressing gender issues, and to identify emerging gender issues and ensure they are brought to the attention of CROP Heads. A key role is to promote gender equality and the empowerment of women in the work of CROP members and intergovernmental processes regionally and internationally.	***Chair:*** SPC and PIFS (co-chairing) ***Members:*** FFA, PASO, PIFS, PPA, SPC, SPREP, and USP
Human Resources Development Working Group (HRDWG)	To coordinate regional activities that support national systems in the areas of formal and non-formal education from pre-school to adult, and across the public and private sector. In addition, the HRDWG in collaboration with the PacREF Steering Committee, is responsible for providing oversight and assisting with the facilitation of the Pacific Regional Education Framework (PacREF) 2018-2030.	***Chair:*** USP ***Members*****:** USP (PacREF Facilitation Unit, Institute of Education, School of Education, Educational Quality & Assessment Programme), UNICEF, MFAT, DFAT, PIFS, APTC, UNESCO
ICT Working Group (ICTWG)	To provide a common platform through which CROP agencies and other relevant stakeholders can address key emerging and on-going regional priorities in the ICT sector in a collective, collaborative and coordinated manner.	***Chair:*** USP ***Members:*** FFA, SPC, PIDP, PIFS, PPA, SPREP, SPTO, USP, PITA, ADB, World Bank, DFAT, MFAT, PIPSO, JICA, WFP, PIF (represented by Forum Troika and SIS). ***Observers:*** Vodafone, Digicel, ITU, CSO Representative, Pacific Island Territories
Marine Sector Working Group (MSWG)	To share information on marine sector issues, particularly with regards to ongoing work programmes and projects, to promote synergies and coordination.	***Chair:*** FFA and SPC (alternating annually) ***Members:*** SPC (Fisheries, Aquaculture and Marine Ecosystems Division and Geoscience, Energy

(continued)

Table 6.1 (continued)

CROP Taskforce/Working Group[a]	Purpose	Chair & Taskforce/Working Group Members
		and Maritime Division), FFA, SPREP, USP, PIFS, OPOC
Communications Working Group	To provide a professional network for CROP communications staff and coordinates and implements CROP communication activities.	***Chair:*** Rotated per meeting between CROP agencies ***Members:*** FFA, PASO, PIFS, PPA, SPC, SPREP, SPTO and USP
Corporate Directors Working Group	To look at CROP harmonisation in all areas and to progress 'quick wins' in corporate harmonisation, as agreed at CROP meeting in June 2019, to roll the responsibilities of CROP Human Resources Group (comprised of HR staff of CROP agencies) into the CROP Corporate Directors Group, and to formalise this Group through a TOR and regular reporting to CROP Deputies.	***Chair:*** PIFS ***Members:*** FFA, PASO, PIFS, PPA, SPC, SPREP, SPTO and USP
Energy Technical Working Group (*Details to be confirmed)*	To guide and coordinate the implementation of the Framework for Energy Security and Resilience in the Pacific (FESRIP): 2021-2030	Chair: SPC Members: PIFS, PPA, SPREP, USP

[a]https://www.forumsec.org/council-ofregional-organisations-of-thepacific/#1507175105152-f00da375-ce46

advice (Government of Fiji. Department of Foreign Affairs, 1972, p. 45). Over time, project funding gained the description "soft money" to indicate the sense of impermanence attached to projects, sometimes even the most successful amongst them. In fact, projects were more likely to gain funding if they were short to medium term, rather than likely to establish long-term dependence on development funds. From UNDP's viewpoint, funding was always provided on the presumption that it was only required until recipient countries attained self-reliance in a field, and not become dependent on external funding in perpetuity. PICs are aware of the gap between knowledge of development opportunities and ability to fund them.

The availability of donor funding has its own history. In the 1950s the World Council of Churches first suggested that the "developed" countries allocate 1% of their GNP to assist "developing" countries. However, this proposition was based on moral grounds rather than economic analysis, and whereas the moral argument took root, the appropriate target for what came to be termed "overseas development assistance" (ODA) took longer to identify. The figure of 0.7% to be reached "by 1975 and in no case later than 1980" was agreed in a UNGA resolution of 24 October 1970. A few Scandinavian countries reached this target for several years, but as at

2016 the average level of ODA, at least for the OECD countries, never exceeded 0.4% (OECD. Development Assistance Committee, 2016). By the 1990s the pattern of Pacific Region's economic performance, and outlook, was clear. Development assistance averaged almost 27% of GDP for the period 1980–1992, with some countries also having high remittances—Kiribati as high as 80%. Aid was contributing to development, but not to growth (Osborne, 1996, p. 4).

The volatility of development assistance was a critical issue well before it gained prominence as MDG 8. In the 1970s UNDP entered the Pacific with high aspirations. But an admission in 1976 that a deterioration in its financial position required reduction in the scope of some of the first regional projects introduced uncertainty into planning for a number of years. Only intensive lobbying to UNDP headquarters and additional financial pledges from Britain's aid program allowed a regional pests and diseases survey and telecommunications projects to proceed as planned (South Pacific Bureau for Economic Cooperation, 1976b, p. 1). SPEC informed UNDP that no additional regional projects would be considered until the issue of adequate financing was resolved, a position UNDP agreed with (South Pacific Bureau for Economic Cooperation, 1976b, p. 2). The same SPEC committee meeting noted the "...continuing need to encourage ESCAP to extend sufficient and effective assistance to the Pacific, in accordance with the needs of the region, and that funds allocated to ESCAP specifically for regional projects in the South Pacific are utilized accordingly." (South Pacific Bureau for Economic Cooperation, 1976b, p. 3).

The unpredictable nature of development funding was further highlighted in 1982, when PICs were surprised to find that UNDP was only planning to fund half of the telecommunications projects that had been put to a visiting assessment mission (South Pacific Bureau for Economic Cooperation, 1982e, p. 1). The EEC's signal at the second ACP Pacific group officials meeting in Suva in April 1984 that the persistence and depth of the world recession and high levels of unemployment in Europe would impact on the quality and quantity of aid available was interpreted by the meeting as the Community's "...intended miserliness" (Pacific Islands Forum Secretariat, 1984a, p. 1). Uncertainty over funding in the telecommunications sector continued into the 1990s. A meeting of the regional telecommunications committee in November 1990 was informed that a decision by UNDP to redirect funds from national programmes to LDCs would result in a 40% reduction in available funding for existing ITU projects run from Suva. The UNDP's representative at that meeting explained that UNDP's purpose was to build up national capacity and create self-reliance. When PIC representatives present indicated a preference for continued support from UNDP/ITU rather than strengthening the Forum's Telecommunications section, they were reminded that UNDP preferred "...recipient countries to consider funding part of the costs of a future telecommunications program through the UNDP/recipient government cost-sharing mechanism." (Pacific Islands Forum Secretariat, 1990c, p. 18). Forum representatives made clear to the meeting that if the ITU projects were discontinued that the Forum had no intention of taking over its roles.

Over time, the expectation has been replaced by an expectation that funding will be continuous—and in larger amounts. UNDP operates four-year funding cycles.

Allocations made by the UNDP governing council for the period 1977–1981 were: Fiji USD 5 million, Gilbert Islands USD 1.1 million; Niue USD 1 million; Solomon Islands USD 2.3 million; Tonga USD 2.0 million; and Western Samoa USD 5.25 million. The Cook Islands, New Caledonia, New Hebrides, Nauru, Tokelau Islands, Tuvalu and Wallis and Futuna shared an additional USD 3.8 million. Papua New Guinea's IPF of USD 8.75 million was administered through a separate UNDP country office and Resident Representative (South Pacific Bureau for Economic Cooperation, 1978, p. 65). In February 1982 UNDP's Governing Council approved a budget of USD 213 million for the whole Asia Pacific for the five years to 1991, within which USD 27 million was allocated to regional projects (ie including at least three countries) in PICs. An additional USD 25 million was allocated for individual PIC country programmes (South Pacific Forum, 1987, p. 150).

UNDP allocated USD 20 million to the Pacific for funding cycle 1982–1986—spread over 33 projects, of which telecommunications—at USD 28 million—was the largest (South Pacific Bureau for Economic Cooperation, 1981, p. 8). Under the fourth cycle, 1987–1991, UNDP's New York office made a list of desirable projects and invited SPEC to comment. At a meeting of planners in Suva in March 1986 to rank the projects in order of regional priority, transport, including inter-island shipping, was not given high priority (Pacific Islands Forum Secretariat, 1990c). UNDP's Suva-based Multi-country office doubled its resource allocation to 10 Pacific Island Countries for 2008–2010 to a total of USD61 million (UNDP contributes more resources to Pacific, 2008). In 2003 the Multi-Partner Trust Fund Office was established within UNDP to facilitate UN coherence and development effectiveness on multifaceted issues—such as humanitarian crises, peacebuilding, recovery and development—and engages in collaborative international efforts on pandemics, climate change and biodiversity conservation.

UN funding is, of course, but one element in the total ODA envelope. Official development assistance to the Pacific Islands doubled between 2005 and 2011. The region is reported as receiving the highest proportions of ODA per capita globally (Pacific Islands Forum, 2015, p. 40). Between 1970 and 2019 official ODA to the Pacific region increased from USD 22,860,000 per year to USD 1.056 billion. As a percentage of gross national income, the increase was from 8.8% in 1979 to 9.9% in 2019.[3] Significant differences are embedded in that regional statistic: Marshall Islands, Tuvalu and Federated States of Micronesia rely on ODA for approximately 40% of their gross national income (Pacific Islands Forum, 2015, p. 40).

The rules for development finance are set by the more powerful countries and agencies (including the UN and the World Bank) and approved in UN consultations and conferences in the form of such mechanisms as the Green Climate Fund (GEF). Whereas AOSIS and the PICs worked hard at Rio and Barbados to set up the GEF, its operational policies were controlled by its major funders (the World Bank, the EU and the wealthy developed countries). GEF works through UNDP, UNEP, the World Bank (implementing agencies of GEF), and other specialised UN agencies,

[3] https://data.worldbank.org/indicator/DT.ODA.ALLD.CD?locations=S2 accessed 27 January 2022

including the Food and Agriculture Organization and regional development banks, including the Asian Development Bank to provide grants and concessional funding for projects and programmes on sustainable development.

At the Thirty-Sixth Pacific Islands Forum meeting in Papua New Guinea in October 2005, leaders agreed the need to "facilitate international financing for sustainable development, biodiversity and environmental protection and climate change in the Pacific including through the Global Environment Fund". (Paragraph 4 36th Pacific Island Forum Official Communiqué) However, this has not proven easy: countries much larger than those of the Pacific—such as Nigeria and South Africa, India, and China—compete for the same funds, as do also some UN agencies (UNDP, FAO, UNEP, etc) seeking to cover the costs of delivering climate programs. In this global context, funding for PICs from the Green Climate Fund has been slow, dissipated, and difficult to access, notwithstanding the fact that PICs played an active role in its creation, and the Pacific Islands Forum and other CROP agencies have been assisting with application processes.

6.4.1 Multi-Partner Funding and Implementation

Development assistance is frequently planned and implemented by multi-agency platforms. The South Pacific Regional Environment programme (SPREP) was jointly administered by SPEC (chair), SPC (facilities), ESCAP and UNEP (both for funding and expertise—although UNEP withdrew after 1988). In the 1990s UNDP and ESCAP's combined contribution to PIF revenue for 1989 (1,104,993) was the second largest behind Australia (Pacific Islands Forum Secretariat, 1990a). In the field of human rights, an EU-PIFS Human Rights Project 2012-2015 used EU funds, PIFS coordination, and UN expertise, to implement human rights consultations at national level. Project goals included ratification and implementation of international Human Rights Treaties and the Rome Statute; and support for Initiative 12.5 of the Pacific Plan which focused on ratification and implementation of international and regional human rights conventions, covenants and agreements. In telecommunications, a regional training coordinator reported to SPEC, under policy established by the South Pacific Regional Telecommunications Meeting (SPECTEL), in close collaboration with UNDP/ITU Pacific Regional Telecommunications Training project.

6.4.2 Regional Development Bank

Proposals for a regional development fund date to an SPC technical meeting on coordination and economic development in 1970. A regional development bank was proposed by UNDP (South Pacific Forum, 1972b, p. 14) but not given region-wide support. Fiji, for one, preferred to establish its own development bank (South Pacific

Forum, 1972a, p. 46). The Cook Islands submitted a paper on the subject at the 1972 Forum that pointed out that many UN proposals had been made to the Cook Islands government which had been not acted upon for lack of financing (South Pacific Forum, 1972a, p. 45). Then, in 1980 a Pacific Islands Development Conference at the East-West Center offered to facilitate the establishment of a regional development fund. A 1981 SPEC/SPC report noted the continuing gap in development finance beyond the support being provided by UN specialized agencies, the European Community, World Bank, and ADB. Control over funding of projects is, of course, the greatest lever that any organization can control, regardless of whether that project is implemented by a single government or agency, or a combination of them "in partnership." The UN might agree in general to providing funds to a disaster preparedness fund, but baulk at the opportunity to donate when specifically requested.

6.4.3 The Monterrey Consensus

In March 2002 the UN convened an international conference on Financing for development in Monterrey, Mexico. Most PICs attended (excepting, it seems, Papua New Guinea, Kiribati, and Vanuatu). Samoa made a statement in the session on Partnership in Financing for Development. Most AOSIS members were present and consulted together during the conference. Forum SG Noel Levy and an adviser were also at present and consulted with PICs while the Monterrey outcomes were negotiated. The conference outcome established several "leading actions" to improve the prospects of development finance. These included mobilizing domestic financial resources; mobilizing international resources, such as foreign direct investment and other private flows; improving the terms of international trade; increasing international financial and technical cooperation; attending to the problem of external debt; and other issues relating to the coherence and consistency of the international monetary, financial, and trading systems (United Nations, 2002).

6.4.4 Case Studies in Multi-Party Project Implementation

Each of the elements essential to the planning and implementation of regional development projects introduced above can be illustrated through the following case studies. Some of the problems that surface in multi-agency projects are also noted—particularly rivalry for agency ascendancy, and the insecurity of development finance. In each case, direct UN-engagement played crucial roles in feasibility studies, leading to small projects which in turn expanded as finance, political will, and administrative cohesion, were brought together. The energy, transport and communications sectors have been highlighted due to there being so little awareness of this early—in some cases continuing—UN engagement in the Pacific Islands context.

6.4.4.1 Communications/Telecommunications

Obtaining adequate funding for telecommunications research, and for the implementation of telecommunications projects, has been a persistent challenge (South Pacific Forum Secretariat, 1976, p. 12). Small populations spread over vast ocean areas combined with low average incomes of PICs, make it difficult to finance such essential telecommunications infrastructure as satellite stations and sub-marine cables. There has always been a presumption that provision of such infrastructure will have significant impact on growth in trade and tourism, and development agencies including the UN, The World Bank, ADB, and private sector partners, have supported these sectors. In 1972 UNDP recruited two experts to report on potential improvements (South Pacific Forum, 1972a, p. 87) and in 1973 Forum Leaders gave top priority to upgrading telecommunications in the South Pacific Region and delegated responsibility to SPEC. The paragraphs below describe a range of initiatives in the sector in the decades that followed.

In 1975 ITU asked each PIC to state its telecommunication plans for the coming years, and SPEC was assessing whether to appoint a regional ITU coordinator. By 1975 UNDP had agreed to pay for more than half of the external costs of a Regional Telecommunications Training School. There were two additional projects underway: a Feasibility Study for a South Pacific Regional Telecommunications Network, and a maritime telecommunications project. By July 1977 the feasibility study and telecommunications training centre were included in a list of priority regional projects submitted by SPEC to the ACP/EEC Lomé Convention for financing by the European Community (South Pacific Bureau for Economic Cooperation, 1977). A training centre was established in Fiji with resources contributed by Australia, New Zealand, and UNDP, under the directorship of an ITU expert. Additional training centres were set up in Tonga, Cook Islands, and Western Samoa.

South Pacific Regional Telecommunications meetings comprising regional organizations and donor were held annually from 1973. A terminal report for the Reginal Telecommunication training program was delivered to the ninth South Pacific Regional Telecommunications Meeting in Niue in 1982 (South Pacific Regional Telecommunications, 1982b). The project had commenced with six countries and expanded to 10 and expanded from an initial period of 4 years to eight. SPEC now proposed that UNDP/ITU shift from short term development consultancies and that telecommunications be planned as regional development project. The eighth meeting, in Sydney in October 1981, reviewed the provisions of a USD 1.5 million 5-year regional program to develop telecommunications starting in 1982 (South Pacific Bureau for Economic Cooperation, 1981, p. 142).

The 32-member Asia-Pacific Telecommunity, comprising governments, telecom service providers, manufactures of communication equipment, and research & development organizations, was established in 1979 through joint initiatives of the United Nations Economic and Social Commission for Asia & the Pacific and the International Telecommunication Union. (ITU).

The Federated States of Micronesia sent representatives to the World Administrative Radio Conference in 1979 to investigate what support it could receive prior termination of its free association with the USA.

The ITU's ICB4PAC project comprehensively surveyed the status of seven essential features of internet governance in the PICs: Numbering; National ICT Policy; International Mobile Roaming; Cybercrime; Licensing; Universal Access and Service; and Interconnection and Cost Modelling. Such policy assessments were accompanied by assistance with policy frameworks, and training. UNDP/UNOPS developed the "e-Pacifika Project" to facilitate the development of National ICT Strategy Plans for 12 Pacific Island countries,

In 1999 the Pacific Islands Forum adopted a Vision for the Pacific Information Economy, and developed a regional approach to telecommunications regulation and tariff levels through the Asia-Pacific Telecommunity, the Pacific Islands Telecommunications Association, and the International Telecommunications Union. A framework for development of ICTs in the Pacific region was outlined in the 2002 regional ICT strategy, the Pacific Islands information and Communications technologies policy and strategic Plan (PIIPP). This "Pacific Umbrella Initiative" brings together the Pacific Island Countries and territories, donor countries, intergovernmental agencies to promote the use of ICTs in support of sustainable development for the people of the Pacific Islands. The project was coordinated by the CROP ICT working group, established in 2001 from the merging of IT-PACNET—a technical cooperation group involving IT managers within the CROP agencies that commenced in 1995, and the Information Sector World Group, which dates to 1998. At its inaugural 2001 meeting the ICT Working Group identified three objectives:

- Review, clarify and provide advice on the development of priorities in the information sector;
- Determine areas of complementarity and overlap and any potential gaps in the coverage of existing and proposed regional activities and initiatives in ICT; and
- Recommend to CROP implementation procedures for enhancing coordination and cooperation, and for ensuring that priority areas are adequately addressed.

This coordinated effort made possible the intergovernmental "Pacific Islands Regional ICT consultation", in Suva in April 2003, in preparation for opportunities to take part in global dialogues under United Nations auspices. The Tokyo Declaration prepared at the Asian regional consultation prior to the World Summit for the Information Society included an "Islands Paragraph" pleading the special circumstances of Small Island Developing states:

> These countries, vulnerable to environmental hazards, and characterized by small, homogenous markets, high costs of access and equipment, human resource constraints exacerbated by the problem of "brain-drain", limited access to networks and remote locations, will require particular attention and tailored solutions to meet their needs. (Para 11, section 2, page four of SWIS/PC-2/DOC/6-E)

The ITU's contribution in the Pacific is widely appreciated, and this appreciation was expressed the President of the Federated States of Micronesia, Joseph Urusemal, when addressing the High-Level Plenary Meeting of the 60th Session of the United Nations General Assembly in 2005:

> In this era of information and communication technological advances, my government applauds the United Nations and the International Telecommunication Union for their commitment to ensuring that information and communication technologies are placed in the service of humankind in the ongoing noble task of economic development. For island countries like the Federated States of Micronesia whose population is spread out over a vast area of the Pacific Ocean, access to reliable and affordable ICT is critical in our nation building. It is for this reason that my Government strongly supports the Declaration of Principles and Plan of Action as endorsed by the first phase of the World Summit on the Information Society (WSIS) and looks forward to the convening of the second phase in Tunisia at the end of this year." (Government of the Federated States of Micronesia, 2005)

6.4.4.2 Energy

Given its pivotal role in all aspects of economic development, one might have presumed that PICs, UN agencies, and other development partners, had given full their full attention to the establishment of successful energy policies at national and regional levels. In reality, the region's quest for appropriate energy policies is marked by institutional rivalries, duplication of effort, and implementation failure.

A UNDP/ESCAP Inter-Country programming Mission for the South Pacific which visited the region in late 1978 solicited government views on the topic of alternative energy sources. Responses were positive and a subsequent joint UNDP/ESCAP/SPC/ESCAP meeting gave the topic of a Regional Advisory Service on Alternative Sources of Energy high priority (South Pacific Bureau for Economic Cooperation, 1979a, p. 2). UNDP proposed the establishment of a regional energy programme to consider energy technology, economics, financing, and institutional arrangements, to be funded through ESCAP and coordinated by UNDAT. A fully costed work program was proposed to a SPEC committee meeting in Honiara in July 1979 (South Pacific Bureau for Economic Cooperation, 1979b).

Then, an UNDP/ ESCAP working group in Apia in June 1980 decided on a regional energy program funded by EEC and UNDP. In 1981 Forum Leaders considered a recommendation that the regional energy program be housed in Suva at ESCAP's cost, alongside SPEC's energy unit (South Pacific Forum Secretariat, 1981, p. 10). This became known as the Pacific Energy Development Programme.

To complicate matters, a 1982 joint UNDP/World Bank project that assessed the energy outlook of six countries recommended bulk purchasing of fuel to decrease cost and guarantee supply—but also concluded that the Pacific Island petroleum market was "basically uncompetitive" (Pacific Islands Forum Secretariat, 1984b, p. 2). ESCAP's Pacific Energy Development Programme (PEDP) reviewed the World Bank's report and concluded that: "The deficiencies of the analytical framework and methodology employed by ESMAP in its assessment of the alternative supply options are so serious as to entirely undermine the conclusions reached in the Report regarding the relative merits of the various options, particularly the relative merits of a regional tender versus national tenders" (Pacific Islands Forum

Secretariat, 1986b, p. 34)—no doubt to the consternation of the officials responsible for selecting future energy policy settings.

A subsequent independent review of the energy sector activities of Forum Secretariat and UN Pacific Energy Department Program requested by the 1988 regional energy meeting recommended that the Forum Secretariat energy division, the regional petroleum unit, and the UN's PEDP be amalgamated into single program under forum auspices (Pre-Forum Officials Committee, M., 1989, p. 4 Attachment: Pacific Energy Programme). The work of PEDP was limited by budget (Pacific Islands Forum Secretariat, 1989), despite having received financial support from UNDP from 1983 to the end of 1991 (Pacific Islands Forum Secretariat, 1990b, p. 2 Agenda Item 7: Report on Twenty-First South Pacific Forum Meeting and Energy Sector Assistance in 1991).

By the early 2000s, and with articulation of the Pacific Plan, the energy sector remained among the region's key policy challenges. The Pacific remained heavily dependent on petroleum, and in 2011 the Forum Leaders approved the *Framework for Action on Energy Security in the Pacific* (FAESP) for the period to 2020. Reviews of the Framework's implementation plan undertaken in 2014 and again in 2019 had consistent findings:

> ...cordination was ineffective, there was limited cooperation or collaboration among the regional agencies. Progress was significant though limited, but with insufficient information to assess overall results; FAESP was often used to guide CROP regional energy activities, but the implementation plan was not used much; governance through the Pacific Energy Oversight Group (PEOG) and a Pacific Energy Advisory Group (PEAG) were useful for networking and information exchange, but not for cooperation or coordination; and, finally, reporting was inaqeuate, and no independent monitoring and evaluation (M&E) were carried out. Following midterm review recommendations, SPC commissioned an enhanced implementation study and considered changes to the terms of reference and membership of the advisory and oversight groups, but no changes were made. (Pacific Community & Pacific Regional Infrastructure Facility & Pacific Community, 2019, p.4)

6.4.4.3 Transport

UNDP has studied the Pacific Regional transport challenges since the 1970s (South Pacific Forum Secretariat. Leader's Meeting, 1971). An inter-country program for 1982–1986, which included a survey of both shipping and aviation in the South Pacific, included UNDEP and ESCAP, and was coordinated by SPEC—to cover both shipping and aviation. A steering committee for the survey first met in Suva in January 1984 and the report completed early in 1985 included a costly shopping list of vessels, port facilities and other infrastructure and technology. The report's findings were considered at a meeting of a working group on maritime affairs in April 1986.

Transport development policy is now considered as an aspect of connectivity, sustainability, and prosperity (Pacific Islands Forum Secretariat, 2022). Marine, air, and land transport are all fundamentally associated with and connectivity, but so are ICTs, and the ability to connect goods and services through information transfer rather than physically can result in less environmental impact, less energy consumption, and greater efficiency.

6.5 Data Collection and Information Management

Several influences in the second half of the twentieth century produced ever-increasing expectations concerning project monitoring and evaluation. The earliest development plans relied on agricultural production data and size of population as indicators of economic development. However, as the value of development assistance funds increased and the numbers of participating development agencies expanded, more focus was placed on program outcomes and not just simple tabulation of "inputs and outputs." Two other significant influences were expansion of the concept of development to include "human development", and increased computing capacity. Departments of statistics emerged as partners with departments of National Planning and Finance. An understanding of how computing hardware, software, and policy settings, can combine to improve government functioning was mentioned as early as 1982 in a UNDAT paper on development processes in the Pacific region:

> The real need is for the evolution of the simplest possible planning and management techniques, compatible with the scale of operations. Even in the smallest countries, the opportunity for retention of planning and management techniques with the desired degree of refinement has been greatly enhanced in the last year or two, with the advent of microprocessors. (South Pacific Bureau for Economic Cooperation, 1982f, p. 2)

Since that statement a great number of initiatives of by UN agencies such as ITU, plus impetus from the World Summit on the Information Society to specific projects of the UNDP, have supported the expansion of digital government in the region. The World Bank and ADB have also supported the installation of infrastructure. Development partners including Australia, China, and Korea, have included ICT components in their development assistance packages.

6.5.1 Monitoring

Development progress by the PICs is documented by multiple domestic, regional, and international bodies. National statistics agencies are the first level of data collection, and much effort has been put into their development. Based on innumerable experiences with such practices, integrity in data collection and results reporting has become increasingly important. The mantra of SPC's Statistics for Development Division's business plan for 2018-2020 was "Trusted Pacific data supporting well informed policy decisions that help improve Pacific people's lives" (Pacific Community. Statistics for Development Division, 2019). Accurate data collection, accompanied by clear reporting of findings, can build confidence in stakeholders and result in increased funding support.

Of course, data collection is only valuable in proportion to its comprehensiveness, timeliness, and accuracy. It is not necessarily a "neutral" exercise, as political ramifications can accompany positive or negative findings. Understanding causes of death in a country depends on making accurate notes on death certificates; keeping track of population growth depends on ensuring that new births are registered; ensuring that education resources are maximized depends on the ability to monitor attendance by both students and teachers. A government whose administration is experiencing high rates of women dying in childbirth might prefer that the causes of death be attributed elsewhere, such as to heart failure; a government that is under-spending due to lack of capacity may seek to shift funds to other budget lines.

Although PICs had been developing statistics and planning capacity since independence or even before, tracking the MDGs during the period 2000-2015 proved daunting. The eight goals had 21 associated targets and 60 indicators, which proved too large a surveillance task for some small states. As assessed in the UN Pacific Strategy document for 2018-2022:

> Most countries have national and sector level development plans, but significant support is needed for their operationalisation, monitoring, and evaluation. Compounding issues include the paucity of data, inadequate links between planning and budgets, uneven sharing of capacity and resources between sectors, and a lack of capacity to effectively implement, monitor, and report on development. (The United Nations, 2018, p. 25)

These capability needs were addressed in the Pacific Statistics Strategy, (2010–2020) coordinated by SPC, which engaged key UN agencies, including the UN Statistics Division, PFTAC, and Australia and New Zealand, to improve statistics capability in all PICs. The strategy focused on building skills for a "core set of Pacific Statistics"—key social, economic, and environmental indicators in particular (Sustineo, 2016), and the ability to monitor population, health, trade, and other vital sectors, and to conduct household surveys and national census.

In 2014, Pacific Ministers responsible for civil registration committed to the Asia Pacific decade for civil registration and vital statistics (CRVS),[4] and the *Regional Action framework for CRVS*, which calls for: (i) universal registration of births, deaths and causes of death (ii) legal certification of all vital events and (iii) production of vital statistics from civil registration data. Associated initiatives were the Ten-Year Pacific Statistics Strategy (TYPSS) and Pacific Vital statistics Action plan (PVSAP), coordinated by SPC with the support of Brisbane Accord Group established in 2010 which comprised the Health Information Systems Knowledge Hub at the University of Queensland, SPC, UNFP, WHO, UNICEF, PHIN, ABS, QUT, UNSW and FNU.

Regional and international agencies possess more information management and communications resources than the small states and use data collected at national

---

[4]https://www.unescap.org/sites/default/files/Asian_and_Pacific_Civil_Registration_and_Vital_Statistics_Decade2015-2024_Booklet.pdf

level to sustain information portals at regional level. SPC's Pacific Data Hub, for instance (https://pacificdata.org/) holds more than 1300 datasets managed by SPC and other Pacific-based organizations, and SPC hosts at least four additional key data portals:

- Pacific Ocean Portal: http://oceanportal.spc.int/portal/ocean.html
- Pacific Regional Data Repository (about energy): https://prdrse4all.spc.int/content/about
- WACOP—Changing Waves and Coasts in the Pacific: https://wacop.gsd.spc.int/
- PDH.Stat: Development Indicator Database: https://stats.pacificdata.org/

SPREP hosts eight key environmental data portals:

- Inform Regional Data Portal: https://pacific-data.sprep.org/
- Invasive Species Battler Resource Base (BRB): https://piln.sprep.org/
- Pacific Climate Change Portal (PCCP) / RTSM / APT / DST: https://www.pacificclimatechange.net/
- Pacific Islands Protected Area Portal (PIPAP): https://pipap.sprep.org/
- Pacific Meteorological Desk & Partnership (PMDP): https://www.pacificmet.net/
- Pacific Network for Environmental Assessment (PNEA): https://pnea.sprep.org/
- Regional Frameworks And Strategies Directory
- Lessons Learned And Best Practices In Environment Management: https://www.sprep.org/lessons-learned-and-best-practices-environment-management

Global agencies also maintain data portals that cover the PICs. For the most part these rely on data provided by national statistics agencies or departments, but they also track development assistance from external sources. The United Nations Development Group's Multi-Partner Trust Fund Office Gateway at https://mptf.undp.org/ provides real time tracking of funds provided by 44 UN agencies, across 140 countries and hundreds of funds. Since 2009 the Lowy Institute has maintained a Pacific Aid Map at https://pacificaidmap.lowyinstitute.org/. Notwithstanding the strength of the portals mentioned above, the UN's Pacific Offices have not developed knowledge portals to make available the extensive array of studies they have produced.

6.5.2 Evaluation

The United Nations has created processes and established agencies to monitor and evaluate results. At macro level these include global conferences and their follow-up conferences held at intervals in subsequent years (see Chap. 9). Program completion reports are also submitted to the initiating organ, such as the General Assembly, the Economic and Social Council, or one the various special purpose entities established to monitor the outcomes of specific treaties and agreements. UNDP established an Evaluation Resource Centre (https://erc.undp.org/) and an Evaluation Group (http://www.uneval.org/) which works with 50 UN agencies to improve the effectiveness, efficiency, impact, and sustainability of their evaluation process. The first country-level independent evaluation of UNDP effectiveness in PICs (excluding Papua New Guinea) published in 2012 (United Nations Development Programme. Evaluation Office, 2012) provided a systematic and comprehensive assessment of the Programme's Pacific context, administrative structure and operation, and contribution to development results. The report reached eight conclusions and made seven recommendations. Overall, these were positive, suggesting that UNDP's Pacific programmes were making significant contributions to development results. It found, on the other hand, that the quality of results varied by country, and that a "differentiated" approach was required for the smaller countries which had thin governance structures and lacked a critical mass of trained people. Most criticism was placed on inefficiencies in UNDP programme management. One of the report's most interesting findings was that UNDP was underutilizing its potential as "...a repository of global knowledge and development experience." (United Nations Development Programme. Evaluation Office, 2012).

6.6 Reporting and Accountability

In 1997 the Forum Economic Ministers Meeting drew on standards developed by the IMF to develop a biennial eight-point stock-take of implementation of leadership codes and principles of accountability:

1. Budget processes, including multi-year frameworks, to ensure Parliament/Congress is sufficiently informed to understand the longer-term implications of appropriation decisions.
2. The accounts of governments, state-owned enterprises and statutory corporations to be promptly and fully audited.
3. Loan agreements or guarantees entered into by governments to be presented to Parliament/Congress.
4. All government and public sector contracts to be competitively awarded, and publicly reported.
5. Contravention of financial regulations to be promptly disciplined.
6. Public Accounts Committees of the legislature to be empowered to require disclosure.
7. Auditor General and Ombudsman to be provided with adequate fiscal resources and independent reporting rights to Parliament/Congress.
8. Central banks with statutory responsibility for non-partisan monitoring and advice, and regular and independent publication of informative reports (Pacific Islands Forum Secretariat. Forum Economic Ministers Meeting, 2000).

These were, in turn, endorsed by the Forum Leaders meeting in 2003. Strengthening accountability has been an important goal for PICs and development agencies alike. But accountability of who to whom, and for what activities? Does accountability lay with individuals, or with organizations? And how is it "exercised"? Public servants face a constant moral dilemma concerning who to give their primary loyalty to:

obviously, it should be 'the public interest'. But who determines what this "public interest" looks like? Should they also meet the interests of their employers? And what of their personal interests? Where does one focus primarily loyalty one is employed on a short-term contract in a small and remote country?

Elected officials are also accountable, both to their electors and to the laws under which they act. But how is accountability exercised by publics that have little knowledge or understanding of what the UN stands for and how it functions? Countries are represented at the UN by the executive branch of government. Some elected officials hold executive power and others legislative power. But representation at the United Nations is through the executive branch of government and there is no formal link with a nation's legislature or parliament. And yet, parliaments have considerable responsibility for the domestication of international obligations and for oversight of executive action in foreign affairs and international organizations. A fourth political group for whom accountability is important are non-government organizations, which justify their participation in UN-related activities through being representatives of interests in civil society. The UN has accepted this assertion and has sought to develop civil society capabilities but doing so raises questions about who civil society organizations represent, and how accountable they are for the impact of their representation.

6.6.1 Paris Principles on Aid Effectiveness, Ownership, and Monitoring and Evaluation

In 2005 the OECD convened a conference in Paris to consult on mechanisms for improving aid effectiveness and accountability. The conference was fully supported by the UN, the World Bank, and other international development agencies, and attended by more than 100 countries. The Pacific was represented by PIF, Fiji, Papua New Guinea, Solomon Islands, Tonga, and Vanuatu. The major Conference outcome, the *Paris Declaration and Aid Effectiveness Principles*, set out five principles to improve the ways in which the work of international development partners aligned with developing country development plans and aspirations:

- OWNERSHIP: Partner countries exercise effective leadership over their development policies, and strategies and co-ordinate development actions
- ALIGNMENT: Donors base their overall support on partner countries' national development strategies, institutions and procedures
- HARMONIZATION: Donors' actions are more harmonized, transparent and collectively effective
- MANAGING FOR RESULTS: Managing resources and improving decision-making for results
- MUTUAL ACCOUNTABILITY: Donors and partners are accountable for development results (Organization for Economic Co-operation and Development, 2005)

The Pacific region has certainly experienced challenges in each of these areas, given the ever-increasing numbers of development agencies and actors in Pacific countries and the Pacific region in addition to UN agencies, including the World Bank, the Asian Development Bank, the European Union, the Commonwealth Secretariat—as well as very many regional and bi-lateral state and non-state actors. Historically, each agency and actor has applied their own funding, assessment and reporting mechanisms, reporting periods, and project cycles. The EU's European Development Fund (EDF), for example offers large financial support, but at the cost of substantial reporting and compliance procedures.

The Paris conference resulted in the establishment by OECD and UNDP of the Global Partnership for Effective Development Cooperation (https://www.effectivecooperation.org) which has, by stages, evolved from the *Paris Declaration on Aid Effectiveness* (2005), to the *Accra Action Agenda* (2008), the *Busan Partnership Agreement* (2011), the *Nairobi Outcome Document* (2016) and in 2019, the first senior-level Meeting at UN HQ in New York. Six Pacific countries (Samoa, Papua New Guinea, Kiribati, Solomon Islands, Tonga and Tuvalu) attended the 2016 Conference in Nairobi. The project monitors five dimensions of development assistance effectiveness: (1) national development planning and results orientation; (2) multi-stakeholder engagement; (3) quality and use of public financial management systems; (4) predictability and forward planning; (5) and mutual accountability mechanisms for all countries receiving development assistance, including 13 PICs.

6.6.1.1 Ownership

Relations between PICs and international organizations such as the United Nations appear to suffer from the classic "Principal-Agent" problem, which the PICs, as the "principal", engage the UN—"the agent"—to provide required expertise. The problem is that the agent has more understanding of the problem—and ability to fix it—than the principal—in much the same manner that a motor mechanic invariably has more knowledge of how an automobile functions than does its driver, or that a medical practitioner generally possesses far more knowledge of a sickness and its remedy than does a patient. Who, then, is to be in charge and take responsibility for decision-making and its outcomes: the principal, who has presented the problem for resolution, or the agent, who is willing to fix it for a price? In the instance of the UN there is the possibility that the UN, although the "agent", may not only fix the problem but also pay the costs (sometimes this agent may even offer to fix problems that haven't been presented yet).

The "political economy" of policy processes must be considered (Jordan & Turnpenny, 2015). Who, for instance, participates in the process? Are the processes authentic or perfunctory? Where is the policy process convened? What are the governance arrangements and where does accountability lay? Are the goals technical, or economic, or political? Who is financing the projects, and with what expectations? What efforts were made, if any, to disseminate the results of meetings, conferences, technical and scientific findings, for public education? Such questions are pertinent when reviewing UN participation development planning and implementation in the Pacific region. If the UN acquires land on one of the capital cites for the purpose of constructing a "UN building", does this imply anything other than long-term engagement with that country? What does it imply for increase or decrease in dependency on UN agency support? Should one assume that the goal is always increased self-reliance—or could it be to reap the benefits of "vulnerability"? Is it possible to seek to maximize "rents" and ODA on the one hand, and yet assert full sovereign autonomy on the other? PICs are known to build ODA into annual budgets, and this relieves the pressure on their governments to fully raise their own revenues. Over time this results in governments relying on externally sourced funding, giving less attention to tax regimes, and potentially losing control over policy, planning, and ownership of results. It has the tendency to generate new forms of clientelism and dependence (Duncan & Hassall, 2011).

Theoretically, the people of the Pacific Islands are the principal beneficiaries of engagement with the United Nations, and this can be accomplished through direct funding of economic and social development; through development of human capacity; and through UN appraisal of government performance. There are other specific beneficiaries of UN engagement: parliamentarians seeking re-election; public servants (local, national, regional, and international) who are climbing professional 'ladders'; experts who garner consulting fees and experience; and academics who have their findings validated, published, and sometimes even remunerated.

The issue of coordination of development assistance has been an ongoing source of tension between Pacific regional organizations and UN bodies. Whereas both sets of parties recognize that sovereign authority rests with PICs and their intergovernmental organizations, the UN agencies have understandably sought to have a say in where and how their resources are being used, and to what effect.

Concern about the influence of foreign consultants and expertise has focused on the limited time that experts take to familiarize themselves with conditions in the region before making recommendations. The Pacific nations have no need to be told about their development challenges—which they are only too familiar with. More beneficial are reports that identify appropriate technologies and strategies, as well as the resources to ensure their sustainability once implemented. "The Pacific Islands countries, relative to population size," a team reported to SPEC in 1976, "... have more foreign specialists than probably any other region in the world. Overseas specialists—whether employed by the international agencies or coming directly on bilateral agreements from various governments, or even those working for regional organizations, are in many cases inadequately prepared. Their task would be much more effectively undertaken if appropriate orientation could be organized." (Finau et al., 1976, p. 21).

Another continuing source of tension has been the cost of consultants and their reports. A UN survey of regional transport took two years and $200,000 to complete, the director of SPEC informed the Forum Leaders meeting in 1974: "The money spent on this report alone, he said, could have gone towards purchasing a vessel." (South Pacific Forum, 1974, pp. 73–74). At a 1974 meeting convened by SPEC to determine a regional strategy for revitalizing the banana industry attended by officials from Cook Islands, Fiji, Niue, Tonga and Western Samoa, at which SPEC reported unsatisfactory aspects of a UNDP/FAO Pests and Diseases Survey: it took some experts away from their other regional projects, it lacked a plan or timetable, and there had been long delay in appointing a team leader (South Pacific Bureau for Economic Co-operation, 1974b, p. 2), p..2. When UNDP allocated 2.4 million for telecommunications projects in its plan for 1982-86, Fiji pointed out that $798,8000 of that amount was consumed by the cost of employing four fulltime experts for a three-year period (South Pacific Bureau for Economic Cooperation, 1981).

There were circumstances, on the other hand, in which consultants were beneficially deployed. The Forum's Telecommunications Division reported to a regional telecommunications meeting in 1990 that the use of consultants "had seemed a good way of satisfying the needs of the FICs, as it provided a flexible and wide range of high-level technical expertise and should be continued for 1991." (Pacific Islands Forum Secretariat, 1990c, p. 2). In more recent years the problem of balancing foreign and local expertise has continued to exist, although increasing numbers of Pacific Islands' consulting businesses have become established and are bidding competitively in their fields.

6.6.2 *The Position of Civil Society*

Whereas the number of sovereign states rose during the twenty century to 193, the number of *international* non-governmental organizations grew from approximately 1000 in 1915 to 37,000 in the year 2000 (Christensen, 2004, p. 50). UN support for NGOs commenced in the 1990s, after UNDP's Governing Council decided to draw civil society organizations into its development activities globally. A 1990-91 survey identified 411 NGOs in ten PICs having attributes that UNDP felt would make them suitable for engagement in development activities: capacity to plan and deliver sustainable economic and social development services through effective grass-roots networks; self-reliance; a special concern for the disadvantaged; a willingness to foster networks and to communicate and cooperate with government while retaining their independence, and a capacity to absorb collaborative inputs from UNDP (Pierce, 1991, p. 7). Having distinguished these organizations from other non-government organizations which lacked a distinctive focus on development, such as trade unions and employer's associations, schools, sports clubs, professional and commercial associations, cooperatives, social clubs and religious bodies, the UNDP review assessed the development needs of those organizations still in scope and identified five categories of need: better management capabilities, network development, more and better qualified staff, funding, and technical assistance (Pierce, 1991, p. 20). Subsequent to this comprehensive report, UNDP established close working relationships with NGOs, which continue to the present time.

Coincidently, the Pacific Islands Association of Non-Governmental Association (PIANGO) was established in 1990-1991. Its inaugural meeting in Pago Pago in August 1991 was funded by AIDAB, Commonwealth Foundation, New Zealand

Aid, and UNDP, and it has participated in major UN events ever since. In 1994 regional NGO peak bodies accepted an opportunity offered by the preparatory process for the International Conference on Population and Development, which was to be held in Cairo that year. With funding from the UNFPA, AIDAB and the South Pacific Alliance for Family Health, PIANGO collaborated with the New Zealand and Fijian peak NGO bodies to bring together 70 NGO delegates from twenty-five PICs to review the *Port Vila Declaration on Population and Sustainable Development*, and the *Cairo Draft Programme of Action* and prepare a Pacific regional NGO statement for PCPD's PREPCOM III (PIANGO, 1994). PIANGO's monthly email digest *Pacific News* was a vital means of keeping civil society across the Pacific informed of UN conferences, prepcoms, seminars, and job opportunities.

PIANGO was part of a vibrant Pacific NGO presence at the Oceans Conference at UN Headquarters in New York in June 2017 and PIANGO's Executive Director. Emele Duituturaga, was for a period the NGO/DPI Executive Committee's Regional Coordinator. PIANGO obtained consultative status with ECOSOC in 2018 but—curiously—neither its *Strategic Plan 2013-2015* (Pacific Islands Association of Non-Governmental Organizations, 2013, p. 14) nor its Strategic Plan for the period 2020-2030[5] refer to UN engagement.

Civil Society organizations have taken active interest in human rights issues. A review of Pacific Islands experiences with the first Universal Periodic Review cycle found that civil society organizations had participated both domestically and during the review process in Geneva:

> NGOs and civil society were involved in all stages of the UPR process from drafting reports, engaging in national consultations, representing themselves in Geneva, lobbying state representatives, conducting side events, reporting on outcomes and holding states accountable to accepted recommendations. The level of engagement differed between states due to technical and financial capacity, as well as access to relevant information. For example it was reported in both Tonga and Tuvalu's national reports that not all civil society organizations were aware of the UPR, with the Tonga national report stating that almost all of the 49 civil society organizations that are members of the Civil Society Forum of Tonga (CSFT) were unaware of the UPR process. (Secretariat of the Pacific Community & Pacific Regional Rights Resource Team, 2012, p. 9)

There is always the possibility that "democratic voice" in Pacific societies will be discretely disciplined and dismissed, as McCarthy suggests occurs elsewhere:

> The public sphere of social-welfare-state democracies is rather a field of competition among conflicting interests, in which organizations representing diverse constituencies negotiate and compromise among themselves and with government officials, while excluding the public from their proceedings. Public opinion is, to be sure, taken into account, but not in the form of unrestricted public discussion. Its character and function are indicated rather by the terms in which it is addressed: "public opinion research", "publicity", "public relations work", and so forth. The press and broadcast media serve less as organs of public information and debate than as technologies or managing consensus and promoting consumer culture. (McCarthy, 1991, p. xii)

[5]http://www.piango.org/home/about-us/strategic-plan/

Pacific civil society interaction with the UN system takes place through dialogue, information sharing, capacity building, and much lobbying and agenda-setting. But civil society support for the United Nations is not well organized within PICs—if the formation of United Nations Associations is a relevant indicator: whereas the World Federation of United Nations Associations has over 100 members,[6] none are (as at 2023) in a Pacific Island Country. Development assistance was provided for the establishment of such Associations in the number of Pacific states, but these initiatives fell away once that funding came to an end. A Fijian association was established in 2014[7] and a Samoan association was established but was short-lived.[8]

6.7 Discussion

Understanding policy networks as they have evolved and currently operate across the Pacific Island countries and region is vital to the success of governance interactions involving governments, development partners, and other non-state actors such as civil society organizations. The term "governance" is itself multi-faceted, and subject to variations in definition and emphasis by agencies wishing to emphasize its democratic, financial, economic, legal, or other dimensions. For UN agencies, the challenge has been to improve the governance performance of their development partners while at the same time improving their internal and inter-agency systems of governance. UN agencies have also established multiple lines of policy dialogue in PICs, both within and outside government, and at both national and regional levels. The complexity of gaining multi-party agreement and sense of common purpose has at times resulted in sub-optimal outcomes as suggested by overviews of the energy and transport sectors. Notwithstanding the range of challenges noted above, the attention to improving governance and policy processes has produced a considerable body of policy literature and expertise relating to all facets of devising, implementing, monitoring, and evaluating, development projects in Pacific Islands context.

References

Asian Development Bank, Secretariat of the Forum of Pacific Island Countries, United Nations Development Programme, United Nations Economic and Social Commission for Asia and the Pacific – Pacific Operations Centre, & United Nations Office for Project Services. (2000).

---

[6]http://www.wfuna.org/who/members/

[7]https://www.unafiji.org/

[8]The United Nations Association of Western Samoa (www.samoa.ws/suna) appears have been established to host a meeting in 2000: the first Global Millennium People's Assembly to Make Peace Stop Wars (April 2-72,000).

Pacific Regional Conference on Governance for Parliamentarians 27-30 March 2000. Conference Report. Asian Development Bank, Secretariat of the Forum of Pacific Island Countries, and the United Nations.

Asian Development Bank. (2009). ADB Support for Public Sector Reforms in the Pacific: Enhance Results through Ownership, Capacity, and Continuity. Asian Development Bank.

Asian Development Bank. (2010). Strengthening Governance and Accountability in Pacific Island Countries (Formerly Strengthening Financial Management and Accountability in PDMCs) (Pacific Islands) Technical Assistance Completion Report. Asian Development Bank. https://www.adb.org/sites/default/files/project-documents//38639-reg-tcr.pdf

Christensen, R. K. (2004). International Nongovernmental Organizations: Globalization, policy learning, and the nation-state. In D. Levi-Faur & E. Vigoda-Gadot (Eds.), *International public policy and management: Policy learning beyond regional, cultural, and political boundaries* (pp. 49–66). Marcel Dekker.

Cumming, L. M., Hammer, M., van Belle, S., & Mayhew, S. H. (2012). *Accountability and engagement of global institutions involved in climate change*. Department for International Development.

Duncan, R., & Hassall, G. (2011). How pervasive is clientelist politics in the Pacific? In R. Duncan (Ed.), *The political economy of economic reform in the Pacific* (pp. 265–276). Asian Development Bank.

Finau, P. P., Craw, C., Morgan, L. R., Crocombe, R. C., Hughes, A. V., Qionibaravi, M., & Spratt, R. G. (1976). *More effective aid: A report to the South Pacific Forum*. South Pacific Bureau for Economic Co-operation.

Government of the Federated States of Micronesia (2005). *Address by H.E. Joseph J. Urusemal, President of the Federated States of Micronesia before the High-Level Plenary Meeting of the 60th Session of the United Nations General Assembly, Friday 16 September.*

Government of Fiji. Department of Foreign Affairs. (1972). *South Pacific Forum Suva Fiji 12-14 September 1972. Summary Record and Final Press Communique*. Suva.

Hassall, G. (2021). *Government and public policy in the Pacific Islands*. Emerald Publishing Limited.

Jordan, A., & Turnpenny, J. R. (Eds.). (2015). *The tools of policy formulation: Actors, capacities, venues and effects*. Edward Elgar Publishing.

Karlsson-Vinkhuyzen, S. I., Groff, M., Tamás, P. A., Dahl, A. L., Harder, M., & Hassall, G. (2017). Entry into force and then? The Paris agreement and state accountability. *Climate Policy, 1-7*. https://doi.org/10.1080/14693062.2017.1331904

Kooiman, J. (2003). *Governing as governance*. SAGE.

McCarthy, T. (1991). *Introduction to Jürgen Habermas, the structural transformation of the public sphere an inquiry into a category of bourgeois society (1st MIT Press pbk ed.)*. MIT Press.

OECD (1988). *Aid Co-ordination: A Review of Country and Regional Experience*. Informal Discussion Group on Regional Economic Issues Meeting, SPEC Headquarters, Suva 26 February 1988. Meeting papers, ((88) MP. 8). Suva.

OECD. Development Assistance Committee. (2016). History of the 0.7% ODA Target. https://www.oecd.org/dac/financing-sustainable-development/development-finance-standards/ODA-history-of-the-0-7-target.pdf.

Organization for Economic Co-operation and Development. (2005). *Paris declaration on aid effectiveness*. OECD.

Osborne, P. (1996). *Trade issues and development prospects of island developing countries of the Pacific*. Paper presented at the United Nations. Economic and Social Council. Commission on Sustainable Development. High-level Panel Meeting on Island Developing Countries, 22-23 April.

Pacific Community. Statistics for Development Division. (2019). *2018–2020 business plan*. Pacific Community.

Pacific Community & Pacific Regional Infrastructure Facility & Pacific Community. (2019). Review of the framework for action on energy security in the Pacific (FAESP) 2010-2020, Sydney & Suva.

Pacific Islands Association of Non-Governmental Organizations. (1994). *Pacific region non-government organizations conference on population and sustainable development, Nadi, Fiji, 2-4 March. Conference Papers*. Suva.

Pacific Islands Association of Non-Governmental Organizations. (2013). *Strategic plan 2013-2015: CSO leaders rethinking development in the Pacific Islands*. Pacific Islands Association of Non-Governmental Organizations.

Pacific Islands Forum. (2015). *2015 Pacific regional MDGs tracking report*. Pacific Islands Forum Secretariat.

Pacific Islands Forum Secretariat. (1983a). National Planning and Statisticis office Republic of Vanuatu. Comments on the Draft Report on Regional Project Criteria and Priorities. Regional Meeting on Criteria and Priorities, SPEc Headquarters, Suva 21-23 February 1983. Meeting papers. *(83) MP. 11*((83) MP. 11).

Pacific Islands Forum Secretariat. (1983b). Project Manager's report. Regional Telecommunications Development Project. South Pacific Regional Telecommunications Meeting, Port Vila, Vanuatu 24 - 28 October 1983. Meeting papers. *(83) MP. 6*.

Pacific Islands Forum Secretariat. (1983c). Regional meeting on criteria and priorities, SPEc Headquarters, Suva 21 - 23 February 1983. Meeting papers. *(83) MP. 11*((83) MP. 11).

Pacific Islands Forum Secretariat. (1983d). United Nations office of the Resident Representative, 13 December 1982, re Draft Report on Regional Project Criteria and Priorities. Regional Meeting on Criteria and Priorities, SPEc Headquarters, Suva 21 - 23 February 1983. Meeting papers. *(83) MP. 11*((83) MP. 11).

Pacific Islands Forum Secretariat. (1984a). ACP Pacific Group of Ministers Third Meeting. Annex 2. Second ACP Pacific Group Officials Meeting, SPEC Headquarters, Suva 24 - 25 April 1984. Meeting papers. *(84) MP. 4*((84) MP. 4).

Pacific Islands Forum Secretariat. (1984b). Agenda item 9: energy petroleum pricing and suppoy: Some potential options. Fifteenth South Pacific Forum, Funafuti, Tuvalu 27 - 28 August 1984. Meeting papers. Folder 1. *(84) MP. 8*((84) MP. 8).

Pacific Islands Forum Secretariat. (1984c). Summary Paper by J. Samy on Negotiations for the Successor to Lome II convention. Annex 4. Second ACP Pacific Group Officials Meeting, SPEC Headquarters, Suva 24 - 25 April 1984. Meeting papers. *(84) MP. 4*((84) MP. 4).

Pacific Islands Forum Secretariat. (1986a). Agenda Iteem 6: UNDP Fourth Inter-Country Programme – Maritime Sector – Annex II: South Pacific Working Group on Maritime Affairs, First Meeting, Suva, FIji, 3 - 4 April. Meeting papers. *(86)MP.12*((86)MP.12).

Pacific Islands Forum Secretariat. (1986b). SPEC/PEDP Regional Petroleum Meeting,, Suva Fiji, 20 – 22 May. Meeting papers. Folder 1. *(86) MP. 8*((86) MP. 8).

Pacific Islands Forum Secretariat. (1989). Regional petroleum unit management committee inaugural meeting, Forum Secretariat, Suva 15 - 16 May 1989. Meeting papers. *(89) MP. 8*.

Pacific Islands Forum Secretariat. (1990a). *Financial statements for the year ended 31 December 1989. Pre-Forum Officials Committee Meeting, Port Vila, Vanuatu 23 - 25 July 1990. Meeting papers, Folder 3*. Suva.

Pacific Islands Forum Secretariat. (1990b). *Regional energy meeting, Nuku'alofa, Tonga 25 - 27 September 1990. Meeting papers*. Suva.

Pacific Islands Forum Secretariat. (1990c). Second regional telecommunications committee meeting. State Legislature Chambers Pohnpei, Federated States of Micronesia, 6-7 November 1990. Agreed Record. *SPFS(90)26*.

Pacific Islands Forum Secretariat. (2004). *An assessment of regional mechanisms and processes in the Pacific*. Pacific Islands Forum Secretariat.

Pacific Islands Forum Secretariat. (2014). *The framework for Pacific regionalism*. Pacific Islands Forum Secretariat.

Pacific Islands Forum Secretariat. (2022). *Second quadrennial Pacific sustainable development: Report 2022*. Pacific Islands Forum Secretariat.

Pacific Islands Forum Secretariat. Forum Economic Ministers Meeting. (2000). *Forum economic ministers meeting, Alofi, Niue, 25-26 July 2000. Session 11 background document: Progress report on the Forum's eight principles of accountability*. Pacific Islands Forum Secretariat.

PIANGO. (1994). *Report of Pacific NGO delegation to the prep comm III meeting of the international conference on population and development (ICPD), 4-22 April 1994, United Nations Headquarters*. Pacific Islands Association of Non-Government Organizations.

Pierce, T. (1991). *United Nations Development Programme Non-Governmental Organization (NGO) Study*. Suva: United Nations Development Programme. office of the Resident Representative for Fiji, KIiribati, Nauru, Solomon Islands, Tonga, Tuvalu, Vanuatu, Federated States of Micronesia, Marshall Islands, Palau and Pacific Regional Programmes.

Pre-Forum Officials Committee, M. (1989). Pre-Forum Officials Committee Meeting, Tarawa, Kiribaiti 3 - 5 July 1989: [meeting papers]. *(89) MP. 1.*

Rhodes, R. A. W. (2001). *Understanding governance: Policy networks, governance, reflexivity and accountability*. Oxford University Press.

Secretariat of the Pacific Community. (2000). *Summary of decisions from the thirtieth meeting of the committee of representatives of governments and administrations. Annex 2*. Nouméa.

Secretariat of the Pacific Community, & Pacific Regional Rights Resource Team. (2012). *Pacific Island states and the universal periodic review: A toolkit for Pacific Island states to measure progress and compliance against universal periodic review recommendations*. Secretariat of the Pacific Community.

South Pacific Bureau for Economic Co-operation. (1974a). *The 5th South Pacific Forum. Rarotonga Cook Islands 20-22 March 1974: Summary Record and Final Press Communique*. Suva.

South Pacific Bureau for Economic Co-operation. (1974b). Technical meeting on the revitalization of the banana industry, Apia, 6–8 August 1974: [meeting papers]. *(74) MP. 1.*

South Pacific Bureau for Economic Co-operation. (1979). *SPEC Committee Meeting, Honiara, Solomon Islands 5-6 July 1979: Agreed report*. South Pacific Bureau for Economic Co-operation.

South Pacific Bureau for Economic Cooperation. (1974). SPEC Committee Meeting, Wellington, N.Z., 3-5 December. Progress Report by the Director. *SPEC(74)55.*

South Pacific Bureau for Economic Cooperation. (1976a). More Effective Aid: A report to the South Pacific Forum. *SPEC(76) 11*(SPEC(76)11).

South Pacific Bureau for Economic Cooperation. (1976b). SPEC Committee Meeting, Nauru 22 - 23 July 1976: UNDP Cutbacks on Regional projects. *SPEC(76)12*(SPEC(76)12).

South Pacific Bureau for Economic Cooperation. (1977). Fourth South Pacific Regional Telecommunications Meeting, Port Moresby, PNG, 11 - 15 /7/77: [meeting papers]. *(77) MP.4.*

South Pacific Bureau for Economic Cooperation. (1978). Directory of Aid Agencies. *(78)FSCOL4.*

South Pacific Bureau for Economic Cooperation. (1979a). Regional advisory services on alternative sources of energy. SPEC Committee Meeting, Honiara, Solomon Islands 5 - 6 July 1979: [meeting papers]. *SPC(79) SCH.1 (Draft).*

South Pacific Bureau for Economic Cooperation. (1979b). United Nations Development Programme. Alternative Sources of Energy. SPEC Committee Meeting, Honiara, Solomon Islands 5 - 6 July 1979: [meeting papers]. *SPC(79) SCH.1 (Draft).*

South Pacific Bureau for Economic Cooperation. (1980). ACP/EEC Pacific Group Ministerial Meeting on Regional Projects to be financed under the LOME II Convention, Apia, Western Samoa 28 - 29 August 1980: [meeting papers]. *(80) MP. 1.*

South Pacific Bureau for Economic Cooperation. (1981). Eighth South Pacific Regional Telecommunications Meeting, Sydney, Australia, 19-23 October 1981. Agreed Record. *(81)FSCOL.15.*

South Pacific Bureau for Economic Cooperation. (1982a). ACP Pacific Group Council of Ministers First Meeting, Suva, Fiji, 28-29 April 1982. Agreed Record. *(82) MP. 12*(SPEC(82)8).

South Pacific Bureau for Economic Cooperation. (1982b). ACP Pacific Group Council of Ministers Meeting, SPEC Headquarters, Suva, Fiji 28 - 29 April 1982. Meeting papers. *(82) MP.1.*

South Pacific Bureau for Economic Cooperation. (1982c). Opening statement by the Director of SPEC, Mr J.P. Sheppard. Regional Conference of Development Planners. SPEC Headquarters, Suva, 7-11 June. *(82) MP. 8.*

South Pacific Bureau for Economic Cooperation. (1982d). Procedural problems affecting the smooth implementation of LOME regional financial and Technical Co-operation Programmes. ACP Pacific Group Council of Ministers Meeting, SPEC Headquarters, Suva, Fiji 28 - 29 April 1982. Meeting papers. *(82) MP.1.*

South Pacific Bureau for Economic Cooperation. (1982e). Regional development project tasks and resources. Ninth South Pacific Regional Telecommunications Meeting, Alofi, Niue 28 October 3 November 1982. Meeting papers. *(82) MP. 12.*

South Pacific Bureau for Economic Cooperation. (1982f). UNDAT – The development process – some thoughts. Regional Conference of Development Planners. SPEC Headquarters, Suva, 7-11 June. *(82) MP. 8.*

South Pacific Bureau for Economic Cooperation, & Forum Secretariat. (1975). Report of the South Pacific Regional Telecommunications Meeting (1975: Nuku'alofa, Tonga). *SPEC(75) 26.*

South Pacific Commission. (1974). Special Meeting between SPEC representatives and SPC Staff: 20-21 May 1974. *(74) MP. 6.*

South Pacific Forum. (1972a). South Pacific Forum, Suva, Fiji 12-14 September 1972. Summary Record and Final Press Communique. *NN (73) NN. 1.*

South Pacific Forum. (1972b). Summary record of proceedings, February 1972. *(72) NN. 1,* 1 v. (various pagings).

South Pacific Forum. (1974). Fifth South Pacific Forum, Rarotonga, Cook Islands: summary record and final press communique. *FSC (74) FSCOL. 1.*

South Pacific Forum. (1987). Eighteenth South Pacific Forum, Apia, Western Samoa 29 - 30 July 1987. Annotated Agenda. *(87) MP. 2.*

South Pacific Forum Secretariat. (1976). *Seventh South Pacific Forum 26 - 28 July 1976 Nauru: summary record.* Suva, Fiji:

South Pacific Forum Secretariat. (1981). Energy co-ordination and energy information dissemination in the South Pacific Region. Twelfth South Pacific Forum, Port Vila, Vanuatu 10-11 August 1981: [meeting papers]. *81(MP)7*(81(MP)7).

South Pacific Forum Secretariat. (1982). South Pacific Regional Environment Programme: Report to South Pacific Forum and South Pacific Conference. Thirteenth South Pacific Forum, Rotorua, New Zealand, 9 - 10 August 1982. Meeting papers. *(82) MP. 11*((82) MP. 11).

South Pacific Forum Secretariat. (1983). SPEC – Civil Aviation. Fourteenth South Pacific Forum, Canberra, Australia 29 - 30 August 1983. Meeting papers. *(83) MP. 1.*

South Pacific Forum Secretariat. (1994). *Review of international support to ECDC: Experiences of developed and developing countries, their problems and prospects.* UNCTAD.

South Pacific Forum Secretariat. Leader's Meeting. (1971). *Communique.* South Pacific Forum Secretariat.

South Pacific Regional Telecommunications, M. (1982a). ITU – Meeting of NDP Aid Coordinators of Asia and the Pacific. Inter-Country Telecommunication Program and Projects 1983-1986. Ninth South Pacific Regional Telecommunications Meeting, Alofi, Niue: 28 October - 3 November 1982: [meeting papers]. *(82) MP. 9.*

South Pacific Regional Telecommunications, M. (1982b). Ninth South Pacific Regional Telecommunications Meeting, Alofi, Niue: 28 October - 3 November 1982: [meeting papers]. *(82) MP. 9.*

SPEC. (1988). Action taken to implement decisions of the Pre-Forum SPEC Committee, Apia, 1987. In *Pre-Forum SPEC Committee Meeting, Nukualofa, Tonga 14 - 16 September 1988. Meeting papers. Folder 1* (Vol. (88) MP. 1 Folder 2).

Sustineo. (2016). *Pacific statistics assessment: A survey of user and producer perspectives.* Government of Australia. Department of Foreign Affairs and Trade.

UNDP contributes more resources to Pacific. (2008, 8 April). *Pacnews.*

The World Bank and The Institute of Social and Administrative Studies. (1993). Making Government More Effective: Pacific Public Sector Reform in the South Pacific: Report on a Four-Day Seminar for Government Ministers with Their Permanent Secretaries. University of the South Pacific.

United Nations. (2002). *Report of the international conference on financing for development, Monterey, Mexico, 18-22 march*. United Nations.

United Nations Development Programme. Evaluation Office. (2012). *Assessment of development results. Evaluation of UNDP contribution Pacific Island countries*. United Nations Development Programme.

United Nations. (2018). *United Nations in the Pacific. United Nations Pacific Strategy 2018-2022. A Multi-Country Sustainable Development Framework in the Pacific Region*. Suva and Apia: Office of the UN Resident Coordinator -Federated States of Micronesia, Fiji, Kiribati, Marshall Islands, Nauru, Palau, Solomon Islands, Tonga, Tuvalu, Vanuatu, and Office of the UN Resident Coordinator - Cook Islands, Niue, Samoa Tokelau.

Part II
Pacific Islands Engagement in the UN

Chapter 7
Membership and Participation in the Major Organs

Chapters 1–6 examined the United Nations-Pacific relationship as it has evolved within the PICs. Chapters 7–12 shift the focus to interactions outside the Pacific, such as at major UN meetings, conferences, and other policy dialogue contexts. In one sense this is a shift in perspective from the Pacific as recipient of UN resources, attention and expertise, to PICs as contributors to UN processes and progress. The goal of attracting additional UN resources remains. But the focus shifts to contributions made *to* the UN, not just contributions *from* the UN. Contributions to meeting the UN's goals include promoting peace and security, helping to define global norms, enhancing international understanding, and participating in international cooperation.

Chapter seven commences with a survey of the PIC presence in the UN through permanent missions in New York, Geneva, and elsewhere. Prior to their gaining independence and becoming members of the UN, the Pacific Island Countries had minimal opportunities to contribute to UN consultations on important issues, particularly the equitable and sustainable use of oceans. At best, they were able to voice their concerns, and receive feedback, through the good offices of Australia and New Zealand. Fiji was also helpful in this way after its admittance in 1970, joining Australian and New Zealand, for instance, as members of the UN Seabed Committee.

Upon attaining independence, PICs faced the issue of when to join the United Nations, and what level of resources to commit to it. Small nations can least afford the costs of membership but have most to gain from it. The decisions concerning when to join the UN, and which agencies and programmes to engage with, took into consideration the membership fee and the costs of maintaining representation in New York and elsewhere. As indicated in Table 7.1, some PICs joined the organization immediately following independence, whereas others deferred for a period of years.

By the late 1960s the challenges of small state participation in the UN began to receive notice. A Committee on United Nations Affairs of the American Bar Association, for instance, reported in 1968:

G. Hassall, *The United Nations and the Pacific Islands*, United Nations University Series on Regionalism 24, https://doi.org/10.1007/978-3-031-34155-7_7

Table 7.1 Pacific Island Country dates of United Nations membership

Country	Date of independence	Date of UN admission	Country	Date of independence	Date
Fiji	10 Oct 1970	13 Oct 1970	Federated States of Micronesia	10 May 1979	17 Sep 1991
Papua New Guinea	16 Sep 1975	10 Oct 1975	Palau	1 Jan 1981	15 Dec 1994
Samoa	1 Jan 1962	15 Dec 1976	Kiribati	12 July 1979	14 Sep 1999
Solomon Islands	7 July 1978	19 Sep 1978	Tonga	4 June 1970	14 Sep 1999
Vanuatu	30 July 1980	15 Sep 1981	Nauru	31 Jan 1968	14 Sep 1999
Marshall Islands	1 May 1979	17 Sep 1991	Tuvalu	1 Oct 1978	5 Sep 2000

> Full membership is not without its cost. Small member states may find onerous the drain on their funds (minimum annual dues of members alone are currently around $50,000) and available trained manpower. There are also drawbacks to having to meet the demands of membership and record voting positions on an imposing number and variety of issues, especially for a small delegation and its home office which may not be equipped to do so effectively. One can expect these difficulties to increase and their burden upon future mini-states to be even heavier than it has been upon small member states to date. (Starr et al., 1968, p. 61)

States pay an annual membership fee calculated by a formula that takes account of a country's economy and population. The UN does not define what constitutes a "small state", since size of population does not determine a state's level of development: there are, for instance, small states with high GDP and HDI, such as Singapore (population 5.6 million), Denmark (5.8 million), Finland (5.5 million), Lichtenstein (38,000), Luxembourg (632,000), Andorra (77,000), Brunei (437,000), Oman (five million), Montenegro 621,000), and others. Small states with equal voting power but less economic development and lower HDIs include not only the PICs, but Sao Tome And Principe (220,000), Bhutan (771,000), and Timor Leste (1.3 million), amongst others.

Whist the membership formula is intended to share the financial burden fairly, small nations have at times struggled to pay. On attaining independence in 1962 Western Samoan Prime Minister Fiame Mata'afa Faumuina Mulinu'u II informed the Trusteeship Council that his country had no immediate intention of seeking U.N. membership but might do so later when its wealth and population had increased; it would, however, seek membership of specialized U.N. agencies such as WHO, UNESCO, and FAO, because their work had particular relevance to Samoan conditions. Trumbull reports a conversation with Mata'afa soon after independence:

> Mata'afa, in his conversation with me, dismissed the question of Western Samoa's joining the United Nations as "too expensive" for a small country with an economy based on coconuts, bananas, and cocoa, and an annual budget of less than ten million dollars. The

> country was, however, the grateful recipient of various services of the United Nations specialized agencies, such as the World Health Organization, and Apia was the site of the Western Pacific regional headquarters of the United Nations Development Program for some years. A call on that organization's offices in the old Casino Hotel, a ramshackle white frame structure whose musty rooms and ceiling fans inevitably reminded visitors of the South Sea tales of W. Somerset Maugham, was like a step into the past. The government-owned establishment, now replaced by a glossy tourist hotel of pseudo-Polynesian design, had been the temporary home of many a high-ranking Allied officer during World War II. (Trumbull, 1977, p. 191).

Samoa carried through this policy, waiting 12 years before seeking UN membership, in 1974 (and being in 1976). Tuvalu acted on a similar sentiment. It gained independence in 1978 but became the UN's 189th member in September 2000. Nauru waited 21 years, and the Marshall Islands 22 years. In contrast, the larger of the small states—Fiji, Papua New Guinea, Solomon Islands, and Vanuatu—joined the United Nations within a year of attaining sovereign independence.

Across the years PICs have struggled to pay their membership fees, let along participate as fully as they would wish in UN activities. In 2015 and again in 2016 **Vanuatu** was among those states whose right to vote in the General Assembly was suspended due to non-payment of its yearly contribution of approximately USD 28,000. Suspension also removes a country's participation in reviewing the UN's budget, appointing non-permanent members to the UN Security Council, receiving reports from other agencies within the UN or contributing resolutions in the General Assembly. The **Marshall Islands**' membership was also suspended for non-payment and resulted in that country calling for reform to the formula. In 2017 **Papua New Guinea** was overdue in paying USD180,000 fees, and news reports in 2021 suggested New York-based members of staff lacked sufficient funds for transport and housing (**Kenneth,** 2021). At the beginning of 2022 Papua New Guinea and Vanuatu were amongst eight countries whose voting rights at the UN were suspended due to fees owing (the others being Iran, Venezuela, Antigua and Barbuda, Republic of Congo, Guinea and Sudan).

7.1 Physical Presence

PIC permanent missions at the UN generally comprise no more than four diplomats, including the Ambassador. In 1984 Western Samoa, Solomon Islands, Vanuatu, and the Seychelles combined resources to open a "joint mission" at 800 Second Avenue. More recently Solomon Islands, Vanuatu, Samoa, Kiribati, and Tuvalu all have offices at 685 Third Avenue, assisted under a Commonwealth programme. Kiribati is the most recent UN member-state to establish a physical presence in New York, in 2013. Some Pacific states also maintain a presence in Europe—either Geneva or Brussels, from where their representatives can participate in activities of UN agencies, the World Trade organization, and UNESCO in Paris. Each of these missions maintains a UN-hosted or government-hosted website, as set out in Table 7.2.

Table 7.2 Location of Pacific Island Country permanent missions to the UN

	New York[a]	Geneva	Vienna	website
Fiji	801 Second Avenue, 20th Floor	Avenue de France 23 1202 Geneva	Serviced from Geneva	https://www.un.int/fiji/
Papua New Guinea	201 East 42nd Street, Suite 2411			https://www.un.int/papuanewguinea/
Solomon Islands	685 Third Avenue, Suite 1105	Avenue Edouard Lacomblé 17 1040 Brussels		https://www.un.int/solomonislands/
Vanuatu	685 Third Avenue, Suite 1103			http://www.un.int/vanuatu
Kiribati	685 Third Avenue, Suite 1109			-
Nauru	300 East 42nd Street, Suite 1601	Avenue de Sécheron 10		https://www.un.int/nauru/
Samoa	685 Third Avenue, 11th Floor Suite 1102			https://www.un.int/samoa/
Tonga	250 East 51st Street			https://www.un.int/tonga/
Tuvalu	685 Third Avenue, Suite 1104			https://www.un.int/tuvalu/
FSM	300 East 42nd Street, Suite 1600			http://www.fsmgov.org/fsmun/
Marshall Islands	800 Second Avenue, 18th Floor,	Rue de Lausanne 37 A1201 Geneva		https://www.un.int/marshallislands/
Palau	880 Third Avenue, 12th Floor			http://www.palauun.org/

[a]https://www.un.org/en/about-us/member-states

The cost of maintaining permanent representation at UN headquarters in New York marks only the start of the expense of UN engagement. Participation in drafting of declarations and treaties can require 6 weeks or more sustenance in a foreign capital—a level of expenditure that few Pacific states have been able or willing to endure—and this has motivated them to collaborate where feasible.

7.2 Participation in the UN's Principal Organs and Agencies

The Introduction to this book explained that the UN Charter establishes six main organs to conduct the UN's four main purposes of providing peace and security, human rights protection, social and economic development, and international coordination. Smaller countries must carefully choose their points of engagement with these organs and their agencies and programmes to ensure adequate return on their investment. How do PICs chose which of UN's 9000 annual conferences and meetings to attend? (McNamara, 2009) During the period 2015–2021 the United Nations Calendar of Conferences and Meetings (https://conf.un.org/admin/byyear) listed approximately 35 meetings per month, convened at Addis Ababa, Bangkok, Bern, Bonn, Chengdu, Doha, Geneva, Guatemala, Hurghada, Istanbul, Kingston, London, Morocco, The Hague, Madrid, Montreal, Nairobi, New York, Paris, Puerto Vallarta, Rome, Vienna, and Washington—and possibly elsewhere. One reason that PICs have not put themselves forward for election to UN bodies is the significant additional cost in human and material resources that this would incur. A Samoan statement to a General Assembly debate on UN Charter reform expressed the view that even if the General Assembly "out of the goodness of its heart", allocated a Security Council seat to a Pacific Islands Forum Group, "...then and only then, will Samoa have a fighting chance of serving on the Security Council. But how can we aspire to such a status when we cannot take up lesser challenges within the United Nations because of our resource constraints?" (Fassbender, 2020, pp. 666–667).

The dynamics of engagement within the UN but outside the Pacific region are quite distinct from engagements within it. First, the costs are inverted. UN programs and visits conducted within PICs are at UN expense, whereas most PIC engagement beyond the region will be at each country's expense. Second, participation in events beyond the region gain minimal press coverage and hence are barely visible to domestic constituencies. Overseas engagement with UN activities was not easily visible to domestic audiences who, even when interested in an issue, had little access to detailed information about the policy detail and dynamics, nor about their representatives' attendance at and reports to routine UN meetings, nor the participation in the daily grind of meetings at UN offices in New York or Geneva. Nor are journalists able to summarize complex issues for public consumption. Third, and more substantially, international engagement generally implies a focus on contributing rather than receiving; Fourth, such contributions could take the form of reporting on actions taken, progress made, and accountability for results. Fifth, and quite seriously, engagement on the international stage implies a shift from being the focus of attention to being a marginal actor amongst a considerable number of other actors who are larger, better resourced, and quite possibly more experienced.

The extent to which PICs have contributed to the operation of the principal organs of the UN is difficult to assess. Table 7.3, which enumerates speeches made in the General Assembly, Economic and Social Council, Trusteeship Council, and Security Council, at least gives an indication of activity.

7.2.1 General Assembly

Presentations to the General Assembly have been the primary means of expressing to other nations the views of Pacific nations as well as the Pacific region's actions toward accomplishment of UN goals. Prime Ministers and Foreign Ministers have

Table 7.3 Pacific Island Country speeches in UN principal organs (to 2021)

	General Assembly	Economic and social council	Trusteeship council	Security Council
Fiji	459	19	2	33
Papua New Guinea	336	8	9	25
Solomon Islands	172	5	1	13
Vanuatu	130	6	1	1
Kiribati	25	0	0	1
Nauru	53	2	0	3
Samoa	113	6	2	2
Tonga	60	1	0	2
Tuvalu	57	2	0	2
FSM	78	0	8	4
Marshall Islands	135	0	31	6
Palau	71	2	22	5

regularly presented details of their government's progress. The issues raised most frequently by PICs in the General Assembly between 2008 and 2013 were climate change, the MDGs and SDGs, Fishing and Ocean preservation; Human rights and empowerment of Women; UN reform; inclusion of Taiwan in the UN system; NCDs; Food security; nuclear testing; peace-keeping, and terrorism. (Hasenkamp, 2016, p. 240).

Each year the country holding the chair of the Pacific Islands Forum submits to the GA the Forum Communique and other regional declarations and sectoral plans. Once submitted, these documents are directed to the relevant agency or initiative. The *Port Vila Declaration on Population and Sustainable Development* of 1993, for instance, was directed to the Preparatory Committee for the then forthcoming International Conference on Population and Development in Cairo. (United Nations General Assembly, 1993) Other documents were transmitted to the General Assembly via "note verbale", such as the *Auckland Declaration* of 2004, which initiated the Forum's program refresh known as the "Pacific Plan" (United Nations General Assembly, 2004). In 2009 Nauru presented a draft resolution on behalf of the PICs concerning the threat to security posed by climate change, which was unanimously adopted as resolution 63/281.

Tables 7.4, 7.5, 7.6 and 7.7 indicate the numbers of speeches made in the General Assembly by PICs between 1971 and 2021. They indicate that only Fiji, Papua New Guinea, and Samoa, contributed in the 1970s—and only to the briefest extent. The remaining PICs participated in subsequent decades. The motivations for addressing the General Assembly, and the content of speeches and their impact or outcomes, deserve their own systematic study.

Table 7.4 Pacific Island Country speeches in the UN General Assembly 1971–1989

	71	72	73	74	75	76	77	78	79	80	81	82	83	84	85	86	87	88	89
FSM														6	9	5	1	1	0
Fiji	1	1	2	8	6	2	1	2	1	2	3	4	11	10	27	32	20	33	25
Kiribati																			
Marshall													5	10	14	17	2	0	0
Nauru																			
Palau															3	7	2	1	3
Papua New Guinea						2	1	2	1	1	1	1	10	25	15	11	14	10	20
FSM																			
Samoa							1	1	1	0	1	0	4	2	5	6	8	3	1
Solomon Islands													2	1	4	8	4	5	2
Tonga																			
Tuvalu																			
Vanuatu											1	0	1	1	1	1	1	1	1

Table 7.5 Pacific Island Country speeches in the UN General Assembly 1990–1999

	1990	1991	1992	1993	1994	1995	1996	1997	1998	1999
FSM	0	5	4	2	5	3	1	1	3	3
Fiji	26	11	8	15	11	11	10	11	17	16
Kiribati										2
Marshall	0	3	1	9	6	20	16	13	9	6
Nauru										1
Palau	3	2	1	2	5	1	1	1	1	2
Papua New Guinea	18	14	14	15	17	26	15	26	11	13
Samoa	2	4	2	6	5	5	7	3	4	7
Solomon Islands	3	4	5	2	3	3	3	8	10	13
Tonga										2
Tuvalu										
Vanuatu	2	1	1	0	0	1	1	1	1	1

Table 7.6 Pacific Island Country speeches in the UN General Assembly 2000–2010

	2000	2001	2002	2003	2004	2005	2006	2007	2008	2009	2010
FSM	8	3	2	2	3	2	1	4	4	7	3
Fiji	15	17	28	19	23	25	17	6	13	4	9
Kiribati	1	0	1	1	1	2	2	1	1	1	1
Marshall	6	3	4	5	5	2	5	4	6	3	2
Nauru	6	12	6	3	3	4	3	1	2	4	5
Palau	2	3	3	2	3	6	5	8	4	3	2
Papua New Guinea	16	8	8	8	10	11	7	7	7	5	11
Samoa	5	4	7	1	8	7	3	6	3	4	4
Solomon Islands	6	6	3	2	2	5	9	9	9	10	8
Tonga	2	2	2	ˋ	ˋ	4	3	6	11	1	1
Tuvalu	3	3	6	5	4	3	4	2	3	3	2
Vanuatu	2	1	1	1	1	2	1	1	1	1	2

7.2.1.1 General Assembly Committees, and Bodies

Full sessions of the General Assembly only occupy a few weeks of each year and are complemented by meetings of its committees and *ad hoc* bodies, in which PICs have opportunity to participate should they decide to. Papua New Guinea and Fiji have been members of the 151-member Special Committee on Peacekeeping Operations, and the 29-member Special Committee on the Implementation of the Declaration on Decolonization. As set out in Table 7.8, some UNGA entities are open to participation by all UN member-states, while others are more optional.

Table 7.7 Pacific Island Country speeches in the UN General Assembly 2011–2021

	2011	2012	2013	2014	2015	2016	2017	2018	2019	2020	2021
FSM	1	5	6	6	3	6	6	5	9	7	0
Fiji	19	3	95	18	11	54	23	17	22	13	5
Kiribati	2	1	2	3	3	1	6	4	2	5	5
Marshall	3	2	4	2	4	2	5	3	5	6	5
Nauru	5	7	5	4	5	6	4	8	3	7	2
Palau	2	2	3	11	14	15	2	5	3	4	1
Papua New Guinea	9	7	11	9	8	17	8	10	7	8	2
Samoa	3	4	5	4	5	3	6	4	6	4	0
Solomon Islands	12	3	14	12	8	2	8	7	2	4	1
Tonga	2	2	11	14	11	7	12	14	4	6	1
Tuvalu	3	2	4	4	4	5	2	3	6	8	1
Vanuatu	1	1	1	1	1	1	1	1	1	1	1

Table 7.8 PIC participation in entities established by the UN General Assembly

Special Committee on Peacekeeping Operations	2005	Fiji
Committee on Information	2005	Solomon Islands
Special Committee on the Charter of the UN and on the Strengthening of the Role of the Organization	2005	All UN member states
Special Committee on the Implementation of the Declaration on Decolonization	2005	Fiji, PNG
UN Commission on International Trade Law	2004	Fiji
Special Committee on Charter of the UN and on the Strengthening of the Role of the Organization		All member states
UN-Open-Ended Informal Consultative Process on Oceans and the Law of the Sea	2000	Open to membership
High-Level Open-Ended Working Group on the Financial Situation of the UN		Open membership
Open-Ended Working Group on the Question of Equitable Representation and Increase in the Membership of the Security Council		Open membership

In the case of the Pacific region's concerns about decolonization, few PICs have sought membership on the General Assembly's Special Committee on Decolonization (C24), Fiji and Papua New Guinea. In 2004 Robert Aisi was elected committee Chair. Fiji, Palau, Papua New Guinea, and Vanuatu have taken an interest in the *Special Committee on Peacekeeping Operations* of the United Nations, or *C34*, first established by the General Assembly's Fourth Committee in 1965 to review and provide recommendations on United Nations Peacekeeping Operations. To date there have been no PIC contributions to the Special Committee on the Charter of the UN and on the Strengthening of the role of the organization.

7.2.2 Economic and Social Council

Scant PIC engagement in deliberations at the Social and Economic Council is surprising given that social and economic development are core objectives of developing countries. The 54-member Economic and Social Council (ECOSOC) makes recommendations to UNGA and relevant UN agencies on development and human rights issues, drafts conventions, convenes global conferences, and arranges consultations with NGOs. ECOSOC also oversees the Committee for Development Policy (CDP), whose decisions have significant impact on the status of PICs. ECOSOC's membership comprises 14 African states, 11 Asian states, 6 Eastern European states, 10 Latin American and Caribbean states, and 13 Western European and other states. Papua New Guinea's only term was 1984–1986. ECOSOC has 9 functional commissions, 5 regional commissions (including one for Asia and the Pacific), 2 standing committees and 12 expert committees and subcommittees.

ECOSOC has oversight of the UN's Human Rights instruments, initially through the Commission on Human Rights (1947–2005) and subsequently through the Human Rights Council. Between 1947 and 2005 no PICs were members of the Commission (by way of comparison, forty-seven African States had served terms on the Commission on Human Rights). The only report by one of the Commissions' Special Rapporteur's concerned a 1995 investigation into extra-judicial killings on Bougainville during the civil war (United Nations. Commission on Human Rights, 1996), although human rights conditions in several other PICs, such as gender disparities in education, freedom of religion, housing conditions, and child exploitation, came to the Commission's attention. In 2006 the Commission was replaced by the 47-member Human Rights Council, on which just two PICs have served to the present time, Fiji in 2019–2021, and the Marshall Islands 2020–2022.

PICs are, however, actively involved with the Economic and Social Council's 53-member Regional Commission for Asia and the Pacific, which is based in Bangkok, Thailand. Government officials routinely participate in seminars and committees on statistics, climate change, ICTs, inter-island shipping, climate change and migration, employment, public-private partnerships, disaster risk management, sustainable agriculture, social sector financing, and other development issues. Fiji (1974–1975, 1981–1983, 2000–2002) and Papua New Guinea (1984–86) are the only two Pacific states to have served on the Economic and Social Council, although Ministers from PIC member states participate in an annual four-day high-level segment of ECOSOC's substantive session.

ECOSOC's Commission on Sustainable Development was established in 1992 to monitor the progress in the implementation of Agenda 21, the outcome of 1992 UN Conference on Environment and Development in Rio de Janeiro. The Commission has also maintained a Small Island Developing States Unit under the Water, Natural Resources and Small Island Developing States Branch. Its substantive meeting is held each July alternately in New York and in Geneva, with an additional organizational session in February, and these require extra travel for PICs which lack representation in Geneva. Pacific members of the CSD have been Fiji (2003–06), Vanuatu (1993–94), and Papua New Guinea (1995–98).

7.2.2.1 Committee for Development Policy

Decisions by ECOSOC's little-known but important Committee for Development Policy (CDP) have had significant impact on the status of PICs within the UN system. The 24-member CDP was established by ECOSOC in 1965 as the "Committee for Development Planning" for a 'consultative group of experts' to provide independent advice to the Economic and Social Council (ECOSOC). During "revitalisation and restructuring" of ECOSOC in the 1990s CDP retained it roles in providing advice on development policy, including categorization of member states as "least developed countries" (LDCs) or "developing countries (Fig. 7.1)."

Whereas the rationale for making this determination has been to identify those countries that most need development assistance, the scheme has also been criticized for inadvertently creating "perverse incentives" for LDC countries which encourage non-graduation rather than graduation. Adam Delaney, who served in Papua New Guinea's delegation in New York, has suggested that the small island developing states have focused on their assessed "vulnerability" as a means of gaining a comparative edge over other states of similar sized economies but which have less volatile physical environments (Delaney, 2006). UNDP's Development Assistance framework for 2008–2012, for instance, allocated 65% of the Pacific region's funds

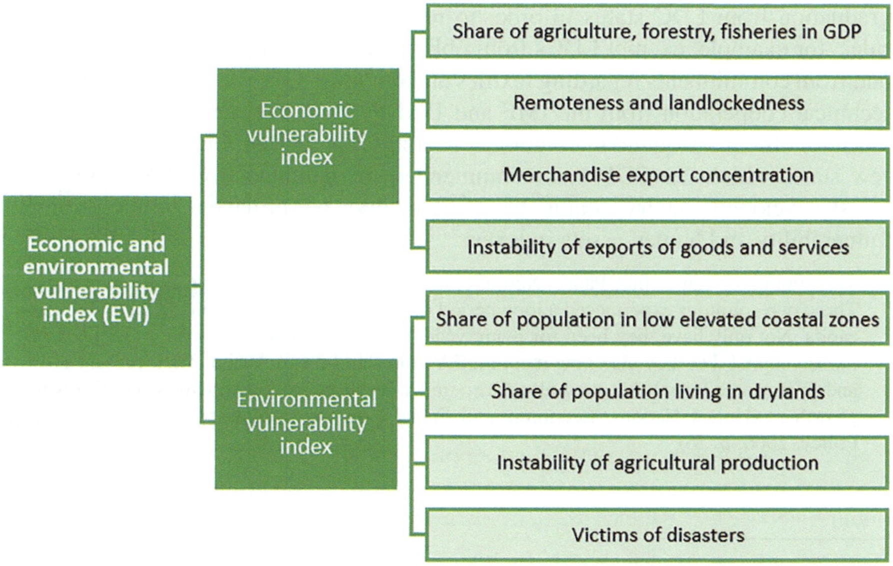

Fig. 7.1 Committee for development policy's economic and environmental vulnerability index

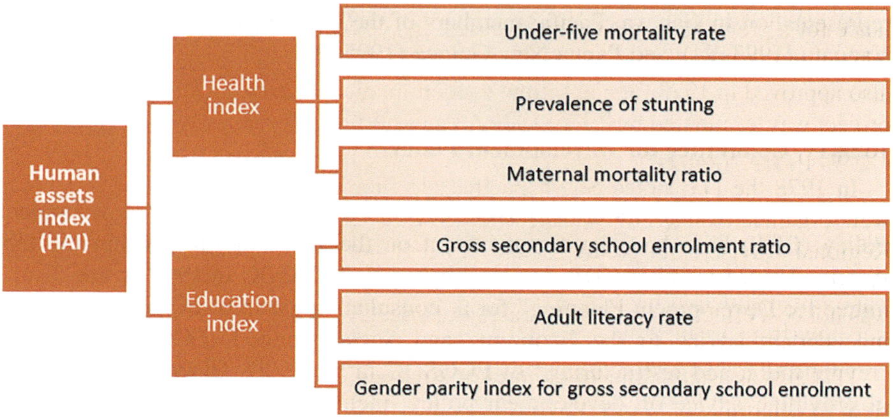

Fig. 7.2 Committee for development policy's human asset index

(USD 309.7 million) to its five LDCs (Kiribati, Solomon Islands, Samoa, Tuvalu, and Vanuatu). (Organization., 2009, p. 12).

The CDP devised a three-dimensional index to assess a country's state of development that considers *Income level* (Gross National Income per capita), a *Human Asset Index*[1] (HAI) and an *Economic Vulnerability Index*[2] (EVI). The EVI, as set out in Fiji 6, is calculated on the basis of four economic and four environmental indicators, while the HAI, as set out in Fig. 7.2, combines three health and three education indicators.

In 2001 49 member states were classified as LDCs, including the PICs Samoa, Vanuatu, Solomon Islands, Tuvalu, and Kiribati. Vanuatu and Samoa sought to defer graduation from LDC status in order to retain benefits of the lower status. WTO rules, for example, exempt LDCs from obligations to reduce barriers in agriculture, and from commitments regarding textiles and clothing. LDCs also receive additional technical cooperation from the IMF and UNDP. They can receive financial assistance for officials to attend international meetings. Papua New Guinea is amongst the few states which the CDP has recommended for demoted from "developing" to "least developed country" status, based on its poor performances on the economic vulnerability and human assets indexes:

> The committee found that Papua New Guinea and Zimbabwe are eligible for inclusion. These two countries experienced a long period of stagnation and/or decline in their GNI per capita. Not only have they been for many years low income countries with a low level of human capital, but they also now meet the EVI threshold for inclusion. Their present HAI and EVI scores reveal that they also face similar levels of severe structural handicaps to growth... (United Nations. Economic and Social Council. Committee for Development Policy, 2006, p. 26)

[1] https://www.un.org/development/desa/dpad/least-developed-country-category/ldc-criteria.html
[2] https://www.un.org/development/desa/dpad/least-developed-country-category/ldc-criteria.html

By 2021 just three PICs had LDC status: Solomon Islands, Kiribati and Tuvalu. Samoa "graduated" in 2014, and Vanuatu in 2020.

7.2.3 *The Security Council*

The Security Council is the smallest of the UN's organs, but arguably one of the most powerful—or at least the most influential in global peace and security issues. I make this distinction because the UNSC, whilst powerful, has major design flaws that flow from its inequitable composition and powers, and render it incapable of providing global collective security. The Pacific States comprise 12 of 67 UN Member countries that have not been elected to the Security Council.[3] The 2019 election of St. Vincent and the Grenadines—population just over 100,000—as the smallest state yet to gain a seat, will test the prevailing view that small states lack the capacity to fully participate on the Security Council. The election of Estonia in 2020, with a population of 1.3 million, is another small state to take up this challenge.

Although no PIC has held an elected seat on the Council, some have engaged in its formal dialogue processes. Papua New Guinea exercised its right to "take a seat at the Council table" to speak on the issue of Bougainville and has asked at times for permission to have Fiji, Japan, and New Zealand present. This intervention resulted in the establishment in 1998 of a United Nations Political Office in Bougainville, replaced in 2004 by the United Nations Observer Mission in Bougainville (UNOMB), which was responsible for supporting the peace process, overseeing weapons destruction, and constitutional processes (see Chap. 8).

The Security Council has not played a major role in resolving other conflicts in the Pacific region, such as those in Solomon Islands and Fiji, apart from receiving status updates. On 21 July 2003, for instance, Fiji transmitted to the President of the Security Council an "outcomes statement" of the Pacific Forum Foreign Affairs ministers, indicating their support for the provision of assistance to Solomon Islands under the terms of the Biketawa agreement. (United Nations. Security Council, 2003).

The one major event that brought the Pacific SIDS into dialogue with the Security Council took place in 2015. On 30th July of that year,[4] when serving a two-year term as an elected member of the Security Council (2015–2016), New Zealand arranged a special session at which the PICs presented to the Council their perspectives on peace and security. (van Bohemen, 2015) Although the session was regarded as a breakthrough by the SIDS because it was their first opportunity to speak directly with the Council, the Council's known reservations about associating the topic of

---

[3] http://www.un.org/en/sc/members/notelected.asp

[4] 7499th meeting, 30th July 2015, *President*: Mr. Mc Cully/Mr. Van Bohemen/Mr. Taula (New Zealand) Maintenance of international peace and security: Peace and security challenges facing small island developing States (S/PV.7499)

climate change with security constrained the possibilities for any outcome statement. (Hawke et al., 2020). Some seventy speakers participated in the debate. Papua New Guinea's Foreign Minister Rimbink Pato urged the Security Council to give greater attention to the traditional and non-traditional security concerns of the Small Island Developing States. He noted the attention that the *UN Framework Convention on Climate Change* was giving to SIDS but suggested that "...more agencies of the United Nations system and others are needed to be engaged", including the General Assembly, the Global Green Growth Institute, the Green Climate Fund, and the Global Environment Facility. (United Nations. Security Council, 2015).

7.2.4 International Court of Justice

The Marshall Islands and Nauru are the only two PICs to have had cases heard by the International Court of Justice. The details of the legal arguments and the judicial process are too intricate to explore in detail here. In 1989 Nauru filed with the Court a claim that Australia had breached its trusteeship obligations regarding implementation of the principle of self-determination and of permanent sovereignty over natural wealth and resources. Australia argued strenuously against the Court's jurisdiction in the matter, on the basis that the Trusteeship Agreement had terminated in 1968 and thus closed the trusteeship period to the satisfaction of all parties. Although the Court denied Australia's objections, the matter did not proceed to a hearing because in 1993 Australia and Nauru reached an out-of-court settlement.

In the case of the Marshall Islands, the Republic filed applications against nine countries in 2014 alleging that they had not fulfilled their obligations under international law to pursue negotiations to end nuclear disarmament. Because six of these countries (China, Korea, France, Israel, the Russian Federation and the United States of America) did not accept the court's compulsory jurisdiction, proceedings only continued concerning India, Pakistan, and the United Kingdom as defendants. The Court ruled in 2016 that no conflict existed between the Marshall Islands and the three states of India, Pakistan, and the United Kingdom, and that it was thus unable to proceed to hearing the merits of the argument. In 2011 Palau announced an intention to seek an ICJ opinion on whether countries have a legal responsibility to ensure that greenhouse gas emitting activities on their territory to not harm other States, and in 2018 Vanuatu first mentioned a campaign to have the ICJ offer an opinion on Climate Change. (Freestone et al., 2022).

7.2.5 Participation in other UN agencies

The preceding section has provided an overview of PIC engagement with the UN's principal organs. The current section continues this survey by outlining their engagement with additional UN agencies established by one of its six organs or which pre-existed the UN. WHO, UNESCO, ILO and ITU for example, became associated with the UN at its founding, and although integrated into the UN system, each maintain unique systems of authority and decision-making. PICs enjoyed beneficial relationships with UN agencies prior to becoming full members of the UN. In 1981, when just Fiji, Papua New Guinea, Solomon Islands and Western Samoa were members of the UN, other PICS were already members of UNDP and members or affiliate members of ESCAP (Cole & Thompson, 1981). Table 7.9 lists 12 major agencies, in which Pacific membership ranges from 4 (UNIDO) to 15 (WHO and UNESCO), including non-members Cook Islands and Niue.[5]

This chapter provides a broad overview of Pacific engagement with these agencies, which later scholarship will no doubt examine in greater detail. As with the major UN organs, PICs must determine which of these associated agencies is most useful to expend resources on. Papua New Guinea, for instance, has pursued its engagement with UNESCO. Theo Zurenuoc, when Papua New Guinea's Minister for Education, remarked at UNESCO's 36th General Conference in Paris in October 2011:

> Madame President, with these remarks, allow me to remind our distinguished delegates that the 17 UNESCO member states in the Pacific has not had a voice at the UNESCO Executive Board over the last two years. You will agree with me that it is unfair to have all 12 seats allocated to the region concentrated in only one part of the region. Therefore the Pacific presents Papua New Guinea as its candidate for a seat at the Executive Board. Colleagues and friend member states, I ask you for your support.

In the case of the Pacific region's concerns about decolonization, few PICs have sought membership on the General Assembly's Special Committee on Decolonization (C24), Fiji and Papua New Guinea. In 2004 Robert Aisi was elected committee Chair. Fiji, Palau, Papua New Guinea, and Vanuatu have taken an interest in the *Special Committee on Peacekeeping Operations* of the United Nations, or *C34*, first established by the General Assembly's Fourth Committee in 1965 to review and provide recommendations on United Nations Peacekeeping Operations. To date there have been no PIC contributions to the Special Committee on the Charter of the UN and on the Strengthening of the role of the organization. The UN Commission on International Trade Law (UNCITRAL) has 60 members, but Fiji has been the only one from the Pacific, serving from 1998 to 2016.[6]

7.2.6 *Pacific Islanders in the United Nations*

Recalling the concept of the "Three UN's" (referred to in the Introduction), the experience of Pacific Islanders at the United Nations includes (a) those who participate as representatives of their country; (c) those who are employed by the United

---

[5] From Un.org

[6] https://uncitral.un.org/en/about/faq/mandate_composition/memberhistory

Table 7.9 Pacific Island Country membership in major UN entities

	Cook Islands	FSM	Fiji	Kiribati	Marshall	Nauru	Niue	Palau	PNG	Samoa	Solomon	Tonga	Tuvalu	Vanuatu
FAO	√		√	√	√	√		√	√	√	√	√	√	√
IAEA			√		√			√	√	√		√		
ICAO	√	√	√	√	√	√		√	√	√	√	√	√	√
ICC	√		√	√	√	√				√				√
IFAD	√		√	√					√	√	√	√		
ILO	√		√	√	√			√	√	√	√	√	√	√
IMO	√		√	√	√	√		√	√	√	√	√	√	√
IPU		√	√		√			√	√	√		√	√	√
ITU			√	√	√	√			√	√	√	√	√	√
OPCW	√	√	√	√	√	√	√	√	√	√	√	√	√	√
UNIDO		√	√		√				√			√	√	√
UNESCO	√		√	√	√	√	√	√	√	√	√	√	√	√
WHO	√	√	√	√	√	√	√	√	√	√	√	√	√	√
WIPO	√		√	√	√	√	√		√	√	√	√	√	√
WMO	√	√		√		√	√		√	√	√	√	√	√

Table 7.10 Pacific Islander UN employees by country

Cook Islands	1
Fiji	293
Kiribati	17
Marshall Islands	1
Federated States of Micronesia	6
Palau	1
Papua New Guinea	108
Samoa	62
Solomon Islands	22
Tonga	12
Tuvalu	2
Vanuatu	28
Total	559

Nations; and (c) those who participate in a non-state capacity, such as through a recognized civil society organization or academia. For those who are employed by the UN there are several pathways to recruitment. Permanent positions in the UN "civil service" require successful completion of an entry exam which is offered periodically. Other positions are contracted for a set term. The UN monitors its workforce to ensure fair representation of the regions as well as gender balance. In 2020 the UN system recorded a total of 550 employees from eleven PICs who were on staff contracts or appointments of at least 1 year. This is approximately 1% of the total UN workforce of 116,388.[7] The numbers of Pacific Islanders employed by UN agencies is set out in Table 7.10.

Current figures indicate that most PIC employees of the UN are Fijian, Papua New Guinean, and Samoan. Those who have held senior positions include Fijian Savenaca Siwatibau, as head of the ESCAP Pacific Operation Centre in Bangkok in the 1990s before becoming Vice Chancellor at the University of the South Pacific; Fijian Naren Prasad as current Head of Training at the International Labour Organization in Geneva; Samoan Malama Meleisea, as UNESCO Director in Bangladesh, Peter Kenilorea Jr who worked in the UN's New York Office of Legal Affairs, DESA, UNOHCHR and the UN Office of the High Representative for Least Developed Countries, Landlocked Developing Countries, and Small Island Developing States between 2000 and 2017, before becoming the Solomon Islands Permanent Secretary of the Ministry of Foreign Affairs and External Trade; in November 2021 Fijian Unaisi Lutu Vuniwaqa was appointed Assistant Secretary-General for Safety and Security; Fekitamoeloa Katoa 'Utoikamanu, formerly Tonga Permanent Representative at the United Nations (2005–2009), was for a time the highest office holding Pacific Islander at the UN, in her role as High Representative for the Least Developed Countries, Landlocked Developing Countries and Small Island Developing States, before returning to Tonga and becoming Minister for Foreign Affairs.

[7]https://unsceb.org/hr-nationality

Among the longest of Pacific Island careers as an international civil servant was that of Tuiloma Neroni Slade, who, amongst other accomplishments, was appointed as Samoa's Attorney-General in 1976, then appointed Samoa's ambassador to the UN in 1993, Slade co-chaired the UN Consultative Process on Oceans and the Law of the Sea, led the Samoan delegation to the Rome Diplomatic Conference on the Establishment of an International Criminal Court, was elected as a judge at the International Criminal Court in 2003, and upon leaving the United Nations, was twice elected Secretary-General of the Pacific Islands Forum (2008–2014).

Many others have held senior appointed positions: Dr. Joe Williams, Cook Islands Minister for Health and Education (1974–1978), as a member of WHO's Executive Board between 1995 and 1997; In 1996 an FAO Sub Regional Office for the Pacific was established in Apia, with a Tongan, Dr. Fuavao, as head; Fijian Ambassador Winston Thompson as vice-president of UNDP's Governing Council in 1978; Fijian Jai Ram Reddy was elected as a judge in the International Criminal Tribunal on Rwanda in 2003; Ni-Vanuatu Ambassador Odo Tevi as Vice-President of the UN's Pledging Conference for Development Activities in 2015; Nazhat Shameem Khan, Permanent Representative of Fiji to the United Nations Office at Geneva, as elected President of the Human Rights Council for 2021 and then as a Deputy Prosecutor at the International Criminal Court. On several occasions PICs have assumed the Presidency of the General Assembly, including Ambassador Peter Thomsen from Fiji in 2016–2017. Yet other PIC representatives holding high UN positions in recent years include:

- Chair of the Alliance of Small States (AOSIS)—Samoa, Nauru.
- Chair of the Group of 77 and China—Fiji.
- President of the Meeting of States Parties to the UN Convention on the Law of the Sea (UNCLOS)—Papua New Guinea.
- President of the Council of the International Seabed Authority –Fiji.
- Chair of the Decolonisation Committee (C24)—Papua New Guinea.
- Secretary-General of the International Seabed Authority—Satya Nandan—Fiji.
- Co-Chair of the UN Consultative Process on Oceans—Samoa.
- Co-coordinator for the text of the Rome Statute of the International Criminal Court—Samoa
- UN Special Envoy for the Ocean—Fiji.
- President of the 23rd annual Conference of the Parties to the *UN Framework Convention on Climate Change*—Fiji.

7.2.7 *Pacific Island Countries on Governing Boards*

The assessment of UN activities within PICs, as well as PIC contributions to UN processes in its principal organs as well as in the operation of its affairs both domestically and internationally, which have been sketched in the preceding sections, are all preliminary findings intended to assist in assessing the role that the UN system allows small states to play in the global multilateral order.

PICs are and have been present on the Governing Council of the UN Environment Program (UNEP). Tuvalu was a member (2004–07). Samoa and the Marshall Islands were members 1996–2003, and Papua New Guinea was a member 1983–1988. Papua New Guinea is the only PIC to have served (in 1994) on the 36-member Executive Board of United Nations Development program (UNDP) and the UN Fund for Population Activities (UNFPA), and twice (1977–85, 93–96) on the Governing Council of UN Human Settlements Programme (UN-HABITAT).

7.3 Commitments to Norms and Treaties

The United Nations system generates norms across many sectors. Some of these remain as recommendations, others as declarations, while yet others become treaties. Each is important. Declarations, for instance, can pave the way for the later establishment of a treaty. A significant body of legal norms has been built, for example, on the *Universal Declaration of Human Rights* agreed by the United Nations in 1948. Although merely a 'declaration' of desirable standards pertaining to human rights, the UDHR has had considerable impact on the ways in which states and citizens understand notions of individual rights and obligations. A *Declaration on the Right to Development* issued in 1986 has generated useful dialogue without becoming formalized by international treaty. The 2007 *Declaration on the Rights of Indigenous People (*UNDRIP) has gained the support of Pacific communities but has drawn little comment from governments. The Melanesian Institute of Philosophy, Science and Technology hosted a Vanuatu Indigenous Peoples' Forum which prepared a statement presented to the sixth session of the United Nations Permanent Forum on Indigenous Issues in New York later that same year, and other meetings have subsequently been held in the region in support of the Declaration.

Norms need not be expressed in laws or declarations to have impact: the concepts "access for all", "leave no-one behind", "information for all", and "universal access", for instance, have influenced policy settings in health, education, social services, communications, and other fields. To give one instance amongst many, Vanuatu based its ICT Services policy on values set out by the ITU Broadband Commission, commencing with goal one "Ensure accessibility and affordability of reliable access to information services to all." (Horne, 2011).

Treaties create commitments to action (or restraints on action). 1966 the UN concluded two "covenants" concerning human rights: the International Covenant on Civil and Political Rights (ICCPR) and the International Covenant on Economic, Social and Cultural Rights (ICESCR). The phenomenon of states freely binding their actions through commitment to international treaties is one of the UN's most successful and fascinating accomplishments. Treaties by which parties accept their rights and obligations in relation to other states are understandable as transactions that facilitate cooperation and avoid conflict. But human rights treaties require states to pledge to the international community the rights and responsibilities they recognize in relation to their own citizens. One might ask why such a contract requires anything more than a domestic charter or constitutional bill of rights, under national law? Clearly, accession to human rights treaties, declarations of agreement with and recognition of trans-national norms of human rights in the full sense of the term—as rights belonging to persons irrespective of their race, nation, colour, or creed. Promising to treat one's own citizens well is part of a state being a "good international citizen."

Member-states are expected to uphold the values espoused by the United Nations, and the most fundamental indicator of this support is through adoption of the core human rights protocols. PICs are making ever-more global commitments through engagement with global governance agencies and policy and diplomatic networks. Trans-national policy engagement on trade, health, customs, security, finance and banking, environment and climate change, transport and communications, science, technology and agriculture, education, and anti-corruption requires ever-growing expectations around preparation, monitoring for compliance, and reporting.

7.3.1 Multi-lateral Treaties

More than 560 multilateral treaties[8] cater to the collective expectations about what Kaul and others have described as "Global Public Goods" (Kaul et al., 2003; Kaul et al., 1999). As expressed by Stone: "When a problem is recognized by nations, the policy tools available are international treaties and conventions. Their effectiveness is problematically reliant on compliance and good international citizenship, and founded upon an implicit assumption that states will act "rationally" and recognize that collective action is to long-term interests" (Stone, 2008, p. 27). Each treaty has been created by, or has established, "trans-national policy communities."

Possible actions concerning multi-lateral treaties include signature, ratification, definitive acceptance, admission to membership, succession, accession, reservation, withdrawal of reservation, and denunciation. There are committees related to human rights treaties, the *Law of the Sea Treaty*, and treaties, and conventions and protocols on the environment. Membership of these bodies is made up of experts (such as lawyers on the Human Rights Committee) or stakeholders (such as on International Seabed Authority). For some, such as the Intergovernmental Panel on Climate Change, membership comprises of all Member States. The Secretariat of the Basel Convention on the Control of Transboundary Movements of Hazardous Wastes and Their Disposal is based in Geneva but works closely with Regional Centres including the South Pacific Regional Environment Program (SPREP).

In 2005 Fiji was the only Pacific country on a treaty-related committee (member of the Council of the International Seabed Authority and a Fiji Lawyer, Alfred

[8]https://treaties.un.org/Pages/Overview.aspx?path=overview/overview/page1_en.xml

Thomas Simpson, was one of the 24 members of the Legal and Technical Commission). All PIC signatories to conventions such as the Basel Convention, or the Convention on biological diversity, are members of the Conference of Parties. All Pacific Countries are members of the Global Environment Facility (GEF), the financial mechanism of the *Framework Convention on Climate Change*, The *Convention on Biological Diversity*, the *Stockholm Convention on Persistent Organic Pollutants*, and the *Convention to Combat Desertification.*

There are yet other ways in which UN norms as expressed in declarations and treaties influence the path of national and regional law. These include "model-laws", which are drafted through UN processes for member countries to adapt to local preferences and circumstances, and "common law", which results from court judgements which draw on United Nations declarations and treaties as relevant legal precedent. The extent to which these two sources of law have been influenced Pacific jurisprudence can only noted here and deserve a separate study, not least through thorough use of the Pacific Islands Legal Information Institute (http://www.paclii.org/).

Many of the fundamental human rights treaties remain unsigned by the PICs, and the depth of their commitment to international law is in many instances yet to be demonstrated. The current situation is summarized in Table 7.11.

The lack of commitment to the International Covenant on Civil and Political Rights and the International Covenant on Economic, Social and Cultural Rights, which have been open for signing since 1966, are particularly concerning. Eight of twelve states have ratified ICESR, and another (Palau), has signed but not yet ratified. Only four of the twelve have ratified ICCPR, while Nauru and Palau have signed but not yet ratified. Palau has ratified just 3 of the nine fundamental HR Covenants, although it signed the remaining six as long ago as 2011. The CRC is the only fundamental treaty that has been signed and ratified by all 12 PIC member states.

Commitment to the human rights norms established by the United Nations can be a demanding process for small states. When a UN treaty signed by a member country requires "domestication" into national law, the relevant government department should provide an information paper to national cabinet, and a process commences to examine both the necessity for the new law and the state's ability to implement it. Once signed, a treaty must be ratified, adopted into national law, implemented, and reported on. Success requires political will, legal capability, public sector support, and sufficient public agreement and cooperation. Opportunity to sign treaties is provided when the UN General Assembly meets each year for member countries.

To be sure, some Pacific states may not yet have developed sufficient legal capacity to ensure domestic compliance with many international protocols, but PICs can receive UPR technical advice from the Office of the United Nations High Commissioner for Human Rights (OHCHR), PIFS, and the SPC's Regional Rights Resource Team (SPC RRRT). (SPC & Pacific Regional Rights Resource Team, 2012).

Table 7.11 Status of Pacific Island Country accession to International Human Rights Treaties. https://treaties.un.org/Pages/Treaties

	ICESR (1966)	ICCPR (1966)	CERD (1966)	CEDAW (1995)	CAT (1984)	CRC (1989)	CRMW	CRPD	CPPED
Cook Is				11 Aug 2006					
Fiji	16 Aug 2018	16 Aug 2018	11 Jan 1973	28 Aug 199	14 Mar 2016	13 Aug 1993	19 Aug 2019	7 Jun 2017	19 Aug 2019
Kiribati				17 Mar 2004	22 Jul 2019	11 Dec 1995		27 Sep 2013	
Marshall	12 Mar 2018	12 Mar 2018		2 Mar 2006	12 Mar 2018	4 Oct 1993		17 Mar 2015	
FSM				1 Sep 2004		5 May 1993		7 Dec 2016	
Nauru		**12 Nov 2001**	**12 Nov 2001**	23 Jun 2011	26 Sep 2012	27 Jul 1994		27 Jun 2012	
Niue						20 Dec 1995			
Palau	**20 Sep 2011**	**20 Sep 2011**	**20 Sep 2011**	**20 Sep 2011**	**20 Sep 2011**	4 Aug 1995	20 Sep 2011	11 Jun 2013	**20 Sep 2011**
PNG	21 Jul 2008	21 Jul 2008	27 Jan 1982	12 Jan 1995		2 Mar 1993		26 Sep 2013	
Samoa		15 Feb 2008		25 Sep 1992	28 Mar 2019	29 Nov 1994		2 Dec 2016	27 Nov 2012
Solomon	17 Mar 1982		17 Mar 1982	6 May 2002		10 Apr 1995	**23 Sep 2008**		
Tonga			16 Feb 1972			6 Nov 1995	**15 Nov 2007**		
Tuvalu				6 Oct 1999		22 Sep 1995		18 Dec 2013	
Vanuatu						7 Jul 1993		23 Oct 2008	**6 Feb 2007**

Bold = signed, but not yet ratified

Key

ICESCR—International Covenant on Economic, Social and Cultural Rights—1966

ICCPR—International Covenant on Civil and Political Rights—1966

CERD—Convention on the Elimination of All Forms of Racial Discrimination—1966

CEDAW—Convention on the Elimination of All Forms of Discrimination Against Women—1979

CAT—Convention Against Torture and Other Cruel, In human or Degrading Treatment or Punishment—1984

CRC—Convention on the Rights of the Child—1989

CRMW—Convention on the Protection of the Rights of All Migrant Workers and Members of Their Families—1990

CRPD—Convention on the Rights of Persons with Disabilities—2006

CPPED—Convention for the Protection of all Persons from Enforced Disappearance—2006

Although a UN Decade for Human Rights Education commenced in 1995, the question of international human rights norms has proved a vexed question in some parts of the Pacific, particularly in circumstances where the "universal" rights advocated by the UN are perceived as being in conflict with culture and tradition. A Tongan Minister expressed the view:

> "...Tongans are not deprived from destitution, but their lives can always be improved in the quality and quantity of services provided for them. These are the very issues I hope will be robustly debated and I encourage all to have dialogue and debate." At the same time, the Honourable Minister cautioned participants, "I would remind you that we are currently the custodians of our cultural values, and the navigators of our future. So I hope that debates and discussions are tempered with a sense of protection of what is Tongan!" (PIFS, 2014)

Several PICs have taken the option of signing the international covenants with reservations. Kiribati, for instance, made reservations to Convention on the Rights of the Child:

> The Republic of Kiribati considers that a child's rights as defined in the Convention, in particular the rights defined in articles 12-16 shall be exercised with respect for parental authority, in accordance with the Kiribati customs and traditions regarding the place of the child within and outside the family.

Reasons for slow update of human rights commitments by PICs include (a) disagreement about their universal applicability; (b) agreement with the norm but lack of legal and administrative capacity to ensure their implementation; (c) belief that domestic laws and norms are protecting human rights adequately; or (d) a concern about the burden of reporting requirements (Baird, 2008). This reluctance to commit to human rights covenants is also expressed in lack of progress toward the establishment of human rights commissions at either national or regional levels, notwithstanding considerable expectation, urging, and development assistance in that direction (Liddacoat, 2007). The Office of the High Commissioner for Human rights has explored the development of Human Rights mechanisms in the Pacific states. Clearly, there are at times tensions between embracing global norms and retaining traditional values and practices. At the level of rhetoric, political leaders are capable of simultaneously espousing the virtues of both stances. At the level of law and implementation, however, choices must be made.

7.3.1.1 Universal Periodic Review

In 2006 the Human Rights Council established the Universal Periodic Review (UPR) Mechanism, which applies to all 193 United Nation Member States. UPRs require preparation by each state of a report on actions it has taken to improve human rights domestically, examination of this report by the Human Rights Council together with recommended actions, and submission of responses to the Council's comments and recommendations by the country under examination. To date, the Universal Periodic Review has completed three "cycles", which yield a clear picture of the state of human rights in each Pacific Islands member-state (Table 7.12).

Table 7.12 Universal periodic review of human rights in Pacific Island countries

	UPR 1	UPR2	UPR 3
Federated States of Micronesia	2010	2015	2021
Fiji	2010	2014	2019
Kiribati	2010	2015	2019
Marshall Islands	2010	2015	2020
Nauru	2010	2015	2019
Palau	2011	2015	2021
PNG	2011	2016	2021
Samoa	2011	2016	2021
Solomon Islands	2011	2015	2021
Tonga	2008	2012	2018
Tuvalu	2008	2013	2018
Vanuatu	2008	2013	2018

https://www.ohchr.org/EN/HRBodies/UPR/Pages/Documentation.aspx

Reviews are extremely thorough and point out to Pacific nations shortfalls in their human rights practices—of which they are most likely already aware. (Baird, 2015) A survey of first cycle reports found that the most pervasive issues concerned gender equality (including violence against women, discrimination in legislation, and low rates of female representation in parliaments), low rates of ratification of international human rights instruments, high numbers of overdue reports to international treaty bodies, failures to establish national human rights bodies, ineffective ombudsman, failures to protect the rights of children, and lack of progress in development outcomes (including food and water security, health and education services, and sanitation). (Secretariat of the Pacific Community & Pacific Regional Rights Resource Team, 2012, p. 11).

Papua New Guinea is a case in point. A review of Papua New Guinea's 2012–17 UNDAF reported a "...disconnect between the adoption and the implementation of recommendations and obligations under human rights treaties and other mechanisms; a lack of government coordination of the human rights mechanisms in Papua New Guinea; and a lack of consistent reporting." (Lund & Lovai, 2016, p. 12) Papua New Guinea's 2016 UPR resulted in 161 recommendations to be considered for potential action by the Government of Papua New Guinea. The country's record of reporting under its human rights treaty commitments, as at 2016, is depicted in Table 7.13.

The UPR process is thorough. Fiji's first review, in 2010, included 103 recommendations, of which the Government accepted 97. The 137 recommendations in Fiji's second report included that it review the 2013 Constitution and review, amend or repeal such restrictive decrees as the Media Industry Development Decree 2010. ("Fiji Yet To Fully Respond To UN Human Rights Review: Repeal various decrees, review Constitution among recommendations", 2014)

Table 7.13 Status of human rights reporting in Papua New Guinea (2016)

Cooperation with treaty bodies (Lund & Lovai, 2016, p. 12)	
Committee on the Elimination of Racial Discrimination (CERD)	Second report overdue since 1984
Committee on Economic, Social and Cultural Rights (CESCR)	Initial report overdue since 2010
Human Rights Committee (HRC)	Initial report overdue since 2009
Committee on the Elimination of Discrimination against Women (CEDAW)	Fourth report overdue since 2014
Committee on the Rights of the Child (CRC)	Combined second and third reports overdue since 2008
Committee on the Rights of Persons with Disabilities (CRPD)	Initial report overdue since 2015

An second example of the thorough assessment is conveyed in the following example drawn from the Human Rights Council's working group's 10-page report within the Federated State of Micronesia's third (2021) UPR:

> II.3. The Committee on the Rights of the Child and the United Nations country team recommended that the Federated States of Micronesia ratify the Optional Protocol to the Convention on the Rights of the Child on a communications procedure. The Committee on the Elimination of Discrimination against Women and the Committee on the Rights of the Child recommended that the Federated States of Micronesia ratify the Convention against Torture and Other Cruel, Inhuman or Degrading Treatment or Punishment, the International Covenant on Civil and Political Rights, the International Covenant on Economic, Social and Cultural Rights, the International Convention on the Elimination of All Forms of Racial Discrimination, the International Convention for the Protection of All Persons from Enforced Disappearance and the International Convention on the Protection of the Rights of All Migrant Workers and Members of Their Families. The United Nations country team recommended that the Federated States of Micronesia ratify the remaining core international human rights treaties. The Committee on the Elimination of Discrimination against Women encouraged the Federated States of Micronesia to ratify the Optional Protocol to the Convention on the Elimination of All Forms of Discrimination against Women. (United Nations General Assembly. Human Rights Council. Working Group on the Universal Periodic Review, 2021, p. 1)

The report subsequently provides feedback on the FSM government's human rights performance in equality and non-discrimination; civil and political rights; economic, and social and cultural rights; with commentary that drills down to the level of adequacy of laws at sub-national level. Concerning the protection of women, for instance, the report states:

> The United Nations country team noted that, while the country had adopted measures to eliminate gender-based violence against women, violence against women remained prevalent. In 2017, the Pohnpei State Legislature passed the Domestic Issues Act, which set out mechanisms and procedures to address incidents of domestic violence. The legislature of Yap had also drafted a domestic violence bill, which was being reviewed. However, federal legislation criminalizing violence against women had yet to be adopted (United Nations General Assembly. Human Rights Council. Working Group on the Universal Periodic Review, 2021, p. 6)

Civil society has also been provided with opportunities to engage with the UPR process, both during the information gathering stage domestically, and when the report is being presented and assessed in Geneva. During the first cycle a side event organized by RRRT, World Young Women's Christian Association (YWCA) and Asia Pacific Women Law and Development (APWLD), featured the strategies of Pacific human rights defenders and activists, while in another side event the International Commission of Jurists (ICJ) and Fijian civil society organizations shared information about the human rights situation in Fiji (Secretariat of the Pacific Community & Pacific Regional Rights Resource Team, 2012, p. 8).

Baird has assessed Pacific responses to the first review cycle as "primarily positive", citing improved rates of treaty ratification and completion of reporting obligations (Baird, 2015, p. 7). The opportunity for civil society to contribute points of view was also beneficial, although the "voice" of civil society was not necessarily balanced, given that a majority were by single-issue advocacy groups, or coalitions led by international NGOs from the global north.

The Convention on the Elimination of all Forms of Discrimination against Women (CEDAW) and the Commission on the Status of Women

In most Pacific Island traditions women enjoy less entitlements than men, and those who seek to continue with tradition view human rights instruments that seek to promote their status as a danger to culture and identity. In such circumstances the decision as to whether or not to sign the Convention on the Elimination of all Forms of Discrimination against Women became a source of tension. The Commission on the Status of Women (CSW) was established in 1946 as a functional commission of the United Nations Economic and Social Council (ECOSOC) (Mishra, 2012, p. 2) and the Convention on the Political Rights of Women—the first international law instrument to recognise and protect the political rights of women—was produced by CSW and adopted by the General Assembly on 20 December 1952.[9]

The Convention on the Elimination of All Forms of Discrimination Against Women was established much later, in 1979. In some PICs, CEDAW was signed at the urging of women's organizations. Fiji ratified with two reservations in 1995, despite a call made by a national conference on the topic to ratify with no reservations (Mishra, 2012). The Government of Tonga refused to sign CEDAW for a considerable period despite intense campaigns mounted by women's organizations, backed by educational campaigns by intergovernmental bodies. The 2015 decision to sign, even if with reservations, was welcomed by Tongan women's rights advocacy groups including the Langa Fonua 'a Fafine Tonga, the National Centre

[9] CSW "Short History of the Commission on the Status of Women" Background note based on the United Nations Blue Book Series on The United Nations and the Advancement of Women, 1945–1996. (available online at http://www.un.org/womenwatch/daw/CSW60YRS/CSWbriefhistory.pdf), 5.

for Women and Children, the National Women and Children's Crisis Centre, and the Ma'a Fafine mo e Famili ("Tonga Lauded For Ratifying Un Women's Discrimination Convention: UN thanked for providing support to promote gender equity", 2015), However, debate continued in parliament as to whether the Convention threatened aspects of Tongan tradition concerning succession to the throne and noble titles, and aversion to same sex marriages and abortion ("Tonga PM: Referendum On UN Women's Convention Possible: Pohiva wants people to be consulted about CEDAW", 2015). This prompted civil society groups to petition parliament to ensure that the government proceed with ratification ("Tongan NGOs Petition For Ratification Of UN Women's Convention: CEDAW can help improve lives and women, girls in Tonga", 2015).

In Vanuatu, CEDAW's provisions have to a certain extent improved gender equality. An Ombudsman Commission report, for instance, found that the *Citizenship Act* (CAP112), in treating male and female applicants for citizenship differently, violated the fundamental right to equality provided in Art 5 of the Constitution and contravened the *Convention on the Elimination of All Forms of Discrimination Against Women (Ratification) Act*, which is binding in Vanuatu. Under the act, males who applied for citizenship after 10 years residence could include their wife and children in their application whereas female applicants could not.

CEDAW's progress in specific member-states is considered at periodic global CSW sessions, and PIFS has been the lead agency for coordinating participation by Pacific Islands' delegates. In 2018 Hilda Heine, President of the Republic of Marshall Islands, delivered a statement on behalf of the Forum at CSW's sixty second session:

> The Pacific Islands Forum's Framework for Pacific Regionalism and the landmark 2012 Pacific Leaders Gender Equality Declaration underpin our collective and country level efforts in support of gender equality. We are committed to supporting rural Pacific women in key areas of gender responsive government policies and programmes; decision making and leadership; economic empowerment; ending violence against women and education and sexual reproductive health services. In addition, we are promoting regional initiatives that advance the rights of rural women through access to health, improved connectivity, appropriate, and accessible ICT, economic and social opportunities and better returns for women in the fisheries and agriculture sector, labour mobility opportunities, and realisation of the rights of rural women and girls with disabilities.

Across the Pacific region, civil society organizations have participated in the CEDAW process by producing "shadow reports." In Solomon Islands, for instance, a report prepared by a group of ten NGOs for the period 2002–2012 raisied ten "critical issues and recommendations":

> Through the process, we realized just how many women, young women and girls in Solomon Islands have not heard of CEDAW and do not know (and therefore do not claim) their human rights. Several of the NGOs that participated in the development of this report have committed to further advocacy and awareness-raising on CEDAW specifically and on human rights more generally. The Solomon Islands government has developed some good policies and initiatives to mainstream gender, but this has largely been concentrated within the Government Women's Machinery. Implementation of these policies is yet to be realised. The government women's machinery itself is lacking in capacity. There is a

> dire need for whole of government to take active steps towards gender equality and undertake law and policy reform and effective measures to stop discrimination faced by women and girls in the country. Civil society organizations encourage government to take action to support service providers through funding. (Christian Care Centre et al., 2012, p. 3)

Similar reports have been generated across the region, not only in relation to CEDAW, but numerous other UN reporting procedures, from UNCAC, involving transparency international (Transparency International & UNCAC Civil Society Coalition, 2021), CRC, involving Cultural Survival (Cultural Survival, 2017), and climate finance, involving OXFAM. (Oxfam., 2020).

Refugees

The treatment of refugees under international law has been an important subject for Nauru and Papua New Guinea, which both entered "offshore processing" agreements with Australia to detain persons who had entered Australian waters illegally, and whose applications for status as refugees were being considered. However, the arrangements did not clarify how these individuals would be treated once that process was complete. In 2016 Nauru's Justice Minister David Adeang requested those present at a UN Summit on Refugees and Migrants to help find final destinations for 924 refugee men, women and children on Nauru. Although the government viewed the processing facility as Nauru's contribution to easing the world's refugee crisis, conditions there attracted the attention of the UN Subcommittee on the Prevention of Torture, which made an inspection visit ("UN Calls For Nauru To Withdraw Recent Amendments: Related to freedom of expression", 2015). Papua New Guinea's facility on Manus Island was closed after a court ruled that it was unconstitutional (Round, 2016). There have been occasional reports of freed detainees given Papua New Guinea residency but unable to travel internationally, and struggling to make their way in an unfamiliar society.

Convention Against Torture

A number of PICs have ratified the Convention against Torture, although those that have military or para-military forces—Papua New Guinea, Vanuatu and Tonga—are not among them. In 2016 Fiji, the other PIC with military forces, signed the convention with reservations, which included querying the definition of torture used by the UN, and querying the jurisdiction of the ICJ. Fiji was probably wary of the potential for action to be taken against its own military, as past military coups had included publicly known acts of violence. The Office for the High Commissioner for Human Rights interpreted Fiji's action as signalling a break with the past, and it encouraged other PICs to also sign, assuring them that it would assist with the technical aspects of periodic reporting required under the convention ("UN Urges Pacific Countries To Sign Anti-Torture Treaty: Fiji ratification, Vanuatu and Nauru only other signees", 2016).

Law of the Sea: UNCLOS

PICs have been vitally interested in the work of the United Nations Committee on the seabed since the 1973 conference on the Law of the Sea. The Pacific states, as archipelagos, have some of the most intricate ocean delimitations of any countries in the world. Since the island countries had more territorial waters than land, their marine resources constituted significant proportions of their economic potential. They sought clear definitions of national jurisdictions. New Zealand put a paper to the 1972 Forum pointing out that the definition of territorial seas, exclusive economic zones and high seas would have significant impact on fishing rights. UN Conferences on the Law of the Sea held in the 1960s took no account of the interests of Pacific Island Countries. (South Pacific Bureau of Economic Cooperation, 1971) The Tongan government noted, for instance, that five definitions for uses of territorial seas were in use, whereas Tonga's Royal Proclamation of 1887 contained a single definition for all such purposes (South Pacific Forum, 1971, p. 3). (South Pacific Bureau of Economic Cooperation, 1971) Fiji was added as a member of the UN Seabed Committee in late 1971 and updated the other Pacific countries at successive leaders' meetings. PIF's first Communique, in 1971, discussed the question of territorial sea claims in the South Pacific:

> ...It was considered that the unique dependence of countries of the South Pacific on marine resources merited special consideration in the recognition of territorial claims. The meeting welcomed the offer of Australia and New Zealand, as members of the United Nations Seabed Committee to keep Island Governments informed of progress of the Committee's deliberations and to draw the attention of the Seabed Committee to the special needs of the South Pacific Islands. (South Pacific Forum Secretariat. Leader's Meeting, 1971)

Fiji was keen to signal its desire to consolidate its control over its seabed, marine and fisheries resources using the "oceanic archipelagos" concept that was not yet fully recognized in international law (Fiji, 1972) (South Pacific Forum, 1972). As part of preparations, PIC law officers from Cook Islands, Fiji, Nauru, Papua New Guinea, Western Samoa, and Tonga met for the first time at the USP's Suva campus to receive an update on the work of the Seabed Committee from Australia, Fiji and New Zealand, and to strategize in preparation for the UNCLOS negotiating session in Caracas in August 1974. In one of the first addresses to an international conference on the Law of the Sea by a Pacific Islander, Cook Islands' Prime Minister Sir Albert Henry expressed the positions of both his own country and the region as a whole:

> From my home in the tiny islands in the huge Pacific Ocean, remote by many hundreds of miles from any neighbours, I have followed as best I could with deep interest the progress that has been made that has led to this great Conference. In my remote corner I have wondered at the influence and strength of the forces that brought this Conference into being, a Conference that could settle a new order in the world and in relationships between states. I have been impressed by the consideration being given to developing states. But at the same time I have worried whether the circumstances of small developing South Pacific Island countries such as mine have been known and fully appreciated by those who have the Influence and strength to decide the matters before this Conference. It is because of this worry, this uncertainty I have left my tiny and remote, but beautiful, homeland to cross the

> world to briefly place before distinguished delegates the reasons leading to my concern. (Henry, 1974)

Henry's mission was to ensure that the Pacific countries obtain the largest possible exclusive economic zone, and that its calculation not be based on landmass or population. This UNCLOS meeting in Caracas was the first major international engagement in which PICs worked together during negotiations and were recognised as the Oceania group of countries. Because the UN had already embarked on major policy initiatives regarding the seas and fishing, the PIC leaders realized the importance of immediate engagement with the dialogue processes. These included the preparatory Sea-Bed Committee, and the Third UN Conference on the Law of the Sea/UNCLOS, with its second 'working' session in 1974.

Issues such as tuna stocks and other highly migratory species, and ocean conservation, are regional as much as national in scope, and so UNCLOS recognises the importance of regional arrangements and has important provisions on, among other matters, global and regional cooperation for the protection and preservation of the marine environment. The Treaty took more than a decade to draft before being open for signing from December 1982. It entered into force at the end of 1994. It established a two-mile territorial sea and a 200-mile exclusive economic zone, recognized coastal states' rights over their continental shelf, set out a detailed code for use of the high seas, rights of passage through territorial seas and straits, and a regime for the exploitation of the seabed—all of which greatly benefited the Pacific Island countries.

The law was of considerable value to the Pacific Islands Region, which has approximately 48 shared boundaries where neighbouring jurisdictions overlap, but not all of which are yet subject to treaty. In 2012, during the 43rd Pacific Islands Forum in Rarotonga, and following several years of effort by technical and legal country teams and regional bodies (SOPAC, SPC, FFA) and other collaborators (the Australian Attorney General's Office and the Commonwealth Secretariat), seven Pacific Island Countries signed and exchanged eight Maritime Boundary Agreements using specifications set out in UNCLOS.[10] This clarification of maritime boundaries provided legal and jurisdictional certainty and a more accurate and secure basis for fishing, marine environmental protection, marine research, and the management and exploration of natural resources.

UNCLOS became an "umbrella" for subsequent agreements on fisheries, environmental protection and even trade. The *Cook Islands Seabed Minerals Act* 2009, for instance, was amended in 2015 to recognize UNCLOS, and the *Marae Moana Act* passed in 2017 to protect and conserve the ecological, biodiversity, and heritage values of the Cook Islands marine environment, was also passed to meet the country's responsibilities under the Conventions Concerning the Protection of the

[10]The seven bilateral treaties were signed by the Cook Islands, Niue, Kiribati, Tokelau, Tuvalu, Nauru and the Marshall Islands. Kiribati, Marshall Islands and Nauru agreed to an additional tri-lateral treaty concerning the determination of a point where the Exclusive Economic Zones (EEZs) of all three (3) countries intersect, referred to as a "Tri-junction point."

World Cultural and Natural Heritage, and on the Conservation of Biological Diversity, as well as UNCLOS.

Additional issues such as biological diversity, the effect of climate change on sea-level rise, and changes in coastal baselines, came later. The 1988 Forum Communique endorsed CCOP/SOPAC's call for PICs to ratify the Convention "...in order to protect jurisdiction of Forum members' marine resources within national exclusive economic zones." (PIF, 1988) It also noted the need for appropriate legislation to protect the region's marine and mining resources in accordance with the United Nations Convention on the Law of the Sea. (PIFS, 1989).

Illegal fishing in the Pacific Ocean became of increasing concern to the region's leaders. The 1989 *Tarawa Declaration* specifically called on Japan and Taiwan to adhere to UNCLOS provisions (particularly Articles 63, 64, 87, 116, 117, 118 and 119) concerning pelagic drift net fishing, which was damaging the region's economies and environment through "indiscriminate, irresponsible and destructive" fishing techniques that threatened the survival of the albacore tuna resource (PIFS, 1989). UNGA Resolutions 44/225 and 45/197 mandated the complete cessation of driftnet fishing in the South Pacific region by 1 July 1991 and the imposition by 30 June 1992 of a worldwide moratoria on all large-scale driftnet fishing on the high seas. In compliance with Resolutions 45/197 Forum members, in consultation with the Forum Fisheries Agency submitted reports to the UN to the Secretary General on the impact of driftnet fishing (South Pacific Forum, 1991).

In 2002 Forum Leaders endorsed the Pacific Islands Regional Oceans Policy that sought to ensure the future sustainable use of the Pacific Oceans and its resources, and to establish high-level leadership on oceans issues. In response to UNGA resolution 59/25 of 2004[11] on the Conservation and Management of Straddling Fish Stocks and Highly Migratory Fish Stocks, the Forum Leaders issued in 2006 a *Declaration on Deep-Sea Bottom Trawling to Protect Biodiversity in the high Seas* which committed PIF members to "... urgently take actions consistent with international law to prevent destructive fishing practices on seamounts in the Western Tropical Pacific Islands Area (WTPIA) and to prevent destructive fishing practices in other areas of high seas in the WTPIA until an appropriate environmental impact assessment has been carried out" (PIF Forum, 2006) the Declaration called for the establishment of effective monitoring, control, surveillance and enforcement measures, and for Forum members to advocate for an interim prohibition on destructive fishing practices.

United Nations Convention against Corruption (UNCAC)

The United Nations Convention Against Corruption (UNCAC) is one contribution to the global challenge of ensuring integrity in government. An Asia-Pacific *Human Development Report* suggests that the costs of corruption include "...eroded

[11] https://undocs.org/en/A/RES/59/25

opportunities, increased marginalization of the disadvantaged and feelings of injustice…" (UNDP, 2008, p. vii) and targets for promoting the rule of law, and substantially reducing corruption and bribery in all their forms are included SDG Goal 16: "Peace, justice and Strong Institutions." The UNCAC establishes an international treaty setting global standards for preventive measures, criminalization and law enforcement, international cooperation, asset recovery, and technical assistance and information exchange. By 2016 all Pacific Island countries had acceded to the treaty and had completed peer review processes (Table 7.14).

UNDP the United Nations Office for Drugs and Corruption (UNODC) are assisting the PICs with the establishment of National Anti-Corruption Strategies and with implementation of anti-corruption initiatives. (Hassall, 2017)

Biological Diversity

At a biodiversity conference in Curitiba, Brazil, in 2006, Palauan President Remengesau announced a plan under the Convention on Biological Diversity to agreement to protect 30% of near-shore marine and 20% of terrestrial resources on the islands by 2020. This pledge is reported to have inspired Grenada to pledge to put 25% of near-shore marine and 25% of terrestrial resources under effective conservation by 2020.

UNESCO Conventions

UNESCO is another U.N. agency that has generated conventions open to adoption, and which establish reporting obligations (Table 7.15).

No PIC has signed the 1954 Hague Convention for the Protection of Cultural Property in the Event of Armed Conflict, nor the 1970 convention on the Means of Prohibiting and Preventing the Illicit Import, Export and Transfer of Ownership of Cultural Property, nor the 2005 Convention for the Protection and Promotion of the Diversity of Cultural Expressions.

7.4 Discussion

Determining which agencies of the United Nations to join, at what cost and for what benefit, has been an on-going task for the small island member-countries of the Pacific region. Nonetheless, they have succeeded in establishing a permanent physical presence in New York, and through collaboration are also increasing their representation in Geneva and elsewhere. Although they do not possess the resources to participate in the UN's full calendar of events, they make strategic choices as to which agencies and programmes have most bearing on their domestic circumstances, and in which they can make the most effective contributions. Similar choices must

Table 7.14 Pacific Island Country participation in UN Convention Against Corruption

Region/country or territory	Signature	Ratification, Acceptance (A), Approval (AA), Accession (a), Succession (d)[a]	Participation in sixth Session (2015)[b]	Participation in fifth session (2013)[c]	Reports
MELANESIA					
1. Fiji		14 May 2008 a	√	√	ES 2012
2. New Caledonia	9 Dec 2003	11 Jul 2005			
3. Papua New Guinea	22 Dec 2004	16 Jul 2007		√	ES 2013
4. Solomon Islands		6 Jan 2012 a	√	√	ES 2014
5. Vanuatu		12 Jul 2011 a	√	√	ES 2013 CR
MICRONESIA					
6. Federated States of Micronesia: participation in UNCAC		21 Mar 2012 a	√	√	
7. Guam	9 Dec 2003	30 Oct 2006			
8. Kiribati		27 Sep 2013 a		√	
9. Marshall Islands		17 Nov 2011	√	√	
10. Nauru		12 Jul 2012 a	√	√	
11. Northern Mariana Islands	9 Dec 2003	30 Oct 2006			
12. Palau		24 Mar 2009 a	√	√	
POLYNESIA					
13. American Samoa	9 Dec 2003	30 Oct 2006			
14. Cook Islands		17 Oct 2011	√	√	2015[d]
15. French Polynesia	9 Dec 2003	11 Jul 2005			
16. Niue	10 Dec 2003	1 Dec 2015			
17. Pitcairn participation in UNCAC	9 Dec 2003	9 Feb 2006			
18. Samoa	?				
19. Tokelau	10 Dec 2003	1 Dec 2015			
20. Tonga	?				
21. Tuvalu		04 Sep 2015 a	√		

(continued)

Table 7.14 (continued)

Region/country or territory	Signature	Ratification, Acceptance (A), Approval (AA), Accession (a), Succession (d)[a]	Participation in sixth Session (2015)[b]	Participation in fifth session (2013)[c]	Reports
22. Wallis and Futuna	9 Dec 2003	11 Jul 2005			

[a]https://www.unodc.org/unodc/en/treaties/CAC/signatories.html
[b]https://www.unodc.org/documents/treaties/UNCAC/COSP/session6/FINAL_LOP_COSP_6.pdf
[c]https://www.unodc.org/documents/treaties/UNCAC/COSP/session5/FINAL_LOP_COSP_5.pdf
[d]http://www.unodc.org/documents/treaties/UNCAC/CountryVisitFinalReports/2015_12_08_Cook_Islands_Final_Review_Report.pdf

Table 7.15 UNESCO Conventions signed by Pacific Island Countries

Country	Joined UNESCO	1972 World Heritage	2001 Underwater Cultural Heritage	2003 intangible cultural heritage	2005 Diversity of Cultural Expressions
Cook Islands	1989			Y	
Federated States of Micronesia	1999	y	y	y	
Fiji	1983	y		y	
Kiribati	1989	y		y	
Marshall Islands	1995	y		y	
Nauru	1996			Y	
New Caledonia	2017 (associate)				
Niue	1993	Y	Y		Y
Palau	1999	Y		Y	
Papua New Guinea	1976	y		y	
Samoa	1981	Y		Y	Y
Solomon Islands	1993	Y		Y	
Tokelau	2001 (associate)				
Tonga	1980	Y		Y	
Tuvalu	1991			Y	
Vanuatu	1994	Y		Y	

be made concerning accession to international treaties, since such commitments invariably create responsibilities regarding domestic law reform and government practice, just as they create resource-intensive reporting obligations. Viewed positively, these domestic and international "burdens" result in capacity development within the small states at the same time that they prompt UN organs and agencies to adjust to accommodate the needs of their smaller members—a theme that will be revised in subsequent chapters.

References

Baird, N. (2008). *Will the universal periodic review make a difference in the Pacific? Conference: Celebrating 60 years of the universal declaration of human rights.* Paper presented at the New Zealand Centre for Public Law, Victoria University of Wellington, Wellington.

Baird, N. (2015). The role of international non-governmental organizations in the universal peridoc review of Pacific Island States: Can 'Doing Good' be done better? *Melbourne Journal of International Law, 16*, 1–37.

Christian Care Centre, Family Support Centre Organization, G., People with Disabilities Solomon Islands, Sistas Savve Group, W. R., Solomon Islands National Council of Women, Solomon Islands Planned Parenthood Association, .World Vision Solomon Islands (2012). *NGO Shadow Report on the Status of Women in Solomon Islands. Initial, Secod and Third report (2002–2012).* Honiara.

Cole, R. V., & Thompson, G. J. (1981). *Proposals for a Pacific Islands Fund: An evaluation: A report to The Secretary General, South Pacific Commission The Director, South Pacific Bureau for Economic Cooperation.* South Pacific Commission.

Cultural Survival. (2017). Convention on the Rights of the Child. In *Shadow Report Submission: Indigenous Children's Rights Violations in The Marshall Islands.* Cultural Survival.

Delaney, A. V. (2006). *The importance of being 'least' developed.* Islands Business.

Fassbender, B. (2020). *Key documents on the reform of the UN security council 1991–2019*: Brill | Nijhoff.

Fiji. (1972). Second South Pacific Forum Canberra, February 1972. *Paper presented by the FIji Delegation for Discussion: Law of the Sea.* (PIFS (72) NN.1). Canberra, .

Fiji Yet To Fully Respond To UN Human Rights Review: Repeal various decrees, review Constitution among recommendations. (2014, 3 November). *Fijilive.*

Freestone, D., Barnes, R., & Akhavan, P. (2022). Agreement for the establishment of the commission of small Island State on climate change and international law (COSIS). *The International Journal of Marine and Coastal Law, 37*, 166–178.

Hasenkamp, O. (2016). The Pacific Island countries and international organizations: Issues, power and strategiees. In A. Holtz, M. Kowasch, & O. Hasenkamp (Eds.), *A region in transition politics and power in the Pacific Island countries.* Saarland University Press.

Hassall, G. (2017). Law, culture, and corruption in the Pacific Islands. *New Zealand Yearbook of International Law, 15*, 9–28.

Hawke, C., Martin, P., & Taula, P. (2020). Keeping our commitments: The small Island developing states open debate. In G. Hassall & N. Partow (Eds.), *A seat at the table: New Zealand and the United Nations Security Council, 2015–2016* (pp. 157–170). Massey University Press.

Henry, A. (1974). *Statement by Sir Albert Henry, Prime Minister of The Cook Islands.* Paper presented at the 3rd United Nations Conference on the Law of the Sea.

Horne, A. (2011). Challenges Facing the South Pacific Islands in Ensuring Universal Access to Communications Sevices: Vanuatu a case history in the making. *2011 Technical Symposium at ITU Telecom World.*

International Labour Organization. (2009). *Decent Work Country Programme Samoa.* International Labour Organization.

Kaul, I., Conceicao, P., Goulven, K. L., & Mendoza, R. U. (Eds.). (2003). *Providing global public goods: Managing globalization: Managing Globalization.* Oxford University Press.

Kaul, I., Grunberg, I., & Stern, M. A. (Eds.). (1999). *Global public goods: International cooperation in the 21st Century*. Oxford University Press.

Kenneth, G. (2021, 30 December). *Head of mission to US has no official vehicle*. Papua New Guinea Post Courier.

Liddacoat, J. (2007). *National human rights institutions: Pathways for Pacific States*. Pacific Islands Forum Secretariat and New Zealand Human RIghts Commission.

Lund, A., & Lovai, B. (2016). *Independent evaluation UN development assistance framework (2012–2017) UN Country Team in* .

McNamara, K. E. (2009). Voices from the margins: Pacific ambassadors and the geopolitics of marginality at the United Nations. *Asia Pacific Viewpoint, 50*(1), 1–12.

Mishra, M. (2012). A history of Fijian Women's Activism (1900–2010). *Journal of Women's History, 24*(2), 115–143. https://doi.org/10.1353/jowh.2012.0018

Oxfam. (2020). *Climate finance shadow report 2020: Assessing progress towards the $100 Billion commitment*. Oxfam.

Pacific Islands Forum. (1988). *Nineteenth South Pacific Forum, Nuku'alofa, Tonga, Forum Communique*. SPEC.

Pacific Islands Forum. (2006). *Declaration on deep-sea bottom trawling to protect biodiversity in the high Seas*. Pacific Islands Forum Secretariat.

Pacific Islands Forum Secretariat. (1989). Twentieth South Pacific Forum, Tarawa, Kiribati. *Forum Communique.. SPFS(89)19*(SPFS(89)19).

Pacific Islands Forum Secretariat. (2014). National human rights consultation a first for the Kingdom [Press release].

Round, S. (2016, 20 September). Pacific Island countries call on UN for help dealing with refugees: Nauru's justice minister defends detention camps as 'innovative model' that saved thousands of lives. *Radio New Zealand International*.

Secretariat of the Pacific Community, & Pacific Regional Rights Resource Team. (2012). *Pacific Island states and the Universal Periodic review: A toolkit for Pacific Island states to measure progress and compliance against Universal Periodic Review recommendations*. Secretariat of the Pacific Community.

South Pacific Bureau of Economic Cooperation. (1971, August). *South Pacific Forum Wellington*. Summary record of proceedings: Part 2. Suva, .

South Pacific Forum. (1971, August 5-7). South Pacific Forum. Wellington, Summary record of proceedings. *(71) NN. 1*.

South Pacific Forum. (1972, February). Summary record of proceedings. *(72) NN. 1*, 1 v. (various pagings).

South Pacific Forum. (1991). *FSM Communiqué*.

South Pacific Forum Secretariat. Leader's Meeting. (1971). *Communique*. South Pacific Forum Secretariat.

Starr, R. I., Gordon, E., Wallace, J., Baker, S., & Rosenstock, R. (1968). The Future Relationship Between Small States and the United Nations. *The International Lawyer, 3*(1), 58–74.

Stone, D. (2008). Global public policy, transnational policy communities, and their networks. *The Policy Studies Journal, 36*(1), 19–38.

Tonga Lauded For Ratifying Un Women's Discrimination Convention: UN thanked for providing support to promote gender equity. (2015, 10 March). Matangi Tonga.

Tonga PM: Referendum On UN Women's Convention Possible: Pohiva wants people to be consulted about CEDAW. (2015, 28 May). *Radio New Zealand International*.

Tongan NGOs Petition For Ratification Of UN Women's Convention: CEDAW can help improve lives and women, girls in Tonga. (2015, 1 June). Matangi Tonga.

Transparency International, & UNCAC Civil Society Coalition. (2021). Papua New Guinea: Civil Society Report by Transparency International Papua New Guinea. In *An input to the UNCAC Impleentation Review Mechanism: Fourth year of review of UNCAC Chapter II*. Transparency International.

Trumbull, R. (1977). *Tin Roofs and Palm Trees: A report on the New South Seas*. Australian National University Press.

UN Calls For Nauru To Withdraw Recent Amendments: Related to freedom of expression. (2015, May 23). *Radio New Zealand International*.

UN Urges Pacific Countries To Sign Anti-Torture Treaty: Fiji ratification, Vanuatu and Nauru only other signees. (2016, March 22). *Radio New Zealand International*.

United Nations Development Programme. (2008). *Tackling Corruption, Transforming Lives: Accelerating Human Development in Asia and the Pacific*. United Nations Development Programme. Regional Centre in Colombo. Human Development Report Unit.

United Nations General Assembly. (1993). *Note verbale dated 93/10/11 from the Permanent Representative of The Marshall Islands to the United Nations addressed to the Secretary-General*. UN.

United Nations General Assembly. (2004). *Note verbale dated 2004/06/04 from the Permanent Mission of New Zealand to the United Nations addressed to the Secretary-General*. UN.

United Nations. Commission on Human Rights. (1996). Report of the Special Rapporteur on extrajudicial, summary or arbitrary executions, submitted pursuant to Commission resolution 1995/66, Addendum: Report by the Special Rapporteur on his mission to Papua New Guinea island of Bougainville from 23 to 28 October 1995. *1996, E/CN.4/1996/4/Add.2*.

United Nations. Economic and Social Council. Committee for Development Policy. (2006, March 20-24). Report on the eighth session. *Economic and Social Council Official Records*.

United Nations General Assembly. Human Rights Council. Working Group on the Universal Periodic Review. (2021). *Compilation on the Federated States of Micronesia*. Report of the Office of the United Nations High Commissioner for Human Rights. *A/HRC/WG.6/37/FSM/2*.

United Nations. Security Council. (2003). *Letter dated 2003/07/21 from the Permanent Representative of Fiji to the United Nations addressed to the President of the Security Council*. UN.

United Nations. Security Council. (2015). *Peace and security challenges facing small Island developing states*. United Nations.

van Bohemen, G. (2015, July). *Letter dated 26 August 2015 from the Permanent Representative of New Zealand to the United Nations addressed to the President of the Security Council: Assessment of the work of the Security Council during the presidency of New Zealand. S/ 2015/621*. http://www.un.org/en/ga/search/view_doc.asp?symbol=S/2015/621

Chapter 8
Regional Peace and Security

8.1 Security

It will be recalled that the first purpose of the United Nations as set out in its Charter concerns the establishment and maintenance of peace and security. Accordingly, this chapter explores the ways in which the UN has assisted the countries of the Pacific region toward these objectives. The Pacific region's security challenges since 1945 included nuclear testing, separatist conflicts, and political unrest in the new and emerging nations. From the 1970s onwards, security issues considered by regional institutions have fallen into two categories: those that originate in the Pacific Islands region and require "Pacific solutions", and those originate outside the region, and involve much larger actors than just the small Pacific nations.

Global geo-politics has certainly affected the region. The Pacific countries arrived on the stage of global affairs through opposition to atmospheric nuclear testing, then through participation in global dialogue processes concerning the Law of the Sea, fisheries, and decolonization. There are also the expanding strategic and economic interests of the super-powers which, if not adroitly handled, could result in considerable political, social, and economic disturbance. The establishment through PIC initiative of a nuclear free zone was a direct challenge to the security doctrines of the nuclear powers. The US, France, Britain, Russia, and China, all wished to maintain close relations with the Pacific Island countries, but their possession of nuclear weapons made it impossible for them to become signatories.

The contest between the People's Republic of China and Taiwan for the loyalty of the Pacific states continues to threaten the stability of the region, and effort is constantly required to satisfy as many parties as possible (regional meetings have expended valuable time, for instance, on whether to refer to Taiwan as Chinese Taipei to appease China, or as Republic of China to please Taiwan). Braithwaite suggests that the UN didn't help Solomon Islands during its period of domestic instability because China would have vetoed assistance. The decision in 2021 by Solomon Islands Prime Minister Manasseh Sogovare to switch recognition from Taiwan to the People's Republic of

G. Hassall, *The United Nations and the Pacific Islands*, United Nations University Series on Regionalism 24, https://doi.org/10.1007/978-3-031-34155-7_8

China sparked riots fuelled by opposing parties that feared the implications for future control of the country's resources (Braithwaite et al., 2010).

This problem of "the two China's" has also had significant impact on Vanuatu. An effort in 2005 by Vanuatu's Prime Minister Serge Vohor to change allegiance from China to Taiwan resulted in his government's rapid downfall.

Despite their ideological differences, China and Taiwan share in common the practice of distributing "development assistance" without levels of transparency and accountability that have become accepted practice in the broader field (Dornan & Pryke, 2017). China does not comply with Forum Compact/Busan Agreement principles on transparency and accountability (on the grounds that their standards were set by the OECD countries without its consent), and refuses to reveal the sources and extent of funding assistance to PICs. The nature of Taiwan's financial assistance to successive governments in Solomon Islands and Nauru has been criticized.

China's desire to forge deeper links with Pacific states was most clearly stated by the attendance by nine Pacific states at the "First Ministerial Meeting of the China-Pacific Island Countries Economic Development and Cooperation Forum" in Fiji in April 2006. The Pacific States' relationship with China is being guided by the *China-Pacific Island Countries Economic Development and Cooperation Guiding Framework*, and Chinese influence in Pacific Islands' economies and trade practices can only increase. In October 2021 China held its first "China-Pacific Island Countries Foreign Ministers' Meeting"—albeit by video link due to the COVID-19 pandemic.[1] China's support for ICT infrastructure in PICs (such as Samoa's National Broadband Highway, and provision of data centres for Papua New Guinea and Vanuatu amongst other places) has prompted Australia to enter the Pacific telecoms market through acquisition of Digicel in Papua New Guinea. Other Western and Asian countries also use aid diplomacy to exert their influence on the Pacific region. Japan's Pacific Islands Leaders Meeting has convened every 3 years since 1997 in the "Pacific Islands Leaders Meeting (PALM). France has hosted five "France-Oceania Summits" commencing in 2003 and most recently in 2021.[2]

8.1.1 Nuclear Testing

The UN's engagement with the Pacific Islands security agenda commenced with the issue of nuclear testing and its consequences, rather than with intrastate conflicts in Papua New Guinea, Solomon Islands, and Fiji. France commenced atmospheric nuclear testing in the Pacific in 1966 and by 1995 had conducted 28 atmospheric and

[1] https://www.fmprc.gov.cn/mfa_eng/wjdt_665385/2649_665393/202110/t20211021_9604831.html

[2] France-Oceania summits have been French Polynesia (2003)'; Paris (2006); New Caledonia (2009); Paris (2015); and, due to the COVID-19 pandemic, virtually (2021).

129 underground tests (United Nations. General Assembly, 1995a, p. 26). The UK and the USA conducted 33 atmospheric nuclear tests at Malden and Kiritimati (Christmas) Islands between 1957 and 1962 (Alexis-Martin et al., 2021). From the outset, PICs were united in voicing their opposition. In addition to voicing their concerns as individual nations, they brought this topic into their core regional policies, and to the UN.

PIC concern about nuclear testing in the Pacific was expressed in annual Forum communiques, beginning with the first, in 1971. Pacific leaders initially endorsed New Zealand's proposal to the 1972 Forum that the best response to France should be diplomatic, through the United Nations, and through environmental and disarmament conferences, the Seabed Committee, and regional groups (South Pacific Forum, 1972, p. 32). In that year, also, New Zealand proposed in the General Assembly a declaration opposing nuclear testing that was endorsed by Fiji and Australia. At their 4th meeting, in 1973, the Pacific leaders expressed satisfaction at the direction of debate within UNGA on the need to halt testing, and in support of establishing a nuclear free zone in the South Pacific:

> Members were once again unanimous in expressing their deep concern at the apparent continuing failure of the French government to comprehend the extent of opposition to the conduct by France of its tests in the Pacific area and to respect the wishes of the peoples of the area. . . .they urged the government of France to heed the call of the United Nations General Assembly and its obligations under international law by bringing about an immediate halt to all testing in the area. (PIFS, 1973, p. 7)

The Forum welcomed UNGA's 1973 resolution condemning atmospheric testing.[3] (South Pacific Bureau for Economic Co-operation, 1973) The topic was raised at the fifth meeting, in Tonga in July 1975:

> Members recalled the expressions of opposition to atmospheric nuclear testing at the last three meetings of the Forum. In particular they recalled the Joint Declaration on such testing made at the meeting last year. They deplored the fact that despite the declaration, France has since carried out a further series of tests in the South Pacific which has resulted in the deposit of radioactive fall-out in a number of countries in the region.
>
> Members once again called on Governments, and in particular the Government of France, the only Government testing in the South Pacific, to heed the views of the Forum, the calls of the United Nations General Assembly and other international bodies and obligations under international law by bringing about an immediate halt to all testing.
>
> Members were unanimous in expressing their concern at the potential health and other hazards to which their people and environment were exposed as a result of radioactive fall-out from such tests.
>
> Members welcomed renewed efforts being made at the international level for a ban on nuclear weapons testing in all environments and agreed to encourage and support any constructive moves that would preserve the human family from the hazards of nuclear weapons testing and further the goal of nuclear arms control and disarmament. (South Pacific Forum, 1974)

The fact that France continued its nuclear tests in the Pacific into 1995, 3 years after the 1992 Rio Conference on Environment and Development, provoked the Forum to

[3] https://www.forumsec.org/south-pacific-forum-apia-western-samoa-17-18-april-1973/

suspend French participation in post-Forum dialogues, and to state its view at UNGA. According to the General Assembly record:

> The delegations on whose behalf (Papua New Guinea Representative Utula Samana) was speaking had reacted with extreme outrage to the resumption of nuclear testing by France in the region. Their call for an immediate halt to the testing had been echoed around the world, and was reflected in a resolution currently before the First Committee of the General Assembly. By resuming nuclear tests France was ignoring its obligations with respect to the precautionary approach explicit in Agenda 21 and threatening the social and economic survival of the countries of the region... (United Nations. General Assembly, 1995b, p. 9)

France used her right of reply to suggest that there was no scientific evidence of the tests being harmful to humans or to the environment, and as usual, Papua New Guinea used her right of reply to suggest that if this was the case then France would surely conduct future tests in the Bay of Biscay or Paris. Recent analysis of previously classified French government records suggests that France underestimated the levels of toxic exposure experience by more than 100,000 residents near Mururoa and has knowingly deceived the public about the impact of the tests on people's health (Philippe & Tomas, 2021).[4]

8.1.2 Nuclear-Free Zone Treaty

Whereas the PICs were unable to halt nuclear testing, they were able to establish a *South Pacific Nuclear-Free Zone Treaty*, also known as the *Treaty of Rarotonga*. The South Pacific Forum's resolution 3477 on a nuclear free zone was endorsed by UNGA in 1975—110-0, with 20 abstentions. Although hiatus set in when conservative governments were subsequently elected in both Australia and New Zealand, the initiative resurfaced when a working group in Suva in November 1984 considered a paper prepared by Australia (SPEC, 1984). The Treaty was opened for signature on 6 August 1985, Hiroshima day. "This South Pacific nuclear free zone Treaty," New Zealand Prime Minster Geoffrey Palmer informed the General Assembly, "...signals in no uncertain terms the deeply anti-nuclear sentiments of our peoples, and their desire to live in peace and independence and to run their own affairs in accordance with their own wishes." (United Nations General Assembly, 1985, p. 61).

The Forum regards the Treaty as the region's contribution to strengthening global security and international non-proliferation of nuclear weapons, and as an indication of its deep commitment to world peace. The Pacific Islands Forum informs the UN the results of periodic conferences of the parties to the treaty. However, the Treaty has not been signed by Palau, Marshall Islands, or Federated States of Micronesia, nor by the three nations with territories within the Treaty's zone: France, the United

[4]Moruroa Files: Investigation into French nuclear tests in the Pacific: https://moruroa-files.org/en/investigation/moruroa-files

Kingdom and the USA. In a vote on UNGA resolution on the Treaty (A/RES/76/34) on 6 December 2021, Papua New Guinea and Solomon Islands voted for, Marshall Islands and Tonga abstained, and the Federated States of Micronesia voted against.[5]

More recently, the PICs have been able to signal their anti-nuclear stance by acceding to the *Treaty on the Prohibition of Nuclear Weapons* which was opened for signature in September 2017. By 2022 there were 9 PIC Parties to the Treaty (Cook Islands, Fiji, Kiribati, Nauru, Niue, Palau, Samoa, Tuvalu, and Vanuatu). Marshall Islands, Federated States of Micronesia, Papua New Guinea, Solomon Islands and Tonga have not signed, indicating that the region does not hold a united position on this aspect of global security.

8.1.3 1992 Forum Honiara Declaration on Law Enforcement Cooperation

Reflection on the Pacific Island Forum's inability to respond adequately to internal crises in Bougainville and Fiji resulted in the establishment of collective guidelines to cope with future situations of a similar nature. The articulation of a comprehensive security framework for the Pacific region included protocols and declarations concerning good governance, a regional security survey, and the deliberations of the Forum Regional Security Committee (FRSC).[6] A discussion paper submitted by Australia to the Committee's first meeting in February 1992 envisaged a closer relationship between the Forum and the UN on security issues:

> At the global level, the United Nations is beginning to assume the collective security role envisaged by its founders. Forum members have a particular interest in the evolution of UN doctrine on the security of small states, an issue which has also been taken up in Commonwealth councils. The existing level of cooperation between Forum members of the UN on environmental and fisheries (driftnetting) issues could be extended to security issues with regional implications, particularly issues pertaining to the security of small states. (Pacific Islands Forum Secretariat, 1992, p. 8 Annex 1: discussion paper)

The *Honiara Declaration on Law Enforcement and Cooperation* of 1992 which established the main principles for law enforcement cooperation was refined and expanded by the Aitutaki Declaration on regional security cooperation (1997), the *Biketawa Declaration* (2000), and the *Nasonini Declaration on Regional Security* (2002). Each of these initiatives provided additional potential entry-points for engagement with relevant UN agencies and processes.

The *Honiara Declaration* identified the key law enforcement cooperation concerns, especially transnational crime, customs, policing and drugs. Both the 1996 Majuro Forum and the 1997 Rarotonga Forum reiterated their commitment to the

---

[5] https://digitallibrary.un.org/record/3950567?ln=en

[6] The FRSC, established in 1992, was preceded by the Committee for Regional Security Information Exchange, which met in 1988 and 1990.

principles contained in the *Honiara Declaration*. The Declaration guided the Secretariat's work with regional specialist bodies such as the Oceania Customs Organization (OCO), the Pacific Immigration Directors Conference (PIDC), the South Pacific Chiefs of Police (SPCPC) and with international organizations such as the Commonwealth Secretariat, the Asia-Pacific Group on Money Laundering Secretariat, the United Nations Office for Drug Control and Crime Prevention (UNODCCP), the United Nations Global Programme Against Money Laundering, the International Monetary Fund (IMF) and the Australian Federal Police, especially through their Law Enforcement Cooperation Programme (LECP).

8.1.4 Light Arms

Pacific Meetings in support of a campaign to limit, and outlaw, small arms and light weapons, commenced in 1997 (United Nations. General Assembly, 2001) and continued for the next decade. The Pacific states participated in a sub-regional meeting in October 2000, an Asia Pacific regional disarmament meeting in March 2001, and a small arms workshop in May 2001. The small arms workshop, convened in Brisbane, addressed draft legislation for the establishment of a common regional approach to weapons control and provided practical advice for legislators and law-enforcement agencies on effective implementation of the model law (2001, p. 8). A regional meeting on program of action was held in Sydney in June 2009, prior to the third biennial meeting of states in New York that year to consider its implementation.

In March 2012, when there were still no regional agreements regulating arms transfers in the Pacific, and most PICs did not have comprehensive and effective national arms control systems, the UN and PIFS convened a session in Brisbane to build Pacific commitment to the 2001 UN Programme of Action that aims to eradicate the illicit trade in small arms and light weapons, and to establish a regional position prior to negotiations being held in July toward establishing a global, legally binding *Arms Trade Treaty* (ATT). There were associated projects, such as PIFs project on control of ammunition in PICs and strengthening of armouries ("Assistance to States for curbing the illicit traffic in small arms and light weapons and collecting them: the illicit trade in small arms and light weapons in all its aspects: report of the Secretary-General," 2009, p. 12).

The *Aitutaki Declaration* of 1997 referred to the region's "vulnerability to natural disasters, environmental damage and unlawful challenges to national integrity and independence" and reaffirmed a commitment to a "...comprehensive, integrated and collaborative approach to maintaining and strengthening current mechanisms for cooperation among members in dealing with threats to the security, broadly defined, of states in the region and of the region as a whole."

The *Biketawa Declaration* of 2000 identified a range of "difficult and sensitive issues including underlying causes of tensions and conflict" (ethnicity, socio-economic disparities, and lack of good governance, land disputes and erosion of cultural values). The framework was first used in 2003, when the Forum responded to a request from the Solomon Islands government to help re-establish the rule of law, which had broken down in the face conflict between militia groups from Malaita and Guadalcanal. The Regional Assistance Mission to the Solomon Islands (RAMSI) stabilized the country within 3 months and peacefully collected almost 4000 weapons. The mission's aims extended beyond the reestablishment of law and order to assisting the processes of rehabilitation and nation-building. Biketawa was also the basis for the Pacific Regional Assistance to Nauru (PRAN).

The *Nasonini Declaration* of 2002 binds the Pacific Islands Forum States to the implementation of internationally agreed anti-terrorism measures such as the UN Security Council Resolution 1373, and the Financial Action Task Force Special Recommendations. Forum countries have undertaken to introduce legislation and other measures to combat terrorism and terrorist financing.

The *Biketawa* and *Nasonini declarations* laid out actions that the Forum Secretary-General can undertake, in consultation with Forum Foreign Ministers:

- A statement representing the view of members regarding the situation
- Creation of a Ministerial Action Group
- A fact-finding or similar mission
- Convening an eminent persons group
- Third party mediation
- Providing support for appropriate institutions or mechanism that would assist a resolution
- Convening of a special high-level meeting of the Forum Regional Security Committee or an ad hoc meeting of Forum Ministers
- Convening of a special meeting of Forum Leaders to consider other options including, if necessary, targeted measures. (van Tongeren & Heemskerk, 2005, p. 74)

In 2003 the Leaders adopted a statement on "nine principles of good leadership"—principles that range from respect for the rule of law, to promotion of "peace, security and harmony."[7] The Forum also established a network of the region's legislative speakers (The Forum Presiding Officers' Conference—FPOC) and hosted talks intended to result in a regional network of national Human Rights Commissions and a Pacific charter of human rights. Unfortunately, none of these initiatives came to fruition. The leadership principles sit unnoticed on the PIFS website, the FPOC network collapsed, and no regional human rights charter or network has been established.

The regional peace and security agenda was set at annual deliberations of the PIFS-administered FRSC, which brought together security sector officials,

[7] The nine principles are: Respect for the law and system of government; Respect for cultural values, customs and traditions; Respect for freedom of religion; Respect for people on whose behalf leaders exercise power; Respect for members of the public; Economy and Efficiency; Diligence; National Peace and Security; and Respect for Office.

development partners, academics, and civil society representatives. National Security Assessments undertaken for the Forum Secretariat 2000–2004 analysed the internal and external security threats facing PICs. The first report, by USP emeritus professor Ron Crocombe, reviewed matters affecting Security in the region generally (Crocombe, 2000). It was followed a year later by a multi-authored report on security in the Southwest Pacific, (Anere et al., 2001) and in 2004 by a single-author assessment of the three Micronesian member-states (Hassall, 2004).

The three reports identify concerns at national and sub-national level but say less about issues at supra-national or regional level. None identified any external military threat, but each suggested a range of domestic issues having a high probability of leading to insecurity: poor governance, crime and violence, socio-economic disparities, ethnic tensions, and land management (Calvert, 2005). "Given Fiji's geopolitical position and role as the regional hub," A report on Fiji's 2006 military coup by the European Commission's Conflict Prevention and Crisis Management Unit noted. "...whatever happens here is more likely to have a bearing on developments and attitudes in the rest of Melanesia than conflict dynamics elsewhere." (Retiere & Schurmann-Zeggel, 2002).

Consequent to these assessments, PIF adopted a "comprehensive security" approach that included real threats to the stability of Pacific Island societies: political instability (frequent changes of government, removal of Ministers, and corruption), economic and social inequities (including denial of human rights and lack of access to services), land pressures, food insecurity, environmental degradation, "new" threats to society such as of HIV/AIDS and drugs, and even poor governance (UNDP, 1994, 1999, 2000, 2004; Westcott, 1996; ADB, 2003, 2004; Holden et al., 2004; Mellor & Jabes, 2004; SOPAC, 2003; World Bank, 2004). More recently, the security agenda has featured, in addition to human security, the "global war on terror", and climate change (Morgan, 2021). This last threat is discussed in more detail in a subsequent chapter.

8.2 Conflict Resolution

Leaving aside the impact of both world wars in the Pacific and global political tensions, the most significant instances of conflict in the region occurred in New Caledonia, Solomon Islands, Vanuatu, Papua New Guinea (Bougainville), and Fiji. The people of Bougainville and Guadalcanal did not imagine their lives being disrupted by violent struggle, just as Fijians hardly imagined their precarious but nonetheless negotiated constitutional order would be overthrown by military coup in 1987 and multiple times again in later years. Long-standing grievances are left unattended to a nation's peril, and peacebuilding requires at least some parties within that society—or outside it—to take the lead in addressing them in ways that prove effective.

8.2.1 Bougainville

The civil war on Bougainville was a different order of conflict for PICs to deal with. This was no longer political and diplomatic conflict with a distant foe, such as France or the United States—it was violent confrontation between Pacific Islanders, about which the rest of the world knew and understood little. Long-standing discontent over the ownership, operation, and impact of the Panguna copper mine, fuelled also by underlying secessionist grievances, culminated in acts of sabotage on capital equipment at the mine in 1987 (these issues were noted in chapter two). Failure to resolve these issues through political or legal means resulted in the declaration of a state of emergency by the national government and a lapse of government authority at provincial level. The Papua New Guinea Defence Force was deployed to Bougainville. When the military established an exclusion zone around the island and a condition of civil war ensued, the U.N. Sub commission on Prevention of Discrimination and Protection of Minorities unanimously adopted a resolution calling on the government of Papua New Guinea "to restore without delay the freedom of movement to the people of Bougainville in the interest of protecting and promoting human rights and fundamental freedoms. The resolution also placed the issue of a peace treaty process between Papua New Guinea and Bougainville under review by the UN Commission's Special Rapporteur on Treaties.

The UN Committee on the Elimination of All Forms of Racial Discrimination requested the Papua New Guinea government to provide it with a full report on human rights violations on Bougainville. A 1993 Amnesty International Report on human rights abuses on Bougainville *Under the Barrel of a Gun* claimed that at least 60 people had been "extra-judicially killed by government soldiers since 1991" and called for United Nations special rapporteurs on extra-judicial executions and torture to visit the island. Whereas the Papua New Guinea government claimed there was no ban on travel into Bougainville, it cancelled the passports of prominent Bougainvilleans attempting to travel overseas to press their claims before UN agencies. In the first months of 1993 the Papua New Guinea Defence Force continued to press for a military solution, engaging the Bougainville Revolutionary Army (BRA) in a battle for Arawa, the Provincial capital. On 10 March 1993 the 49th Session of the UN Sub-commission on Human Rights passed a resolution calling on the government of Papua New Guinea to lift its blockade of medical supplies and to recommence negotiations with all parties to the conflict.

Peace agreements were achieved by stages, beginning with a truce (1997–98). Following accords facilitated by New Zealand in 1997 a permanent cease-fire was established in April 1998. The conflicting parties agreed to involve Australia, New Zealand, Fiji, and Vanuatu, in a peace monitoring group (PMG) to report on compliance on all aspects of the ceasefire, and to promote and instil confidence in the peace process through its presence at ground level to assist with democratic resolution of the situation. The parties also agreed to establish an all-party Peace Process Consultative Committee (PPCC), and for the United Nations to be requested to provide an observer mission. Papua New Guinea put this request to the Security

Council in 1998 (United Nations. Security Council, 1998), and the peace process was eventually overseen by two groups, the Peace Monitoring Group (PMG) and the United Nations Observer Mission to Bougainville (UNOMB).

UNOMB was established in August 1998 to monitor disarmament and weapons disposal. The mission had to build trust with government, with the population and the peace-keeping mission, and keep the UN Security Council's attention (United Nations. Department for Disarmament Affairs. Conventional Arms Branch, 2002). Its personnel established good relations with former combatants from all sides, including the BRA and the supporters of Francis Ona, and its mandate was extended several times.[8] UNOMB set up subcommittees on weapons disposal in 12 districts and collaborated with each of the armed factions as well as with government officials. A comprehensive peace agreement was signed on 30th August 2001 (United Nations. Security Council, 2001). 2014 weapons were collected by 2005, and 17,000 combatants applied for or received compensation. One of the last groups to lay down arms was the Me'ekamui Defense Force (MDF), some of whose members continued to support Francis Ona's declaration of an independent kingdom, and an estimated 4000 people continued to live in a "No Go Zone" near the copper mine. On 22 March 2005 some of these joined Ona in a march through Arawa to presented UNOMB representative Tor Stenbeck with a three-page petition demanding that the UN recognize Bougainville as the "independent nation of Mekamui islands.

In accordance with agreements in the peace process, changes to Bougainville's constitutional status required the approval and assent of the Papua New Guinea Parliament. After UNOMB verified completion of the second of three stages in the weapons disposal scheme the Papua New Guinea parliament, in March 2002, passed a constitutional law that gave effect to the Agreement,[9] which in turn came into effect in August 2003.

Establishment of the autonomous province of Bougainville provided the legal basis for the establishment of the Bougainville Constitutional Commission, which had the task of preparing a draft constitution for presentation to an elected constituent Assembly. Constitutional Amendment 23 (CA23), which became law on August 7, 2003, incorporated terms of the Bougainville Peace Agreement, and amended sections 276(2) and 296(2)(a) of the national constitution to give the Bougainville legislation an overriding effect where it is inconsistent with a national legislation. The constitution of the Autonomous province of Bougainville was promulgated in 2004.[10] With a peaceful transition to self-government in sight, and sufficient levels of reconciliation between previously antagonistic parties in place,

[8] At the end of 2004 the Security Council agreed to a request for extension of the Mission's presence for a further 6 months to June 30 2005; it subsequently approved an extension to December 31st, with an expectation that general elections would have been held by then.

[9] (*Chapter XIV of the* Papua New Guinea *Constitution* and the *Organic Law on Peace-Building in Bougainville—Autonomous Bougainville Government and Bougainville Referendum*)

[10] For background, see (Ghai & Regan, 2002).

the UNSG informed the Security Council on 28 March 2005, that UMOMB would wind up on 30th June, after which time the UN's political function could give way to its development and humanitarian agencies (United Nations. Security Council, 2005).

Elections were held 20th May to 2nd June, 2005 for the position of President and a 33-member legislature. The United Nations, the Commonwealth Secretariat, and PIFS organized and funded a ten-person team observer team. A referendum held in November 2019 found 97.8% of voters in favour of independence—although post-referendum consultations between the Autonomous Bougainville Government and the national Government on Bougainville's on the region's future political status, is yet to put a decision to the national parliament for ratification.

8.2.2 Solomon Islands

Through the 1990s, Solomon Islands experienced ever-increasing tension between ethnic groups based on their perception of unjust distribution of resources and opportunities between the provinces. Tensions between ethnic groups which had simmered since independence in 1978 were fanned into conflict 2000–2001, ostensibly over pressures on land-use on the main island of Guadalcanal. and a fragile peace was established through peace talks in Townsville late in 2000.

After the Ulufalu government was ousted by militia groups, public buildings and resources ransacked, and opposing militias brought the state to the brink of failure, the Pacific states (including Australia and New Zealand) invoked the provisions of the *Biketawa Declaration* and established in 2003 the Regional Assistance Mission to the Solomon Islands (RAMSI) to re-establish the rule of law and work towards the rebuilding of the state institutions (Kabutaulaka, 2002). Weapons were removed and a peace treaty brokered before the Mission shifted to rehabilitation of the nation's "machinery of government." These steps were vital to the country and to the region, but on such small scale compared to similar conflicts elsewhere in the world as to barely rate a mention within the UN. In August 2003 the UNSG commended "countries in the Pacific Islands region" for their collective support to restore law and order in the Solomon Islands and the unanimous support for the "concerted regional initiative" under the framework of the *Biketawa Declaration* (Graham & Felício, 2006, p. 139).

LeRoy describes the over-long and successive efforts made in Solomon Islands from 2001 to undertake constitutional reform, commencing with UNDP-supported technical assistance, through to the long-awaited publication of a draft federal constitution. The Kemakeza government obtained UNDP support for constitutional review in 2002, and this exercise continued through the Sogovare and Sikua governments. A 32-person Constitutional Congress representing the country's 9 provinces, plus 17 "distinguished Solomon Islanders" was launched by then PM Sogovare in June 2007 to consider the draft, with expectations that the new constitution would be in place in time for general elections in 2010. Although the Congress and its secretariat laboured through 2008 and halted toward the end of 2009 through lack of operational funds, its process remained incomplete when the term of the Sikua government expired in 2010. The manifesto of the incoming Philip Government referred to the need for constitutional reform without acknowledging the existence of a process almost one decade old (Government of Solomon Islands. Office of the Prime Minister, 2010). Had the parliament considered the 1987 Constitutional Review Committee's first recommendation "That the Solomon Islands become a Federal Republic in which:—(a) there is a democratic union of states having equal status under the Federal Constitution..." (Solomon Islands, 1987, p. 1) the subsequent path of Solomon Islands political history may have been different.

Although ethnic conflict has subsequently been contained in Solomon Islands, it cannot be said that the underlying problems have been addressed, or that the potential for further conflict has been removed. To the contrary, although constitutional reform exercises have on several occasions (and with considerable UNDP support) proposed federal arrangements to meet regional aspirations for greater autonomy, the national parliament has not found the conviction to depart from the system of government established at independence, leaving many political and economic issues unresolved.

8.2.3 Fiji

In the case of Fiji, a third PIC to experience prolonged ethnic conflict and periodic political upheaval, the situation has drawn the United Nations into political dialogue, but little more. Efforts by Pacific regional institutions and intergovernmental bodies to help resolve political conflict in Fiji following a military *coup d'etat* in December 2006 were protracted and had indeterminate results. Fiji was subsequently suspended from the Commonwealth (for the third time, following suspensions in 1987 and 2000 that followed coups in those years). A joint working group established by PIFS comprising Australia, Fiji, Kiribati, Nauru, New Zealand, Papua New Guinea and Tuvalu met 28 times between January 2007 and by May 2008, without making any substantial progress (Hassall, 2009). In February 2009 a joint UN-Commonwealth Secretariat technical team visited Fiji to consult with a broad range of political forces on the national, regional and local levels and in April 2009 the UNSC called for restoration of democracy. On 2 May 2009 Fiji was suspended from the Pacific Islands Forum. The country subsequently returned by stages from military rule to parliamentary democracy, under a constitution promulgated in 2013. That the United Nations played so little role in assisting the country to restore parliamentary democracy is due to its mandate to address inter-state conflicts rather than the intra-state conflicts which have occurred in Fiji.

8.3 Peace-Keeping

Participation by Pacific Islanders in peace-keeping operations in other parts of the world is a larger larger story than UN engagement in conflict resolution within the Pacific region. Fijian Prime Minister Ratu Mara decided in the 1970s that participation in Peace-keeping would provide employment, provide troops with active service experience, and promote the training of youth (Mara, 1997, p. 140). Fiji has subsequently been deployed on more than 30 missions, the largest PIC engagement by far (Sharland & Feely, 2020). Palau, Papua New Guinea, Samoa, Solomon Islands, Tonga, and Vanuatu have also engaged in peace-keeping operations overseen by the Security Council. Vanuatu's Mobile Force and Police Force have served in UN Peacekeeping missions in East Timor, Haiti, Sudan, and Bosnia, in the Peace Monitoring Group and Transition Team in Bougainville, and in the Regional Assistance Mission to Solomon Islands. Nauru, Fiji, and Samoa have contributed UN policing missions internationally.

Peace-keeping is very expensive, although costs are shared between the participating countries who pay salaries—and the UN-which contributes an additional daily rate. In 2015 Fiji‘s shadow foreign minister, Ratu Isoa Tikoca, claimed that Fiji had outlaid US$194 million dollars for peacekeeping operations but had been repaid less than a third of that, and that the UN owed Fiji US$138 million dollars for peace keeping duties, going back a decade ("Demand In Fiji For UN To Settle Peacekeeping Repayment Has been repaid less than 1/3 of peacekeeping," 2015).

8.4 Conflict Prevention and Peace Building

The concepts and practices of "conflict prevention" and "peace-building" are related to, but identical to those of conflict resolution. Peacebuilding often requires a third party to assist parties in dispute to reach agreement. The inability at first of the Forum's regional security framework to assist member countries was due to the primacy given to the doctrine of national sovereignty. While this core value has inhibited neighbouring states from interfering with each other's "internal affairs", it has also rendered the Pacific Islands Forum ineffective throughout a decade of conflict in Papua New Guinea's North Solomons Province. The same doctrine prevented the Forum from mitigating the effects of a military coup in Fiji in 1987 even though that coup, and those that followed, occurred in Suva, the nation's capital and the location of Pacific Islands Forum secretariat and the offices of some like 20 intergovernmental organizations, including the United Nations.

8.4.1 UNDP's "Peace Facility"

In the context of political upheaval in Papua New Guinea, Fiji, and Solomon Islands, UNDP's Pacific Regional office established a "Peace Facility" in 2005 to cultivate conflict prevention capacities within the region. In April of that year UN agencies

(UNDP and DPA) and Pacific regional organizations (PIFS and USP) convened a workshop on Peacebuilding to identify specific actions that could be taken to prevent conflict and to strengthen peacebuilding activities (PIF, 2005). The meeting noted the benefits of restorative justice processes in post-conflict Bougainville and called for "Track II processes to engender broader understanding of peace building and conflict prevention", beginning with a pilot dialogue process engaging Fiji's squatter settlements. Other initiatives focused on land management, and women in peacebuilding. Partners in these projects included UNHabitat, UNICEF, ECREA, USP, and AusAID. Resulting workshop recommendations included ensuring that the Pacific Plan encompass peace building and conflict prevention issues; that coordination between international, regional and bilateral development assistance agencies be strengthened; and that greater attention, research and assessment to be given to the issue of people movement within the region. Also in 2005, UNDP's Suva-based Pacific "Peace Facility" conducted "Peace, Stability and Development" assessments for Papua New Guinea, Solomon Islands, and Fiji to identify opportunities, initiatives and strategies linking peace with development planning.

8.5 Discussion

Although peace, security, and development, are central goals of both the United Nations Organization and its Pacific Island member-countries, they have been difficult aspirations to realize through collaborative effort. In the first instance, the PICs constitute but a small voice in matters of global security. Their concerns regarding the consequences of nuclear testing in the Pacific region, for example, received scant consideration in the UN Security Council, and weak support in findings of the International Court of Justice. Even when Pacific conflicts such as Bougainville reached the table of the UN Security Council, they were regarded as small-scale when compared to the conflicts occurring in Africa, Asia, and elsewhere.

Secondly, the PICs embraced the concept of human security—including its associated concerns security of food, energy, housing, and energy—at a time when the UNSC maintained a focus on traditional concepts of "hard security" associated with cross-border military incursions by one state into the territory of another. The post-9/11 "global war on terror" required the PICs to address global security concerns rather than domestic ones, and to meet "unrealistic standards" with insufficient resources. As the permanent representative of Samoa expressed in a statement on behalf of the Forum Group to the Security Council, small developing states were not consulted on international Counter-Terrorism standards and therefore such standards do not reflect challenges of implementation. For example, although Kiribati signed and acceded to Counter Terrorism Conventions and passed legislation to implement these and major transnational crime treaties, it expressed its inability to enforce them effectively. In sum, the UN framework for dispute resolution, together with the small size and influence of the PICs, have resulted in UN agencies and processes contributing relatively little to solving conflicts or improving security in the small island countries of the Pacific region.

References

Alexis-Martin, B., Bolton, M. B., Hawkins, D., Tisch, S., & Mangioni, T. L. (2021). Addressing the humanitarian and environmental consequences of atmospheric nuclear weapon tests: A case study of UK and US test programs at Kiritimati (Christmas) and Malden Islands, Republic of Kiribati. *Global Policy, 12*(1), 106–121. https://doi.org/10.1111/1758-5899.12913

Anere, R., Crocombe, R., Huffer, E., Horoi, R., Tuimaleali'ifano, M., Trease, H. V., & Vurobaravu, N. (2001). *Security in Melanesia: Fiji, Papua New Guinea, Solomon Islands & Vanuatu*. Pacific Islands Forum Secretariat.

Asian Development Bank. (2003). Millennium development goals in the Pacific: Relevance and progress.

Asian Development Bank. (2004). *Responding to the priorities of the poor: A Pacific strategy for the Asian Development Bank 2005–2009*. Asian Development Bank.

Braithwaite, J., Dinnen, S., Allen, M., Braithwaite, V., & Charlesworth, H. (2010). *Pillars and shadows. Statebuilding as peacebuilding in Solomon Islands*. ANU E Press.

Calvert, I. (2005). Forum Secretariat: What assistance does it provide for regional peace building? *GLobal partnership consultation on the prevention of armed conflicts, lagoon resort, 18–22 April.*

Crocombe, R. (2000). *Enhancing Pacific security*. Forum Regional Security Committee.

Demand In Fiji For UN To Settle Peacekeeping Repayment Has been repaid less than 1/3 of peacekeeping. (2015). *Radio New Zealand International.*

Dornan, M., & Pryke, J. (2017). Foreign aid to the Pacific: Trends and developments in the twenty-first century. *Asia & the Pacific Policy Studies, 4*(3), 386–404. https://doi.org/10.1002/app5.185

Ghai, Y., & Regan, A. J. (2002). Constitutional accommodation and conflict prevention. In A. Carl & S. L. Garasu (Eds.), Weaving consensus: The Papua New Guinea - Bougainville peace process. *Conciliation Resources Accord Series*. Accessed December 18, 2005, from http://www.c-r.org/accord/boug/accord12/constitutional.shtml

Government of Solomon Islands. Office of the Prime Minister. (2010). *The National Coalition for reform and advancement (NCRA) government. Policy statement.* Government of Solomon Islands.

Graham, K., & Felício, T. N. (2006). *Regional security and global governance: A study of interaction between regional agencies and the UN Security Council with a proposal for a regional-global security mechanism*. VUB University Press.

Hassall, G. (2004). *Micronesian security assessment: The Marshall Islands, the Federated States of Micronesia, and the Republic of Palau*. Pacific Islands Forum Secretariat.

Hassall, G. (2009). The mediation of public policy disputes in Pacific Islands: The case of Fiji. In D. Bagshaw & E. Porter (Eds.), *Mediation in the Asia-Pacific region: Transforming conflicts and building peace*. Routledge.

Holden, P., Bale, M., & Holden, S. (2004). *Swimming against the tide? An assessment of the private sector in the Pacific*. Asian Development Bank.

Kabutaulaka, T. T. (2002). A weak state and the Solomon Islands peace process. *Pacific Islands Development Series, 14*.

Mara, K., Ratu Sir. (1997). *The Pacific way a memoir* Honolulu: Center for Pacific Islands Studies, University of Hawai'i; Pacific Islands Development Program, East-West Center; University of Hawai'i Press.

Mellor, T., & Jabes, J. (2004). *Governance in the Pacific: Focus for action 2005–2009*. Asian Development Bank.

Morgan, W. (2021). Large ocean states: Pacific regionalism and climate security in a new era of geostrategic competition. East Asia (Piscataway, N.J.). https://doi.org/10.1007/s12140-021-09377-8
Pacific Islands Forum. (2005). Regional Workshop on Peace Building and Conflict prevention - Outcomes Statement.
Pacific Islands Forum Secretariat. (1973). Fourth South Pacific Forum, Apia, Western Samoa, 17-18 April. Proceedings. (73)NN.1.
Pacific Islands Forum Secretariat. (1992). Forum Regional Security Committee Meeting, Forum Secretariat Headquarters, Suva 18–21 February 1992. Meeting papers. Folder 1. (92) MP. 4 Folder 1.
Philippe, S., & Tomas, S. (2021). *Toxique: enquête sur les essais nucléaires français en Polynésie*. PUF.
Retiere, A., & Schurmann-Zeggel, H. (2002). Conflict prevention and peace consolidation in the South Pacific: Papua New Guinea, Solomon Islands, Fiji Islands.
Sharland, L., & Feely, G. (2020). *Mapping Pacific contributions to UN peacekeeping*. Australian Strategic Policy Institute.
Solomon Islands. (1987). *Constitutional review committee. Report. Vol. 3 - recommendations*. Honiara.
South Pacific Applied Geoscience Commission, S. (2003). Pacific regional action plan on sustainable water management. Suva: South Pacific Applied Geoscience Commission (SOPAC) and Asian Development Bank (ADB).
South Pacific Bureau for Economic Co-operation. (1973). *Fourth South Pacific Forum. Apia, Western Samoa 17–18* April 1973. Proceedings. Suva: South Pacific Bureau for Economic Co-operation.
South Pacific Bureau for Economic Cooperation. (1984). Working Group on Nuclear Free Zone in the Region First Meeting, SPEC Headquarters, Suva 13–16 November 1984: informal summary record. SPEC(84) 21(SPEC(84) 21), 46 p.
South Pacific Forum. (1972). South Pacific Forum, Suva, Fiji 12-14 September 1972. Summary record and final press communique. *NN (73)* NN. 1.
South Pacific Forum. (1974). Fifth South Pacific Forum, Rarotonga, Cook Islands, 20–22 March 1974. Forum Communique.
UNDP. (1999). *Pacific human development report 1999: Creating opportunities*.
UNDP. (2000). Pacific regional conference on governance for parliamentarians 27–30 March 2000.
UNDP. (2004). Support to demobilization of special constables in the Solomon Islands. Inception Report, Status of Field Activities.
United nations development programme (UNDP). (1994). *Pacific human development report: 'Putting people first'*. United nations development programme.
United Nations General Assembly. (1985). Provisional verbatim record of the 7th meeting: Held on Tuesday 24 September 1985, New York, general assembly, 40th session. UN.
United Nations. General Assembly. (1995a). General assembly official records, 50th session: 29th plenary meeting, Wednesday, 11 October 1995, New York. UN.
United Nations. Department for Disarmament Affairs. Conventional Arms Branch. (2002). Mission report. Bougainville (Papua New Gineaa) 1–10 May. United Nations.
United Nations. General Assembly. (1995b). Summary record of the 21st meeting: 2nd committee, held on Wednesday, 1 November 1995, New York, general assembly, 50th session. UN.
United Nations. General Assembly. (2001). *Letter dated 2001/01/09 from the Chairman of the Pacific Islands Forum group of countries in New York for January 2001 addressed to Jayantha Dhanapala, Under-Secretary-General for Disarmament Affairs, transmitting a paper entitled "Small arms and light weapons: the interests and involvement of Pacific Islands Forum Countries."* UN.
United Nations. Secretary General. (2001). *Illicit traffic in small arms and light weapons: Report of the secretary-general*. UN.

United Nations. Secretary General. (2009). *Assistance to states for curbing the illicit traffic in small arms and light weapons and collecting them: The illicit trade in small arms and light weapons in all its aspects: Report of the secretary-general*. UN.

United Nations. Security Council. (1998). *Letter dated 2 June 1998 from thee secretary-general addressed to the president of the security council* (Vol. S/1998/506). New York.

United Nations. Security Council. (2001). Letter dated 31 October 2001 from the president of the security council addressed to the secretary-general (Vol. S/2001/1028). New York.

United Nations. Security Council. (2005). Report of the Secretary-General on the United Nations Observer Mission in Bougainvllle (Papua New Guinea." S/2005/204.

van Tongeren, P., & Heemskerk, R. (Eds.). (2005). *Reader: UN, Regional Organizations and Civil Society Organizations. Global ConferenceL From Reaction to Prevention: Civil Society Forging Partnerships to Prevent Violent Conflict and Build Peace. 19–21 July 2005* Global Partnership for the Prevention of Armed Conflict.

Westcott, C. (1996). UNDP. Preparatory assistance phase. Governance for sustainable human development in the Pacific Islands. First quarter report. 19 Sept - 31 December 1996.

World Bank. (2004). Pacific regional strategy.

Part III
Small Island Developing Countries, Regionalism, and Global Multilateralism

The first two sections of this book have examined the role of the UN in the self-determination processes and development of new nations in the Pacific Islands, the engagement of the new nation-states within the United Nations system, and the effort that was required to have the UN engage with the Pacific's regional organizations. Part III examines the role of Small Island Developing States in global policy networks and global summitry. These activities further extend their resources but are essential activities for all nation-states in the international order. Chapter 9 introduces the idea of global policy networks and their integration with global summitry. The role that small states can play in global affairs is the subject of growing literature (Súilleabháin, 2014). A 2017 report on the State of Pacific Regionalism noted that some PICs were joining up to as many as thirty international organizations and that while "both the range of spaces and actors related to Pacific regionalism provides greater opportunities, it also poses challenges of ensuring a consistent Forum voice across groupings and relevant political processes." (PIFS, 2017, p. 9)

After reviewing the key global conferences of the past five decades and the elements required for successful pre- and post-conference engagement, the chapter considers current examples of PIC engagement: commitments to adaptation measures under the UNFCCC climate change agreement. Chapter 10 examines implementation of multi-partner platforms developed under global development plans—the MDGs and the more expansive Global Partnership for Sustainable Development, otherwise known as the SDGs. Chapter 11 traces the emergence of the UN's Small Island Developing States (SIDS) initiative, and Chap. 12 discusses the idea of small states as "good states" in the global era.

References

Pacific Islands Forum Secretariat. (2017). *State of Pacific Regionalism Report 2017*. Pacific Islands Forum Secretariat.

Súilleabháin, Ó. (2014). *Small States at the United Nations: Diverse perspectives, shared opportunities*. International Peace Institute.

Chapter 9
Global Policy Networks and Global Summitry

9.1 Global Policy Networks

Although networks have been discussed in Chap. 6, they are considered once again here, this time more specifically in relation to global public policy processes. Networks generally comprise a set of "... relatively stable relationships which are of non-hierarchical and interdependent nature linking a variety of actors, who share common interests with regard to a policy and who exchange resources to pursue these shared interests acknowledging that co-operation is the best way to achieve common goals" (Borzel, 2002, p. 254). Networks bear a variety of names, including alliance,[1] partnership,[2] forum, initiative, panel, platform, campaign, and coalition. Under the auspices of UNHRRLLS (the UN Office of the High Representative for the Least Developed Countries, Landlocked Developing Counties, and Small Island Developing Countries), the third SIDS conference, held in Apia, Samoa, resulted in the establishment of the "SIDS Action Platform", which convenes 306 "SIDS Partnerships" having sustainable development objectives.[3] Their most common purposes are knowledge sharing; standard setting; issue advocacy; support mobilization, and compliance monitoring (Streck, 2002).

Although often distinguished from *institutions* by their fluid and impermanent existence, *networks* nonetheless require administrative and financial capacity to operate, and vary greatly in their expectations of members. Although they need not have a permanent secretariat, those that do have dedicated secretariats tend to operate most effectively. The United Nations Environment Programme (UNEP), for instance, provides the secretariat for REN21, a global renewable energy policy multi-stakeholder network of governments, nongovernmental organizations, research and academic institutions, international organizations and industry,

[1] http://www.who.int/alliance-hpsr/en/

[2] http://www.globalpartnership.org/, http://www.ngowgsc.org/

[3] http://www.sids2014.org/partnerships/

G. Hassall, *The United Nations and the Pacific Islands*, United Nations University Series on Regionalism 24, https://doi.org/10.1007/978-3-031-34155-7_9

established to "...facilitate knowledge exchange, policy development and joint action towards a rapid global transition to renewable energy."[4]

Stone points out that "Networks ... enable actors to operate beyond their domestic context, and networks are the means by which organizations individually and in coalition can project their ideas into policy thinking across states and within global or regional forums" (Stone, 2001). Global Public Policy Networks have thus emerged to address complex transnational and supranational policy problems in a manner that works with, and supplements, the efforts of national governments. They can be useful in bringing a broad spectrum of knowledge and experience to the table and can also help improve transparency and accountability to stakeholders, and communication with the public. Whereas some networks only admit states as members, others are more open, and include non-state actors. GPNs provide both an opportunity and a stress for small states, which must choose which are most advantageous for them to associate with.

A considerable number of global policy networks have emerged through global agency patronage, including under UN auspices. Wiess has characterized the Intergovernmental Panel on Climate Change, for instance,[5] as a powerful recent illustration of "...the role of intellectuals in creating ideas, of technical experts in diffusing them and making them more concrete and scientifically grounded, and of all sorts of people in influencing the positions adopted by a wide range of actors, especially governments," an influence felt because "... the network of world-class volunteer scientists from several disciplines translate scientific findings into the language comprehensible by policymakers." (Weiss, 2010, p. 6) The Intergovernmental Platform on Biodiversity and Ecosystem Services (https://ipbes.net/) was based on the success of the IPCC model.

Networks can range from low entry-level commitment (such as the Internet Society), to those requiring subscription, or specified levels of participation and commitment, such as the anti-corruption "networks" convened by the IMF. The Global Health Council established a Global Health Action Network. The Global Water Partnership and the World Water Council collaborated in the Integrated Water Resource Management (IWRM) movement (Kramer & Pahl-Wostl, 2014). Other issue-specific global networks include Safer Cities, Global Water Quality Data and Statistics (GEMStat), the Global Compact (for corporate social responsibility), Energy for Sustainable Development (GNESD), Coral Reef Monitoring, Promoting Digital Technologies for Sustainable Urbanization, and Monitoring and Evaluation for Disability-inclusive Development.

Networks established by the World Health Organization to enlist multi-sector support for global health objectives include the Global Noncommunicable Disease Network, Global Alliance against Chronic Respiratory Diseases,[6] Global Network

[4] http://www.ren21.net/about-ren21/about-us/

[5] https://www.ipcc.ch/

[6] http://www.who.int/gard/en/

of Age-friendly Cities and Communities,[7] and Global Health Workforce Alliance.[8] Whereas some networks dissolve when their objectives are met, others evolve in new directions, or even transform into more permanent institutions. WHO's Global Health Workforce Alliance, for example, created in 2006 with a 10-year mandate to coordinate engagement of multi-sectoral stakeholders to advocate for Human Resources for Health, issued a "legacy report" at its conclusion in 2016. The G77, a network of developing countries established at a meeting of the United Nations Conference on Trade and Development, on the other hand, continues to exert influence within the UN system.

Global networks may or may not be associated with global summitry. Within the context of the UN, not all policy networks are global in scope, but even those that exist at country level invariably feed into policy dialogue at regional and then global levels. Networks can be useful for:

1. Country level agreements
2. Pacific Region agreements—UNDAF for the Pacific (2013–17 and 2018–2022)
3. UN Organ functioning (General Assembly, Economic and Social Council/ ESCAP, Secretariat/Security Council
4. Treaty creation, monitoring, and reporting
5. Global Conferences and their "prep-coms" and post-conference dialogues
6. Other Specialist Conferences and their "prep-coms" and post-conference dialogues

9.1.1 Alliance of Small Island States (AOSIS)

The Alliance of Small Island States (AOSIS) is an instance of a highly successful advocacy network, a "voluntary negotiating alliance", (Laatikainen, 2020, pp. 137–138) of low-lying coastal and small Island countries, that includes 14 PICs. AOSIS was established during the Second World Climate Conference in 1990 to assist with small island states' advocacy on global climate change. It has at times been coordinated by a bureau comprising three permanent representatives to the United Nations but does not have a permanent secretariat or charter. As Betzold explains:

> Even before formal negotiations began in 1991, island micro-states recognised that they could only meaningfully participate in the negotiation process and thus contribute to the outcome if they came together as a cohesive group. Led skilfully by some key AOSIS representatives, including Robert Van Lierop of Vanuatu, Tuiloma Neroni Slade of Samoa and John Ashe of Antigua and Barbuda, island states thus 'acted as though they were members of *one* negotiating team' (McMahon, 1993, p. 11, emphasis added). Together, AOSIS countries were among the most active participants, and could ensure 'that the positions adopted by the group would be represented fully not only within the INC as a

[7] http://www.agefriendlyworld.org/

[8] http://www.who.int/workforcealliance/en/

> whole, but also within the respective regional groupings and within the larger Group of 77'. (Ashe et al., 1999, p. 212; Betzold, 2010)

Just before COP1, Nauru (on behalf of AOSIS) was the first state to submit a draft protocol, in legal language and containing a specific emission target (a 20% CO_2 reduction target, to below 1990 levels by 2005), and this initiative exerted considerable influence on the Kyoto process. Although this target was not adopted by that first conference, the initiative served as a point of reference for the Kyoto negotiations:

> AOSIS uses dialogue and negotiation to achieve its aims. In its efforts, the group is guided by clearly stated principles, including the principle of preventative action, the precautionary principle, the "polluter pays" principle, the duty to cooperate and the principle of equity. AOSIS pushes for strong emission-reduction commitments, strong monitoring and compliance measures in the UNFCCC and Kyoto Protocol and assistance for adaptation to the adverse effects of climate change. (Barnett, 2005, p. 211)

AOSIS has argued in UNFCCC negotiations, firstly, that small states contribute only small amounts per capita to global greenhouse gas concentrations and that the international community must take urgent action to significantly reduce emissions, and second, that its members face significant impacts of climate change inversely proportional to their contribution to emissions. These arguments have had some impact. In the estimation of Betzold, "...the Alliance of Small Island States (AOSIS) became one of the most vocal participants in the negotiations for the *UN Framework Convention on Climate Change* (UNFCCC) of 1992 and its 1997 Kyoto Protocol and is now widely recognised as one of the key players in the climate change regime. The literature... however, mentions AOSIS only in passing; no attention is paid to how island micro-states were able to gain such a strong position in the climate change debate. Yet their recognition and influence on the negotiation process, as well as its outcome, are remarkable, given AOSIS's limited size and almost negligible political clout, and merit further analysis." (Betzold, 2016, p. 131)

At the Copenhagen Conference of Parties (COP 15) in 2009, the Pacific delegations successfully convey their islands' vulnerability to sea-level rise when pleading for a commitment to a maximum 1.58 Celsius global average temperature increase. Although the Copenhagen Accord fell far short of their call, the effectiveness of AOSIS' strategy had gained additional attention and respect:

> AOSIS's interventions drew first and foremost on the context by appealing to norms and principles, and linked these appeals to target-based strategies highlighting the common interest all participants should have in combating climate change, as scientific findings (third party) underscore. Supported by NGOs (third party), AOSIS could as a bloc fully participate in and shape the negotiation process, gaining visibility and political clout. (Betzold, 2016, p. 138)

For the first time a seat was specifically reserved on the Bureau of the negotiations for a group other than a UN regional group, and this helped AOSIS affect the co-ordination of the negotiations and to ensure its voice was heard in the process. Betzold considered this Bureau seat "AOSIS's main accomplishment and direct measure of success. This is all the more remarkable as OPEC countries similarly fought to obtain a specific seat on the board, but failed." (Betzold, 2016, p. 139) At the United Nations Climate Change Conference COP17 in Durban, South Africa in 2011, which attracted some 20,000 delegates, the voices of 14 Pacific Island countries were again represented by the Alliance of Small Island States (AOSIS).

Laatikainen suggests that AOSIS has produced statements about climate change consistently for three decades, and that it has been more "effective and active" in conference diplomacy than in the standing bodies of the UN (referring to the General Assembly and its committees)—conferences under the *Framework Convention on Climate Change* in particular—but that it has been less effective at the three SIDS conferences (discussed in Chap. 11), where small states are united on the issue of climate change but differ in their prioritization of other issues such as nuclear testing and marine dumping versus chemical and oil spills, or agriculture and fisheries versus tourism and offshore banking (Laatikainen, 2020, p. 145). Perhaps due to the difficulty of establishing a unified position, AOSIS has not contributed to debate on UN reform, or on such basic issues as disarmament, reduction in nuclear arsenals, or strengthening the authority of the UN's General Assembly or Courts.

9.2 Global Summitry and Conferences

States have participated in major summits at important points throughout history to discuss war and peace, and international cooperation and development. Alexandroff and Brean suggest that Global Summitry involves:

> . . . the architecture, institutions and, most critically, the political and policy behaviour of the actors engaged in the influence of outcomes of common concern in the international system . . . . [it] includes all actors—international organizations, trans-governmental networks, states and non-state entities whether individuals, corporations or associations—that influence the agenda, the organization and the execution of global politics and policy. (Alexandroff & Brean, 2015)

Global summitry includes several types of meeting: (1) those that seek the participation of all nations for the purpose of decision-making on global public policy settings; (2) those that include fewer participants, but which make decisions having global implications; and (3) those about issues or problems that are pervasive at national level across most nations. More broadly, global summits and conferences are held to raise consciousness on global issues, to define and agree on shared values and standards, and build commitment to uphold them (Schechter, 2005). Despite their limitations in capacity and capability, Pacific SIDS have had no option but to engage with all other states in such emerging global policy dialogue spaces. In the 1960s and 1970s Pacific perspectives were presented on decolonization, environment, globalization, and fishing rights; and to these were added in the 1980s trade, and the law of the law. Pacific perspectives were also offered on the potential impact of global warming on the small island states in negotiation that produced the *United Nations Framework Convention on Climate Change*.

Whereas issues of climate change, sustainability, inequality, and security, require solution at global level; issues of health, housing, education (which can of course benefit from global standards setting and experience sharing), nonetheless require solution at the level of nations, provinces, and local communities. From the 1980s, global environmental conferences drew attention to the human impact on natural resources and eco-systems; from 1990 UNDP *Human Development Reports* drew attention to the significance of health and education data, and from the 1990s, too, UN conferences drew attention to the benefits of "south-south" dialogue. The 1994 Barbados Programme of Action (BPOA) exemplifies a global summit of a sub-group of nations, as it was formulated specifically to tailor an Agenda 21 programme for Small Island Developing States. Beyond summits convened under United Nations auspices are others of smaller size but nonetheless global impact: these include summits sponsored by the World Trade Organization, the World Bank, the G7, the G27, and the Non-Aligned Movement.

Global summits do not need to bear that name to qualify as such: for climate change they are called "Conferences of the Parties". At other times "World Conference" is attached to a theme such as "Environment", "Women", or "Disaster Risk Reduction", just as there have been "World Summits" for "Social Development", "the Information Society", "Children", "Food Security", etc. The *Treaty on the Non-Proliferation of Nuclear Weapons* requires a periodic "Review Conference of the Parties." Global conferences about human settlements organized by UN Habitat commenced as Habitats I (1976), II (1996), and III (2016)[9] and are now known as the World Urban Forums.[10]

In 1999 UNGA (resolution 54/33 of 24 November) established an Open-ended Informal Consultative Process on Oceans and Law of the Sea, to facilitate its annual review of developments in ocean affairs. A first meeting convened in 2000 by Ambassador Tuiloma Neroni Slade (Samoa) and Mr. Alan Simcock (United Kingdom of Great Britain and Northern Ireland), included detailed consultation on the theme "Responsible Fisheries and Illegal, Unregulated and Unreported Fisheries: Moving from principles to implementation" and "Economic and Social Impacts of Marine Pollution and Degradation, Especially in Coastal Areas: International aspects of combating them." Such meetings have been held each year, with 20 states (Federated States of Micronesia, Papua New Guinea and Tonga from the Pacific) participating virtually in 2021.[11]

Key vehicles for the development of global public goods have been UN-initiated global conferences, and the many policy networks that these conferences have generated. Between 1990 and 2015 some 45 such global summits, including those listed in Table 9.1, played a leading role in cultivating the "global summitry" that has shaped collective values about issues from sustainable development to the

[9] http://habitat3.org/

[10] http://wuf9.org/

[11] https://www.un.org/Depts/los/consultative_process/icp21/statement21.htm

Table 9.1 Selected Global United Nations Conferences

Year	Event	Pacific Island Country participation
1972	UN Conference on the Human Environment (Stockholm)[a]	Fiji
1975	World Conference on Women (Mexico City)[b]	133 States, including Fiji and PNG (observing)
1976	Habitat I (Vancouver)[c]	Fiji, Papua New Guinea
1980	2nd World Conference on Women (Copenhagen)[d]	Fiji, Papua New Guinea, Samoa,
1985	World Conference on Women (Nairobi)	Fiji, Kiribati, Papua New Guinea, Solomon Islands Vanuatu
1990	World Summit for Children (New York)[e]	n/a
1992	United Nations Conference on Environment and Development (UNCED) (Rio de Janeiro)[f]	Cook Islands, Fiji, Kiribati, Marshall Islands, FSM, Niue, Palau, Papua New Guinea, Solomon Islands, Tokelau, Tonga, Tuvalu, Vanuatu, Western Samoa
1993	World Conference on Human Rights (Vienna)[g]	Fiji, FSM, PNG
1994	Global Conference on the Sustainable Development of Small Island Developing States (Bridgetown, Barbados)	Cook Islands, FSM, Fiji, Kiribati, Marshall Islands, Nauru, Papua New Guinea, Solomon Islands, Tonga, Vanuatu (Observers: American Samoa, French Polynesia, Guam, New Caledonia, Niue, Northern Mariana Islands, Palau; South Pacific Forum SPREP SOPAC, USP)
1994	International Conference on Population and Development (Cairo)	Cook Islands, FSM, Fiji, Kiribati, Marshall Islands, Niue, Papua New Guinea, Samoa, Solomon Islands, Tonga, Tuvalu, Vanuatu (plus SPC, Pacific Islands Development Program, South Pacific Forum Secretariat)
1995	World Summit for Social Development (Copenhagen)[h]	110 States, including Cook Islands, FSM, Fiji, Marshall Islands, Papua New Guinea, Vanuatu (plus SPC)
1995	Fourth World Conference on Women (Beijing)	189 Countries
1996	Second UN Conference on Human Settlements (Habitat II - Istanbul)	Fiji, Kiribati, Marshall Islands, Niue, Palau, Papua New Guinea, Samoa, Solomon Islands, Tonga, Vanuatu
2000	Millennium Summit: "The role of the United Nations in the Twenty-First Century" (part of UNGA Fifty-Fifth Regular Session, New York)	Kiribati, Marshall Islands, FSM, Nauru, Palau, Papua New Guinea, Samoa, Solomon Islands, Tonga, Tuvalu, Vanuatu
2002	International Conference on Financing for Development (Monterrey)	Cook Islands, Fiji, Marshall Islands, FSM, Nauru, Palau, Samoa, Solomon Islands, Tonga, Tuvalu
2002	World Summit on Sustainable Development (Johannesburg)	Fiji, Kiribati, Solomon Islands, Tonga, Cook Islands, Marshal islands, FSM, Niue, Palau, Papua New Guinea, Samoa, Tuvalu, Vanuatu

(continued)

Table 9.1 (continued)

Year	Event	Pacific Island Country participation
2003	World Summit on the Information Society (Geneva)	Fiji, Tonga, FSM, Samoa
2005	World Summit on the Information Society (Tunis)	Samoa (was on credentials committee), Marshall Islands, FSM, Tonga, PNG
2011	Fourth UN Conference on the Least Developed Countries (Istanbul)	Kiribati, Samoa, Solomon Islands, Tuvalu, Vanuatu
2012	Rio + 20 Sustainable Development (Rio de Janeiro)	Fiji, Kiribati, FSM, Marshall Islands, Nauru, Palau, Papua New Guinea, Samoa, Solomon Islands, Tonga, Tuvalu, Vanuatu, PIF, Permanent Commission for the South Pacific[i]
2015	UN Conference on Disaster Risk Reduction—Sendai, Japan	Cook Islands Fiji, FSM, Kiribati, Marshall Islands, Nauru, Palau, Papua New Guinea, Samoa, Solomon Islands, Tonga, Tuvalu, Vanuatu
2016	Habitat III (Quito)	Fiji, Papua New Guinea,
2016	World Humanitarian Summit (Istanbul)	180 countries, including Cook Islands, Fiji, FSM, Marshall Islands, Nauru, Palau, Samoa, Solomon Islands, Tonga, Tuvalu, Vanuatu Fiji Commerce and Employers Foundation, National Youth Council of Fiji, Pacific Conference of Churches, Pacific Islands Climate Action Network, Pacific Disability forum, Pacific Islands private Sector Organization, PIANGO, Palau National Emergency Management Office, USP, Radiation Exposure Crusaders for Humanity—Marshall Islands; Women United Together Marshall Islands; Samoa Umbrella NGO, People with disability Solomon Islands, Solomon Islands Development Trust, Tonga Community Development Trust, Tonga National Youth Congress, Ekalesia Kelisiano Tuvalu, Tuvalu Association of NGOs, Tuvalu National Council of Women, Vanuatu Association of NGOs, Vanuatu National Youth Council, Vanuatu Red Cross Society, Papua New Guinea Civil Society Organization Forum, University of Papua New Guinea

[a]https://www.un.org/en/conferences/environment/stockholm1972
[b]https://www.un.org/womenwatch/daw/beijing/mexico.html
[c]https://www.un.org/en/conferences/habitat/vancouver1976
[d]https://www.un.org/womenwatch/daw/beijing/copenhagen.html
[e]https://www.un.org/en/conferences/children/newyork1990
[f]https://www.un.org/en/conferences/environment/rio1992
[g]https://www.un.org/en/conferences/human-rights/vienna1993
[h]https://www.un.org/development/desa/dspd/world-summit-for-social-development-1995.html
[i]The speaking rights of Cook Islands and Niue were removed by UNGA Resolution 66/197, which limited participation to UN member states

status of women, human rights, population growth, urbanization, the internet, and climate change (Farrell, 2017, p. 227).

The Pacific's regional organizations have participated in global summits as they deemed feasible and appropriate, but not to the same extent as member states. In the early years, Fiji and Papua New Guinea, and then Samoa, were the lone PICs at these UN Global Conferences. Some major events including, it seems, the first World Climate Conference in Geneva in 1979, had no Pacific representation. Just three PICs (Fiji, Federated States of Micronesia and Papua New Guinea) attended the World Conference on Human Rights convened in Vienna in 1993 to review global progress in advancing human rights (United Nations, 1993).

Pacific delegates to the 1980 World Conference on the United Nations Decade for Women succeeded in embedding in the conference report a subsection on "Health and well-being of the women of the Pacific" that focused specifically on the impact of nuclear testing:

> 37. Health and well-being of the women of the Pacific.
>
> The World Conference of the United Nations Decade for Women: Equality, Development and Peace,
>
> Recalling that the subtheme of the Conference is "Employment, Health and Education",
>
> Gravely concerned that the continued testing of nuclear weapons constitutes a serious hazard to the environment and the health of present and future generations of mankind,
>
> Reiterating its conviction that the cessation of nuclear-weapon testing in all environments would be a major step towards the attainment of world peace,
>
> Considering that the continued nuclear testing in the Pacific Ocean constitutes a serious hazard to the health and well-being of the peoples of the Pacific,
>
> Recognizing that the health of women and children is of primary importance for the achievement of the aims and objectives of the Decade,
>
> Calls upon all nuclear-weapon States to refrain from conducting any testing of nuclear weapons and other nuclear explosive devices in the Pacific and calls up on all States to support the conclusion of the comprehensive test ban treaty to bring about the cessation of all nuclear testing. (United Nations, 1980, pp. 98–99)

In 2012 the Pacific Islands Forum issued the *Waiheke Declaration* as a contribution to the Rio+20 conference held to assess sustainable development two decades after the seminal "earth summit" in Rio de Janeiro.

UN global conferences are not without their detractors, who focus on the cost of attending the event, the "carbon footprint" created by delegates' travel, the bloated size of delegations, an impression that they are "all talk and no action", and in many cases, a paucity of outcomes. In my experience there are two main criticisms of delegations to international meetings: the first is that leaders and officials attend too many international meetings, in unnecessary numbers, and at exorbitant cost; the second is that leaders and officials fail to attend important international meetings or attend them with inadequate numbers and preparation!

Papua New Guinea's delegation to COP26 in Glasgow, Scotland, in 2021, received considerable social media scrutiny for its size and composition and quite possibly epitomised the delegation comprising officials with no direct authority for decision-making in the meeting subject matter, in larger numbers than required. The larger picture concerns, nonetheless, the advances that flow from successfully convened global events. Results of the 1992 Earth Summit, for instance included the Rio Declaration's 27 universal principles; the *United Nations Framework Convention on Climate Change (UNFCCC)*; the *Convention on Biological Diversity*; the *Declaration on the Principles of Forest Management*; negotiations for the establishment of the *Agreement on Straddling Stocks and Highly Migratory Fish Stocks*; establishment of the Commission on Sustainable Development; and the follow-up first world conference on the sustainable development of small island developing States in Barbados in 1994.

UN global conferences have some broad procedural features in common but may differ according to the composition and mandate of the lead organization. Because the United Nations is an inter-governmental body, its initial global conferences only recognized delegates who represented member-states. Over time, however, non-government delegates were allowed into delegations in some instances. The preparatory process for the International Conference on Population and Development in Cairo in September 1994 included a sizeable conference for NGOs in the Pacific region. Conferences on the most pressing global issues – especially climate change, the environment, the status of women, and urbanization - eventually required the convening of parallel conference programs to accommodate non-government participants, who far outnumbered official government delegates. Global Summits have thus become the hub of global discussion for governments, civil society, business, academia, and the media. Unfortunately, with the exception of those leaders, officials, and activists able to participate directly in the planning and convening of global summit events and processes, only a small portion of the public has taken note. The potential for addressing global public goods has for the most part escaped the public's attention in PICs, and even the largest summits have garnered only passing attention in the region's press.

Whereas global UN conferences have often been presented in the media as isolated events, they have generally been the most visible component of ongoing policy dialogue on policy, law, and implementation by UN member countries. The UN's global calendar is busier in some years than in others. It not only sets the pace for global meetings but influences meetings and dialogue processes in the years proceeding and following them. Protocols have evolved to guide these pre-summit, mid-summit, and post-summit decision-making, implementation and verification processes, as set out in Table 9.2 (United Nations Research Institute for Social Development, 2006, p. 2).

Successful PIC engagement with these UN global conferences and policy dialogue processes have at least four facets: the ability to attend; adequate preparation for attendance; meaningful participation during sessions; and effective post-conference impact. These facets and some examples are considered in more detail below, and the policy dialogue process used for SIDS III is illustrated in Fig. 11.1.

1. The ability to attend

PICs and regional intergovernmental organizations receive official invitations to a considerable number of global conferences and associated meetings. In some instances non-governmental organizations are also invited. Factors influencing a

Table 9.2 Policy cycle for UN summits

Various Processes of UN Summits and Civil Society Organization Participation		
Time frame (years)	Stages of Preparation	Nature of CSO involvement
3 years prior	General Assembly agrees to hold a UN Summit. Preparatory Committee 1	
2 years prior	Expert Seminars Regional Conferences	Inputs from CSOs sought by summit organizers
1 year prior to the summit	Summit secretariat synthesizes the outcomes of regional conferences and expert seminars Preparatory Committee 2 Preparatory Committee 3	
Year of the summit	Holding of UN world summit Action oriented summit: Plan of Action and Declaration with preamble — or — Rule-making summit: Legal Document	CSOs are most active
	National follow-up process UN follow-up process	CSOS are expected to participate
5 or 10 years after the summit	+5 or +10 meeting	CSOs are usually involved

government's decision to attend or not include its prioritization of policy issues, its ability and willingness to cover the costs, and/or the availability of external funds. The official Pacific delegation to UNCED, for example, numbered 111, and preparations coordinated by SPREP amounted to an "investment" of USD 2 million, provided by ADB, UNDP, Commonwealth Secretariat, Australia, and New Zealand. (PIFS, 1992a, p. 2 Energy and Environment: the new Global Agenda for sustainable development and its implications for energy use in the Pacific).

In their first post-independence decades few PICs were sufficiently resourced to allow expansive participation in global conferences. Fiji was the only PIC to attend the first UN Conference on the Human Environment, held in Stockholm in 1972 (United Nations, 1972, p. 43). No PIC attended the first World Climate Conference, held in Geneva in February 1979 (Hare & Patrick, 1979). In fact, the Pacific Islands were not mentioned in the conference proceedings. Fiji and Papua New Guinea attended the UN Conference on Human Settlements (Habitat I) in Vancouver in June 1976 but did not submit country reports.[12]

The first global conferences with substantial regional representation by PICs were the 1994 conferences on the Sustainable Development of Small Island Developing States in Barbados, which was also attended by the South Pacific Forum Secretariat SPREP, SOPAC, and USP; and on Population and Development in Cairo, which was attended by the SPC, Pacific Islands Development Program, and South Pacific

[12] https://habitat.scarp.ubc.ca/habitat-i-document-archive/

Forum Secretariat. The challenge of funding delegations to the World Conference for Social Development in Copenhagen in March 1995 were set out clearly in the Pacific Technical Team's post conference report:

> During initial briefing meetings with Pacific Island delegations that attended, there was concern that Pacific regional issues should have been more visible in the final global document. Comments were also presented on the coordination of the Pacific Island countries concerns into the various preparatory committee meetings and the low numbers of Pacific Island countries attending preparatory meetings and the Summit itself. . . .Funding Support from the Forum Secretariat for Forum Island members only became available at the end of February, while numerous proposals submitted by the SPC to development partners including the UNDP were not successful. As a result, the SPC has had to draw on its own reserves to ensure the availably of facilities, a secretariat and the technical support team to Pacific Island delegations. In this regard the financial support received from the Forum Secretariat and UNFPA is acknowledged. In previous global meetings for environment and population, UNDP, through its specialised agencies, has provided full funding support to the Pacific regional efforts. (WSSD. Pacific Technical Team. Report on the Summit Pacific Islands Forum Secretariat, 1995, p. 8)

2. Adequate preparation for attendance

Meaningful participation in a global conference can only occur on the basis of adequate preparation. One or several government departments will be tasked with preparing background documents for the delegates, and a country report for the country if that is required. By 2000 UNDP's Pacific Operations Centre in Suva had begun assisting PICs prepare for attendance at UN international conferences (United Nations. Economic and Social Council, 2001, p. 19).

Global conferences are generally preceded by regional meetings and other preparatory committee meetings which provide opportunities for regional perspectives to enter into the conference's formal agenda, dialogue channels, and final reports. Preparatory sessions at regional level are inter-woven with those at higher levels. The 1994 Barbados conference, for instance, was preceded by a Pacific Regional Preparatory Meeting in Apia Samoa parallel with similar meetings held in other parts of the world.

Considerable preparation occurred at regional level prior to the World Summit for Social Development in Copenhagen in 1995. Pacific reports were prepared by UNESCAP and a report on *Pacific Islands Social and Human Development*. (SPC, 1995) by a 'South Pacific Organizations' Coordinating Committee' comprising representatives of SPC and USP. ESCAP's ministerial-level preparatory meeting in Manila in October 1994 produced a declaration that, despite being tabled at PREPCOM III and promoted vigorously by the PICs present, "met with significant resistance" and was not widely discussed in plenary or working groups (SPC, 1995, p. 20 WSSD. Pacific Technical Team. Report on the Summit).

In the case of the World Summit on the Information Society (WSIS), convened in 2003 in Geneva and 2005 in Tunis, very few PICs appear to have prepared well. "Have just been wandering around a series of web sites about the World Summit on the Information Society (to be held in Dec 2003, and 2005)", read one communication circulated at SPC in June 2002:

> . . . It's all totally new to me, and full of the usual impenetrable "UnitedNationese" jargon, but it seems to be something that we (Global Islands Network) MUST be represented at, and which is not [yet] very sympathetic or attuned to Global Islands Network issues, such as: Access, Access, Access. and "Breaking the Back of the Cable and Wireless Troll to achieve Access." [I'm joking, but if you check at <http://unicttaskforce.org/groups/members/public.asp?cod_tema_menu=34>, you will see the result of the UN Task Force on ICT, as it wrestles with access issues.] Way at the bottom of this message I have quoted some info about UNESCO's involvement with NGOs in preparing for the WSIS (Potter, 2002a). Since this WSIS is being organized out of Geneva by the International Telecommunications Union, maybe Arthur Dahl or Iain Orr could come up with suggestions about how to stay well informed about this. In any event, another topic for discussion at the International Small Islands Study Association. (Potter, 2002b)

In preparation for the Geneva portion of the conference the Forum Secretariat convened a Pacific Islands Regional ICT Consultation in April 2003 at which 80 representatives of governments, ICT providers, and NGOs from 12 countries and 17 regional organizations assisted with preparation of the Pacific Islands report delivered to the conference. (PIFS, 2003). Apart from this useful regional survey of existing ICT policies and plans, few PICs spoke at the WSIS conferences. In subsequent years, however, all PICs realized the significance of ICT development and have worked individually and collectively toward digital transformation (Cullen & Hassall, 2017). This has included significant engagement with the ITU (International Telecommunication Union, 2014, p. 20).

3. Meaningful participation

Meaningful participation in a global UN conference results from the quality of preparation referred to above, as well as from the quality of formal and informal engagement of delegates during the event. The size of a delegation can also figure in meaningful participation, with larger delegations required when conference programs have multiple program streams. Again, depending on the conference purpose and format, meaningful contributions can include shaping of draft resolutions, coalition-building, and securing commitments to outcomes documents. Because formal presentations by government leaders gain most press coverage and are usually the main country report included in the conference proceedings, they should include substantial content rather than political rhetoric.

Participation at global conferences also bring PICs into contact with small island states from other regions of the world, with whom they collaborate in voicing common concerns. At the Third World Conference on Women in Nairobi in 1985, for instance, Fiji, Samoa and Solomon Islands worked with Caribbean states to propose a resolution on support for women in island developing countries (United Nations, 1986, pp. 198–199). Participation in AOSIS, noted above, is another good example of such collaboration.

Experience of the Pacific delegation at the 1995 World Summit for Social Development resulted in insights and recommendations being presented to an SPC/PIFS meeting of development planners:

> During initial briefing meetings with Pacific Island delegations that attended, there was concern that Pacific regional issues should have been more visible in the final global

> document. Comments were also presented on the coordination of the Pacific Island countries concerns into the various preparatory committee meetings and the low numbers of Pacific Island countries attending preparatory meetings and the Summit itself. ...Some of the delegations attending the World Summit were not involved in the processes leading up to the Summit and were only able to rely on the regional reports disseminated through their Foreign Affairs Ministries. Therefore briefing sessions and notes prepared by the Pacific Technical Team assisted the delegations comprehend the various levels of negotiation and redrafting proceeding during the course of the week. However, actual participation at the negotiation sessions relied heavily on the Pacific Island delegations that had been involved particularly at the preparatory committee meetings in New York. This situation engendered Pacific NGOs and a few delegations to comment on the level of Pacific input and influence at the Summit as compared to Cairo and Rio. (WSSD. Pacific Technical Team. Report on the Summit, SPC, 1995, p. 9)

4. Effective post-conference impact

Each of the preceding phases of conference participation come to naught unless steps are taken to ensure post-conference impact domestically. These begin with submission of end-of-mission reports to the appropriate government departments, and briefings for all stakeholders, including the media. If new commitments have been entered into, follow-up must reach those responsible for legislative reform. The Fourth World Conference on Women, for example, produced a 12-point "Beijing Platform for Action"[13] which required extensive legislative reform to be effective.

On returning from the World Summit for Social Development in Copenhagen in March 1995 the Technical Committee that represented the Pacific recommended to the CROP agencies that PICs be invited to prepare or consolidate national social development programs in line with the *Copenhagen Declaration and programme of Action for Social Development*, liaise with the relevant regional agencies, and highlight social development in future regional conferences. The committee also recommended that SPC disseminate their report and the conference outcome document widely to PICs for reference and feedback and consider establishing a regional social development coordinating group to support implementation of the *Copenhagen Declaration and programme of Action* (WSSD, Pacific Technical Team. Report on the Summit SPC, 1995, p. 10).

While all these steps may appear obvious, one cannot assume that they are all carried through. Every conference takes place in the midst of multiple additional responsibilities and pressures on delegates' time and competes with other critical policy issues for the attention of a nation's leaders. In addition to UN-related global summitry, the WTO, World Bank, G7, G27, and other organizations also convene events that require attention. Indeed, competition for the attention of leaders and the public is such that major events invest in their unique "visual identity", such as for

[13]The Beijing Platform for Action's 12 critical platforms for action: Women and Poverty; Education and Training of Women; Women and Health; Violence against Women; Women and Armed Conflict; Women and the Economy; Women in Power and Decision-making; Institutional Mechanism for the Advancement of Women; Human Rights of Women' Women and the Media; Women and the Environment; and The Girl-child.

Habitat III, the third global conference on urbanization convened by UN Habitat in Quito in 2016 (UNHABITAT, 2016). These processes of pre-conference preparation, conference participation, and post-conference advocacy are illustrated below by the cases of the 1992 UN Conference on Environment and Development, and the global conferences on climate change.

9.2.1 1992 United Nations Conference on Environment and Development

The Forum Leaders recognized the value of the 1992 UN Conference on Environment and Development (UNCED—the "Earth Summit") to the Pacific region, and encouraged regional preparation and full participation in the conference's preparatory meetings:

> The Forum agreed that its members be collectively represented, at ministerial level where possible, with co-ordination and support from SPREP, at the Preparatory Committees leading up to the Conference as these will take many of the key decisions. The third and fourth Preparatory Committees were of particular importance in this regard. At the UNCED Conference itself, participation at the highest possible political level would be desirable to focus attention on the region's interest. Given the complexity and size of the UNCED process, the Forum considered that issues could be usefully engaged in concert with other like-minded countries, especially members of the Alliance of Small Island States (AOSIS). Consideration should be given to a co-ordinated strategy whereby Forum representatives might "specialise" as spokespersons for the region on specific issues. Support from SPREP would be vital. The Forum urged international support to facilitate the effective participation of Forum Island Countries at the Preparatory Committees and the UNCED itself. (South Pacific Forum, 1991, p. 4)

SPREP's Intergovernmental Meeting of 8–9 July 1991 issued a *Ministerial Declaration on Environment and Development*, which the Forum endorsed in its annual communique later that same month (South Pacific Forum, 1991). There was regional participation in PrepComs I, II, III, and IV. UNCED was preceded by PrepComs in Nairobi in 1990, twice in Geneva in 1991, and in New York from March to April 1992. Fourteen PICs collaborated on a joint Pacific report to the Conference, with SPREP coordination and ADB/UNDP funding (South Pacific Regional Environment Programme, 1992b). The report, named *The Pacific Way*, identified 14 "issues and constraints" and five "priorities for further action" (The Earth Charter, Social and Economic Dimensions, Conservation and Management of resources for development, strengthening the role of major groups, and means of implementation). In addition to the Pacific regional report, nine PICs presented individual country reports (UN, 1992a, b). As the Forum did not yet have general consultative status with the UN, PIFS' Secretary General applied directly to Maurice Strong, the conference's Secretary-General for permission to attend. (PIFS, 1992b Ieremia T Tabai to Maurice Strong, 27 February 1992) SPREP also prepared an 86-page brief for the summit, which included two pages backgrounding, the conference, plus a summary of the Agenda 21 draft highlighting the themes most relevant to the Pacific region, and updates on the status of negotiations on the Biodiversity and Climate Change conventions (SPREP, 1992a).

Coordination of the Pacific's regional position at UNCED presented challenges because the multifaceted agenda embraced the focus areas of not just SPREP and SOPAC, but also AOSIS. UNCED already had the G77 group of countries, and SOPAC and SPREP were the official intergovernmental bodies for the Pacific region, and G77 was reportedly sensitive about special interest groups established by developing countries weaking its own position (South Pacific Regional Environment Programme, 1992a, p. 23).

UNCED attracted representatives of 178 nations, including 108 heads of State, and 35,000 participants from approximately 1000 NGOs. It adopted *Agenda 21*, a global blueprint for sustainable development that identified the major groups essential to successful implementation: women, children, and youth; indigenous communities; NGOs; local governments; workers and trade unions; business and industry; and the scientific and technological community. Agenda 21 articulated, furthermore, means of implementation, such as raising finances, promotion of education and capacity building, and strengthening institutions, legal frameworks, and information systems. Chapter 17G of Agenda 21 specifically recognized Small Island Developing States (SIDS) as a distinct group of developing countries facing specific social, economic, and environmental vulnerabilities.

The conference resulted in new international treaties on climate change, biological diversity, desertification, and high-seas fishing. It also led to establishment of the UN Commission on Sustainable Development to monitor implementation of the conference's agreements and commitments. Over 1800 cities and towns created their own "local Agenda 21" and more than 150 countries set up national advisory councils to promote dialogue on sustainable development policies. Following the Barbados 1994 conference, for instance, Samoa translated Agenda 21 into a national program of action by initiating policy review in all sectors to ensure consistency with the objective of sustainable development, paying particular attention to forests, fresh water, fisheries, and agriculture. It formulated new policies for population, biodiversity protection, and waste management, and revised legislation to make environmental impact assessments mandatory (United Nations. General Assembly, 1995, p. 13 Ms. Williams-Stewart (Samoa)).

The extent of response at national level was not always transparent, as not all countries were reporting. Meetings of the Apia and SPREP conventions in October 1995 found that the PICs were yet to submit national reports to the Commission on Sustainable Development, making it difficult to identify action that had been taken at national level (Miles et al., 1995, p. 126). Part of the problem was the small staff in PIC environmental units.

Follow-up conferences included Rio+5 in 1997, Rio+10 in 2002, and Rio+20 in 2012. The UNGA special session in 1997 (Rio +5) reported that the global environment continued to deteriorate—especially access to fresh water. Rio+20, which attracted 45,000 participants from 188 countries—12,000 official delegates and 32,000 representatives from 10,000 NGOs and other organizations (United Nations. Department of Public Information, 2012)—continued to report mixed results and uncertain resolve at national level, as reported in *The Future We Want* (United Nations, 2012). One month following the 2012 conference PIFS convened a meeting in Suva with the Pacific Region's development partners to assess its key outcomes and to develop "a coordinated approach at both national and regional levels"—an assessment that included, in addition, preparatory dialogue on the "post-2015 development agenda" (i.e., the Sustainable Development Goals) (PIFS, 2012).

9.2.2 2002 World Summit on Sustainable Development

The World Summit on Sustainable Development in Johannesburg in 2002 built on the structures and accomplishments of UNCED, whilst also extending the agenda and further elaborating platforms for national action. The *Johannesburg Declaration on Sustainable Development* and the *WSSD Plan of Implementation* also outlined the challenges that "Small Island Developing States"" face in achieving sustainable development. The conference also established "type-II" umbrella initiatives to network interested agencies in diverse regions of the world, across a range of critical sectoral challenges (the term "type-II" was used to distinguish these networks from state-centred networks, referred to as "type-I"). These initiatives, set out in Table 9.3, were viewed as "an opportunity for countries and interested parties to develop flexible and voluntary partnerships that would allow implementation of the summit plan of action in their own countries and regions, as appropriate." Not all networks became active for the Pacific region, nor were those with activities necessarily effective. It was envisaged that each activity area would comprise three phases, commencing in 2002 with a programming phase, followed by a design phase in 2003–4, and implementation spanning 2004–08. Guiding principles for these partnerships provided in a 2002 "explanatory note" referred to:

- partnership objectives
- their voluntary nature
- linkage with globally agreed outcomes
- integrated approach to sustainable development
- a multi-stake-holder approach
- transparency and accountability
- tangible results
- funding arrangements
- new/value added partnerships
- local involvement and international impact, and
- follow-up processes.

The Water network was one of the best coordinated. It raised awareness of issues through a regular "Island Climate update newsletter", found funding for training of hydrological technicians; developed an inter-SIDS Pacific-Caribbean programme for action on water and climate which included scavenger well trials in Majuro and some Caribbean Islands. The Energy network focused on building institutional and

Table 9.3 World Summit on Sustainable Development "type II" Networks

"type II" Network	Convening organization
capacity building and distance education	SPREP
land resources	
education and community development	
mainstreaming conservation	
adaptation	
waste management	
land resources	SPC
health	
water	SOPAC
oceans	
vulnerability and disaster management	
energy	
tourism	SPTO
information communication technology	PIFS
governance	PIFS and USP

professional capacity to assist PIC's develop and implement their National Energy Policies and Plans. Japan assisted the waste management network devise a waste management strategy for the region (PIFS, 2004). Each of the other networks identified similar projects, although these were implemented with varying degrees of success.

9.2.3 Framework Convention on Climate Change

Climate change is a global public policy challenge that requires responses at local, national, and global levels. As a "policy problem", it does not respect bureaucratic boundaries, and its impact is felt by all sectors, and by all communities. At global level, nations have agreed the *United Nations Framework Convention on Climate Change* (UNFCCC), which they review and revise at regular conferences. Whereas much attention has been given to the policy responsiveness of the larger nations, the efforts of small states to take their share of responsibility for mitigation and adaptation efforts are also important. Indeed, the challenges small states face in addressing the political, legal, financial, administrative, and socio-ethical dimensions to climate change, given their resource and capability constraints, are considerable.

In official documentation, the implications of global climate change for the Pacific region were first brought to the attention of Pacific leaders in a one-page brief presented to their annual meeting, held in Nuku'alofa, Tonga, in 1988. The brief highlighted the possibility that an increase of global temperature by as much as 4°C by 2025 could result in changed rainfall pattens, sea-level rise of between 20 and 140 cm, salination of groundwater, increased frequency of cyclonic storms, the impact of changed ocean currents on fish stocks, and impacts on native and agricultural plants—all of which threatened significant economic and social impact. Australia and New Zealand were already collaborating on research aimed at reducing carbon emissions and could play a useful role in monitoring sea-level change. The brief concluded by proposing the establishment of a network of measuring instruments, providing training in measurement, and "assessing the physical stability of the islands under conditions of rising sea level." (SPEC, 1988).

A Regional meeting convened in the Marshall Islands in July 1989 by SPREP, UNEP, and the Association of South Pacific Environment Institutions, discussed implications of climatic changes and rises in sea levels. Forum Leaders were kept informed of the structure of the Intergovernmental Panel on Climate Change (IPCC) established in November 1989, and first referred to climate change in their 1989 annual communique:

> (2) Climatic Changes and Sea Levels
>
> 20. The Forum expressed concern about the possible effects on Island countries of rising sea levels, resulting from global warming, and emphasised the importance of a regional approach to environmental matters. It agreed that Forum members should take decisive action to draw world attention to the way the environmental problem affected the South Pacific, and to represent regional views at appropriate international gatherings, possibly including by way of a resolution in the United Nations General Assembly. The Forum supported the candidature of New Zealand to the Governing Council of the United Nations Environment Programme as one means of achieving this.
> 21. The Forum welcomed confirmation by the Prime Minister of Australia that his Government would fund a core project over the next five years to establish a network of monitoring stations in the region, to research such areas as changes in sea levels, atmospheric pressures and rainfall patterns. The total cost, including the design phase, will be A$6.25 m, with implementation to begin early in 1990. The Prime Minister emphasised that the results of this research would be available to all Forum members, and he suggested that this be done in the context of a programme of information exchange. He also emphasised Australia's commitment to a long-term programme in this area (Pacific Islands Forum Secretariat, 1989).

In the following decades individual PICs as well as Pacific regional organizations recognized climate change as the most pressing issue before them, and it is the policy challenge that propelled Pacific voices onto the world stage. Engagement in UN activities was no longer merely an exercise in maximising development assistance but a global campaign to alter the industrial practices of the world's largest economies. UN platforms became essential to the Pacific's quest for long-term viability. PICs focused on policy processes relating to climate change with more unity of purpose than they had they had achieved in any other sector. The Pacific Islands position has been based on six principles:

1. Immediate and significant cuts in the emission of carbon dioxide and other greenhouse gases in the atmosphere;
2. A preventive approach based on the precautionary principle;
3. New and equitable funding mechanisms, including institutional arrangements in which the small island countries are recognized as a special geo-political grouping;

4. The transfer on an equitable basis of appropriate and environmentally sound technologies, as well as appropriate and necessary amendment to international intellectual property laws;
5. Recognition of the applicability of the polluter-pays principle to the consequences of climate change;
6. Commitment to binding energy conservation and efficiency requirements and the development of renewable energy sources. (South Pacific Regional Environment Programme, 1992a, p. 22)

Since the *United Nations Framework Convention on Climate Change* was established in 1992 PICs and regional organizations have placed reports with, spoken at, and attended no less than 27 COPs—the most recent (at the time of writing) in Glasgow in 2021 during the unsettling times of the COVID-19 global pandemic. A new language emerged around adaptation and mitigation, sustainability, and "climate change diplomacy." UNFCCC seeks (Article 2) "stabilization of greenhouse gas concentrations in the atmosphere at a level that would prevent dangerous anthropogenic interference with the climate system" in a time frame "sufficient to allow ecosystems to adapt naturally, and to enable economic development to proceed in a sustainable manner."

As the small states produce a very small component of global carbon emissions their diplomatic challenge has been to convince other states to reduce emissions, whilst their domestic challenge has been to implement adaptation strategies that promote the future well-being of their peoples. Publicity surrounding the participation of small island states in global climate change meetings focused on their efforts to influence the commitments of large carbon-producing states to effective reduction targets (McNamara & Gibson, 2009). As put by Goldsmith: "Why has Tuvalu achieved such prominence in the global warming debate? In part, because of some very effective telling of its story. Ever since Bikenibeu Paeniu went to the 1992 United Nations Conference on the Environment and Development (UNCED), also known as the Rio Earth Summit, each successive Prime Minister has assumed the mantle of spokesperson for small nations victimised by the policies and actions of the industrialised world." (Goldsmith, 2012, p. 143).

In 2005 Forum Leaders endorsed the *Pacific Islands Framework Action on Climate Change 2006–2015*, which provided for implementation of national adaptation measures and for improving awareness of the issues. The subsequent 2008 *Niue Declaration on Climate Change* expressed Forum Leaders' "deep concern" at the "current impacts of and growing threat posed by climate change to the economic, social, cultural and environmental well-being and security of Pacific Island Countries", recalled, recognized, welcomed, and stressed the importance of existing international and regional decisions, protocols, compacts and treaties concerning climate change, committed the Forum members, amongst other things, to:

> . . .continue to advocate and support the recognition, in all international fora, of the urgent social, economic and security threats caused by the adverse impacts of climate change and sea level rise to our territorial integrity and continued existence as viable dynamic communities, and of the potential for climate change to impact on intranational and international security. (Pacific Islands Forum, 2008)

From this time on, momentum for regional collaboration on climate advocacy has continued to grow. The Pacific Adaptation to Climate Change (PACC) Project was established 2009 through collaboration between SPREP, the Global Environment Facility, Australia, UNDP and UNITAR to initiate practical adaptation schemes in 14 participating countries and to generate and distribute knowledge of best practice.[14] By 2012 the PICs had established a regional mechanism for coordinating engagement in the UNFCCC (*United Nations Framework Convention on Climate Change*) process. It was coordinated by SPREP, with political leadership and resourcing coordinated by the Forum Secretariat. The CROP Executives Subcommittee on Climate Change (CES-CCC) set out the respective roles of other CROP agencies (USP: research and development and human resource development; FFA: pelagic fisheries; SPTO: tourism; PPA: utilities; FSMed: health) (Council of Regional Organizations in the Pacific, n.d.).

9.3 Discussion

Global summitry has provided a foundation for collaboration by nation-states and other entities including regional organizations, in the formulation of global public policy. Global summits have facilitated their joint efforts to explore issues and seek common action, develop mechanisms for global decision followed by national implementation, monitoring, reporting, and evaluation. The PICs have at times struggled to fully engage with global summitry due to high transaction costs involved as well as due to scarcity of the necessary human resources and expertise. However, the PICs have drawn on their regional cohesion and institutions to supplement their individual efforts. While some reservations have been expressed about the effectiveness of small-state engagement in global summits and policy discourse, non-engagement has been similarly criticised. While Pacific engagement has been strongest, and arguably most effective, on such crucial global challenges as climate change and sea-level rise, it has extended to a wide range of other pressing policy challenges.

References

Alexandroff, A. S., & Brean, D. (2015). Global summitry: Its meaning and scope Part One. *Global Summitry, 1*(1), 1–26.

Ashe, J., Van Lierop, R., & Cherian, A. (1999). The role of the Alliance of Small Island States (AOSIS) in the negotiation of the United Nations Framework Convention on Climate Change (UNFCCC). *Natural Resources Forum, 23*(3), 209–220.

[14]https://www.sprep.org/pacc

Barnett, J. (2005). Titanic States? Impacts and responses to climate change in the Pacific Islands. *Journal of International Affairs, 59*(1), 203–219.

Betzold, C. (2010). 'Borrowing' power to influence international negotiations: AISOS in the climate change regime, 1990–1997. *Politics, 30*(3), 131–148.

Betzold, C. (2016). Fuelling the Pacific: Aid for renewable energy across Pacific Island countries. *Renewable and Sustainable Energy Reviews, 58*, 311–318. https://doi.org/10.1016/j.rser.2015.12.156

Borzel, T. A. (2002). Organizing Babylon: On the different concepts of policy networks. *Public Administration, 76*(2), 253–273.

Cullen, R., & Hassall, G. (Eds.). (2017). *Achieving sustainable E-Government in Pacific Island States* (Vol. 27). Springer International Publishing.

Farrell, M. (2017). Group politics in global development policy: From the millennium development goals to the post-2015 development agenda. *The Hague Journal of Diplomacy, 12*(2–3), 221–248. https://doi.org/10.1163/1871191X-12341367

Goldsmith, M. (2012). The Colonial and Postcolonial Roots of Ethnonationalism in Tuvalu. *Journal of the Polynesian Society, 121*(2), 129–150.

Hare, F. K. M., & Patrick, J. (1979). *Proceedings of the World Climate Conference: A conference of experts on climate and mankind, Geneva, 12–23 February 1979.* World Meteorological Organization.

International telecommunication Union. (2014). *Final WSIS targets review: Achievements, challenges and the way forward. Partnership on Measuring ICT for Development.* International Telecommunciation Union.

Kramer, A., & Pahl-Wostl, C. (2014). The global policy network behind integrated water resources management: Is it an effective norm diffusor? *Ecology and Society, 19*(4).

Laatikainen, K. V. (2020). Chapter 8 The Alliance of Small Island States at the UN: The promise and pitfalls of single-issue groups in multilateral negotiations government of Papua New Guinea. Climate Change (Management) Act, (2015). Group Politics in UN Multilateralism. In K. Smith & K. Laatikainen (Eds.), (pp. 135–155). Brill | Nijhoff.

McMahon, V. M. (1993). Environmental Nongovernmental Organizations at Intergovernmental Negotiations. In L.E. Susskind, W.R. Moomaw & A. Najam (Eds.), Papers on International Environmental Negotiation, Vol. III, Cambridge, MA: The Program on Negotiation at Harvard Law School, pp. 1–21.

McNamara, K. E., & Gibson, C. (2009). 'We do not want to leave our land': Pacific ambassadors at the United Nations resist the category of 'climate refugees'. *Geoforum, 40*, 475–483.

Miles, G., Fuavao, V., & Smith, A. (1995). Implementing Agenda 21: Oceans, coasts and the Barbados outcomes in the Pacific region. *Ocean & Coastal Management, 29*(1), 125–138. https://doi.org/10.1016/0964-5691(96)00014-2

Súilleabháin, Ó. (2014). *Small States at the United Nations: Diverse perspectives, shared opportunities*. International Peace Institute.

Pacific Islands Forum. (2008). *Niue Declaration on Climate Change.* Pacific Islands Forum Secretariat.

Pacific Islands Forum Secretariat. (1989). Twentieth South Pacific Forum, Tarawa, Kiribati, 10–11 July. Forum Communique. SPFS(89)19(SPFS(89)19).

Pacific Islands Forum Secretariat. (1992a). Energy Ministers' Meeting, Forum Secretariat Headquarters, Suva Fiji 29 September 1992. meeting papers. (92) MP. 13.

Pacific Islands Forum Secretariat. (1992b). UNCED Earth Summit Rio de Janiero, Brazil, 3–14 June 1992. Briefing Notes for Mr. Ieremia Tabai, Secretary General. (92)NN.11.

Pacific Islands Forum Secretariat. (1995). Joint SPC/Forum Secretariat Regional Development Planners Fifth Meeting, Forum Secretariat Headquarters, Suva 23–24 May 1995. Meeting papers. (SPFS(95)6).

Pacific Islands Forum Secretariat. (2003). *Pacific Islands input to the World Summit on the Information Society.* Pacific Islands Forum Secretariat.

Pacific Islands Forum Secretariat. (2004). *An assessment of regional mechanisms and processes in the Pacific.* : Pacific Islands Forum Secretariat.
Pacific Islands Forum Secretariat (2012). Pacific Island Countries/Development Partners Meeting. Annotated Agenda, Suva 19–20 July. PIFS(12) PICS.01.
Potter, B. (2002a). World Summit on the Information Society *SPC records.*
Potter, B. (2002b). World Summit on the Information Society: To Discuss at PEI... *SPC records.*
Schechter, M. G. (2005). *United Nations global conferences.* Routledge.
South Pacific Commission. (1995). *Pacific Islands social & human development.* South Pacific Commission.
South Pacific Forum. (1991). *FSM Communiqué.*
South Pacific Regional Environment Programme. (1992a). Earth Summit Brief: Regional Brief for the United Nations Conference on Environment & Development Rio de Janeiro, 3–14 June 1992. 86.
South Pacific Regional Environment Programme. (1992b). The Pacific way: Pacific Island developing countries' report to the United Nations Conference on Environment & Development.
SPEC. (1988). Global Climate Changes - Implications for the South Pacific. Nineteenth South Pacific Forum, Nukualofa, Tonga 20–21 September 1988. Meeting papers. Folder 2. (88) MP. 3 Folder 2.
Stone, D. (2001). Learning Lessons, Policy Transfer and the International Diffusion of Policy Ideas, Centre for the Study of Globalisation and Regionalisation (CSGR), 69/01.
Streck, C. (2002). Global public policy networks as coalitions for change. In D. Esty & M. Ivanova (Eds.), *Global environmental GovernanceL options and opportunities.* New Haven.
UNHABITAT. (2016). *Habitat III visual identity.* UNHabitat.
United Nations. (1972). *Report of the United Nations Conference on the Human Environment, Stockholm, 5–16 June.* New York.
United Nations. (1980). *Report of the World Conference of the United Natoins Decade for Women: Equality, Development and Peace: Copenhagen, 14 to 30 July 1980.* The United Nations.
United Nations. (1986). *Report of the World Conference to Review and Appraise the Achivements of the United Nations Decade for Women: Equality, Development and Peace. Nairobi, 15–26 July.* United Nations.
United Nations. (1992a). *United Nations Conference on Environment and Development: National Reports Summaries. Nations of the Earth report. Volume 1.* UNCED.
United Nations. (1992b). *United Nations Conference on Environment and Development: National Reports Summaries. Nations of the Earth report. Volume 2.* UNCED.
United Nations. (1993). World Conference on Human Rights Vienna, 14–25 June 1993, Vienna Declaration and Programme of Action. A/Conf.157/23.
United Nations. (2012). *The future we want.*
United Nations Research Institute for Social Development. (2006). UN World Summits and Civil Society Engagement. *UNRISD Research and Policy Brief, 6.*
United Nations. Department of Public Information. (2012). *Rio+20 in Numbers.* United Nations.
United Nations. Economic and Social Council. (2001). *Regional cooperation in the economic, social and related fields: report of the Secretary-General: addendum.* UN.
United Nations. General Assembly. (1995). Summary record of the 21st meeting: 2nd Committee, held on Wednesday, 1 November 1995, New York, General Assembly, 50th session. UN.
Weiss, T. (2010). How United Nations ideas change history. *Review of International Studies, 36*(Suppl 1), 3–23.

Chapter 10
Global Development Plans

The establishment of the MDGs for the period 2000–2015 and the SDGs for the period 2015–2030 introduced a new level of program harmonization to development assistance. Each country still pursued its own development goals, but all countries were now pursuing common objectives. The MDGs exposed the gap between capacity of the UN as a global agency providing advice, resources and training, and the PICs as "small island developing states" with insufficient capacity and capability to implement. The emergence of "development communications" and more attention to branding made the MDGs more visible to the public than did previous UN plans. Icons, symbols, logos, graphs, photographs, press releases and policy briefs—all available online for download—replaced old-style typed technical proposals, work programs, and completion reports duplicated in small numbers and distributed to selected government offices, development partners, and possibly the parliamentary library.

10.1 Human Development Index

Among the first of the global development initiatives to establish successful online bringing and visibility was the Human Development Index (HDI) and its associated *Human Development Report* (HDR). The HDI was not a "global development plan" as such. However, since its establishment in 1990, UNDP's annual HDR and HDI have influenced planning processes on a global scale and provided some of the most useful comparative indicators of development outcomes.[1] The index is built on a concept of human development that considers economic development as a means toward improvement of individual and collective life, rather than as an end-in-itself. It combines data for life expectancy, quality of education, and means of income, to

[1] https://hdr.undp.org/

G. Hassall, *The United Nations and the Pacific Islands*, United Nations University Series on Regionalism 24, https://doi.org/10.1007/978-3-031-34155-7_10

produce an "indicator" of change that can be compared over time against indicators for comparable countries or against a country's own past performance.

As explained by one of the concept's founders, Pakistani economist Mahbub ul Haq, development cannot be measured simply in terms of income or growth, since people's well-being extends beyond economic data to include their opportunities for education, nutrition, security, and political and cultural freedoms. Development is related to enlarging peoples' opportunities to exercise choice, and to "...enjoy long, healthy and creative lives."[2]

The second major contribution of the HDI has been stimulation of standardization of data collection and analysis. The HDI's requirement for economic and social data across a spectrum of indicators that can be updated annually has been a major driver of improving statistical competency in the PICs. By adhering to standardized definitions of literacy, numeracy, poverty, water quality, etc., a country can gain a reasonable assessment of its pace of development in relation to comparator countries. HDI data is only useful to the extent that it provides feedback on policy effectiveness, and on how policy can be improved. UNDP These skills have subsequently been applied in other sectors, including gender equity, democratic institutions, and digital government.

A third contribution made by the HDI concept and programme has been its impact on government thought and practice. Any government that aspires to improving its HDI score will necessarily prioritize effective expenditure on health, education, and welfare. Although the HDI is useful for the comparisons it makes concerning levels of human development across countries, it also assists countries compare their individual performance over time.

For small island countries with agricultural and natural resource economies, small financial sectors, and rapidly expanding populations, a development index that takes into consideration economic growth without being wholly reliant on it, is appreciated. However, the task of collecting data on a regular and systematic basis across an ever-expanding number of sectors has been a substantial one for the small island nations. The accompanying chart shows the Human Development Index for Pacific Island Countries for the earlier years 2000, then 2017, and 2020, plus the current HDI global ranking (Table 10.1).

These figures place Palau, Fiji, Tonga and Samoa in the high human development category; Federated States of Micronesia, Vanuatu and Kiribati in the medium category, and Solomon Islands in the low human development category. Palau is ranked highest (50) and Papua New Guinea and Solomon Islands amongst the lowest ranked (155 and 151) PICs in global ranking—at the same level of human development as such African nations as Cameroon, Zimbabwe, and Rwanda.[3]

Six of the ten countries for which the HDI is available dropped down the global ranking between 2017 and 2021. In plain language, the citizens of these countries

---

[2] *http://hdr.undp.org/hd/default.cfm*

[3] See http://hdr.undp.org/en/countries. The HDI for 12 PICs is not known, either because it is not calculated, or because it is embedded in the HDI for a metropolitan country.

Table 10.1 Human development index ranking of Pacific Island countries

Region/country or territory	First HDI	HDI 2017	2017 RANK	HDI 2020	2020 RANK
MELANESIA					
Fiji	0.641 (1990)	0.736	91	.743	93
New Caledonia	n.a.				
Papua New Guinea	0.471	0.516	154	.555	155
Solomon Islands	0.442 (2000)	0.515	156	.567	151
Vanuatu	0.591 (2010)	0.597	134	.609	140
MICRONESIA					
Federated States of Micronesia	0.604 (2000)	0.638	127	.620	136
Guam					
Kiribati	0.585 (2010)	0.588	137	.630	134
Marshall Islands	n.a.			.704	117
Nauru	n.a.				
Northern Mariana Islands	n.a.				
Palau	0.741 (2000)	0.788	60	.826	50
POLYNESIA					
American Samoa	n.a.				
Cook Islands	n.a.				
French Polynesia	n.a.				
Niue	n.a.				
Pitcairn	n.a.				
Samoa	0.598 (1990)	0.704	104	.715	111
Tokelau	n.a.				
Tonga	0.648 (1990)	0.721	101	.725	104
Tuvalu	n.a.				
Wallis and Futuna	n.a.				

live shorter lifespans and enjoy fewer life-choices on average, than do citizens of most other countries in the world. While some of this disadvantage might still be attributed to the injustices of the colonial era, decisions concerning policies for the contemporary period emerge from the current governance capacities within these countries. Whereas PICs are making progress, they are doing so at a slower pace than countries on other continents and are slipping down the global rankings.

In the Pacific context, UNDP and associated development agencies have placed considerable effort in communicating the HDI's utility to decision-makers. UNDP convened a regional conference in 1994 that coincided with publication of the first

Human Development Report for the Pacific Islands region (UNDP & PIFS, 1994; UNDP, 1994), and it has already been noted that in 1994 the Forum gained Observer Status at the United Nations and endorsed the *Suva Declaration on Sustainable Human Development in the Pacific*.

UNDP convened a meeting of PIC representatives in Nadi in May 1997 to review its "Governance for Sustainable Human Development in the Pacific" (GSHD) regional programme for the period 1997–2001. This programme focused on key leveraging institutions at country level (parliament, the judiciary, media, office of public prosecution and ombudsman, public sector reform committees or non-government organizations). The aspiration was to improve participatory, transparent, accountable, effective, equitable and rule-based governance practices, through training on strategy and systems, convening action-planning workshops, facilitating access to knowledge on best practices partnerships, and consolidating links to professional networks.

In their consultation on these prospective activities, PIC representatives made several interesting recommendations concerning implementation. First was the need to promote greater congruence between traditional and western systems of governance by improving accountability, grassroots participation and rule of law. Additional recommendations included educating about rights and responsibilities; improving enforcement of existing laws and regulations; assisting judicial development; working with civil society organizations; enhancing principles of accountability; improving donor collaboration to ensure effective use of allocated funds; working with well-respected champions to gain the support from the people; building sustainable capacity in financial management, accountability, rule of law, investigative journalism, and elections; convening workshops on national rather than regional basis to include more participants for the same cost; and increasing use of email and internet for communication and access to knowledge.

10.1.1 National, Regional, and Sub-Regional Human Development Reports

In addition to annual calculation of an HDI for each PIC for publication in the UNDP's global Human Development Report,[4] UNDP has produced Human Development Reports at National and Regional levels, as listed in Table 10.2.

Whereas the Asia-Pacific reports at times made effort to include proportionate coverage of PICs and the Pacific region, the sheer size of the Asian countries commanded most attention in these texts. Human Development Reports for the sub-region have proven difficult to prepare, and their impact on public policy has similarly been hard to discern.

[4] http://hdr.undp.org/

Table 10.2 Pacific National and Regional Human Development Reports

National	Asia-Pacific	Pacific Sub-Region
Fiji—1997 Palau—1999 Papua New Guinea—1998, 2014 Samoa—1998, 2006 Solomon Islands—2002 Vanuatu—1996	2005—Promoting ICT for human development in Asia: Realizing the millennium development goals 2006—Trade on human terms 2008—Tackling corruption, transforming lives 2010—Power, voices and rights: A turning point for gender 2012—One planet to share: Sustaining human Progress in a changing climate 2014—Trade and human development 2016—Shaping the future: How demographics can power human development	1994—Putting people first 1999—Creating opportunities 2014—Vulnerability and exclusion in a time of rapid change

10.2 Millennium Development Goals

A decade after the commencement of the HDI, 191 countries at the Millennium Development Summit in New York in 2000 issued a declaration that was both a stocktake of the condition of the United Nations at the turn of the century and a commitment to spending the next 15 years addressing the most critical needs. The Secretary General's report *We the Peoples* identified four areas of concern: (a) The challenge of abject and dehumanizing poverty, in which more than 1 billion people were currently confined; (b) The challenge of the scourge of war and especially the violence of civil conflict and the fear of weapons of mass destruction; (c) The ultimate concern of life on a planet irredeemably spoilt by human activities, and whose resources may no longer provide for human needs; (d) The immediate concern of a United Nations lacking effective instruments for confronting all three of the above. (Annan, 2000, pp. 77–80).

At that global meeting in 2000 all member-states committed themselves to achieving eight MDG goals by 2015: (1) eradicate extreme poverty and hunger; (2) achieve universal primary education; (3) promote gender equality and empowering women; (4) reduce child mortality; (5) improve maternal health; (6) combat HIV/AIDS, malaria and other diseases; (7) ensure environmental sustainability; and (8) establish a global partnership for development (United Nations General Assembly, 2000) (Fig. 10.1).

These eight over-arching goals, which are depicted in Fig. 10.1, included more specific targets:

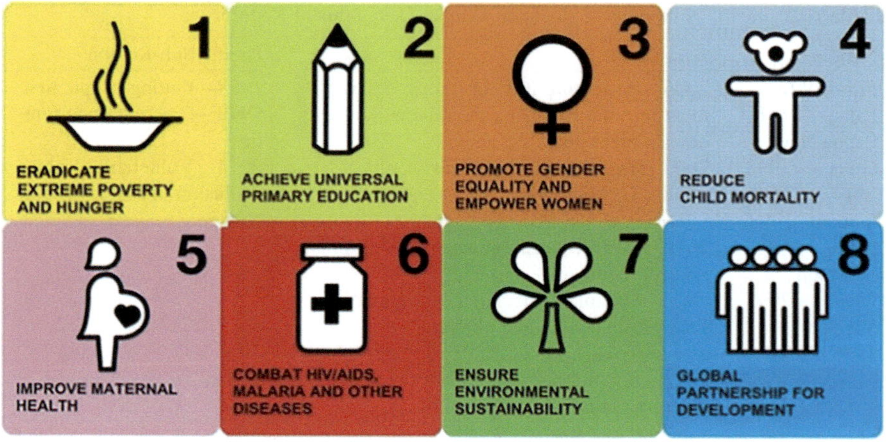

Fig. 10.1 Millennium development goals

1.1.1	Halve, between 1990 and 2015, the proportion of people whose income is less than US$1 a day.
1.1.2	Halve, between 1990 and 2015, the proportion of people who suffer from hunger.
1.1.3	Ensure that, by 2015, children everywhere, boys and girls alike, will be able to complete a full course of primary schooling.
1.1.4	Eliminate gender disparity in primary and secondary education, preferably by 2015 and to all levels of education no later than 2015.
1.1.5	Reduce by two thirds, between 1990 and 2015, the under-five mortality rate.
1.1.6	Reduce by three-quarters, between 1990 and 2015, the maternal mortality rate.
1.1.7	Have halted by 2015, and begin to reverse, the spread of HIV/AIDS.
1.1.8	Have halted by 2015, and begun to reverse, the incidence of malaria and other major diseases.
1.1.9	Integrate the principles of sustainable development into country policies and programmes and reverse the loss of environmental resources.
1.1.10	Halve, by 2015, the proportion of people without sustainable access to safe drinking water.
1.1.11	By 2020, to have achieved a significant improvement in the lives of at least 100 million slum dwellers.
1.1.12	Develop further an open, rule-based, predictable, non-discriminatory trading and financial system.
1.1.13	Address the special needs of least-developed countries.
1.1.14	Address the special needs of land-locked countries and Small Island developing states.
1.1.15	Deal comprehensively with the debt problems of developing countries through national and international measures in order to make debt sustainable in the longer term.
1.1.16	In cooperation with developing countries, develop and implement strategies for decent and productive work for youth.
1.1.17	In cooperation with pharmaceutical companies, provide access to affordable, essential drugs in developing countries.
1.1.18	In cooperation with the private sector, make available the benefits of new technologies, especially information and communications.

Some commentaries suggest that global programs, standards, and timetables, have been imposed on developing countries. In the case of the MDGs, the claim is that the 8 goals were designed in New York with little input from the countries in which they were to be applied. Some PICs expressed reservations about the ability of the PICs to meet 5 of the MDG targets by 2015 without significant support from the international community. However, many of these goals were not new. They had origins in the Barbados Programme of Action (BPoA) for the Sustainable Development of Small Island Developing States (SIDS) and its Mauritius Strategy for Implementation (MSI). They also drew on analysis by the OECD's Development Assistance Committee, which combined goals for poverty alleviation and human development into a package with environmental goals, and mechanisms for financing development (goal 8) (Farrell, 2017). What was new was shared commitment to a global effort to reducing extreme poverty by setting time-bound targets and establishing a framework for the international community to collaborate towards a common end. Whereas goal setting had always been an aspect of development planning, the specification of indicators and targets required considerable improvement in monitoring and evaluation processes beyond those that had been built to meet the requirements of HDI reporting.

Pacific governments and development agencies alike made a slow start to the MDGs. The Smaller Island States meeting in Palau in October 1999 discussed black pearl farming, mineral prospecting, aviation revenues from use of upper flyover airspace, and the logistics and frequency of their meetings—but not the MDGs. (PIFS, 1999) The Forum Leaders meeting in 1999, also in Palau, consulted on aviation, economics, communications, the situations in New Caledonia and in East Timor, the *UN Framework Convention on Climate Change*, the UN Special Session on Small Island Developing States, UNCLOSS, and nuclear issues—but not the pending Millennium Development Goals (South Pacific Forum, 1999). Some PIC development plans initiated during the period of the MDGs pursued goals in health, education, and environmental sustainability, as well as strengthening monitoring and evaluation capabilities—all without referencing the Millennium Development Goals. (Republic of Kiribati & Secretariat of the Pacific Community, 2008, p. 1 Annex 1: Port Vila Declaration on Accelerating Progress on the Achievement of the Millennium Development Goals).

To address this slow start, UNDP invited participants from 19 PICs to a Pacific Regional Workshop in March 2003 to determine how best to integrate MDG Goals and targets into national planning processes. The first target, "eradicate extreme poverty and hunger" elicited strong responses from some Pacific leaders, who objected to the imposition of "foreign" definitions of poverty. Their argument was that Pacific lifestyles were not totally determined by size of income, since they were traditionally self-sufficient in food and used barter and exchange to meet their needs: they preferred the term "hardship." There is sufficient evidence, on the other hand, that low incomes are a cause of poverty in the Pacific, particularly in urban areas. A compromise position was to agree that "poverty of opportunity" should be addressed through pro-poor policies and good governance.

The Forum Leaders made few references to the MDGs throughout the 15-year programme. In August 2003 they recognized the "broad relevance of the MDGs to the Pacific", in their annual communique: ". . .particularly if modified to better reflect Pacific circumstances, and their usefulness in focusing and improving the integration of planning for sustainable development, and in the monitoring of progress." (PIF, 2003) The Micronesian President's Summit Communique that same month was focused on acquisition of fibre-optic cable, new anti-terrorism obligations, and financial negotiations with the United States—but not the MDGs. (Micronesian Chief Executives Summit, 2003).

Pacific leaders were prepared to attribute failure in achieving MDG targets to the international community rather than to national or regional efforts. "After five years since we committed ourselves to the UN Millennium Declaration," the President of the Federated States of Micronesia informed a High-level plenary meeting at UNGA's 60th Session in 2005,

> . . .the MDGs remain relevant as the basis of our sustainable development. Unfortunately, it is the experience of my country, and our general observation in Micronesia that we face an uphill battle in meeting and implementing the Millennium Development Goals. It is widely recognized that Oceania remains one of the identified regions of the world that is worst off in meeting its Millennium Development Goals. Even with this recognition, we find more often than not, that the international community has not followed up with fair and proportionate allocation of resources to these more vulnerable small island developing states. For too long our islands of the Pacific have been overlooked. We, therefore, appeal to the United Nations and the international donor community to apply a fair and more proportionate allocation of development assistance to our region if we are to achieve these goals. (Federated States of Micronesia, 2005)

By the midpoint of the MDGs when the shortfall in progress toward the goals had become obvious, UNDP's Fiji Office doubled its support for ten PICs for the period 2009–2010 to a total of USD 61 million. Much of this additional finance was to be spent on "strengthening national institutions and developing capacities." ("UNDP Contributes more resources to Pacific", 2008) On reviewing its challenges with reaching its MDG targets, Papua New Guinea launched its 2005–2010 *Medium Term Development Strategy* to provide a roadmap for implementation. A first progress report toward the MDGs, issued in 2004, and set out in Table 10.3, pointed out that the government considered most MDG targets to be "over-ambitious, unrealistic and therefore out of reach, at least for the year 2015", and that it had therefore built "tailored or customised" targets and indicators into its medium-term development plan to reflect Papua New Guinea's "realities and priorities. (Government of Papua New Guinea & United Nations Development Programme, 2004, p. 41) Even with this localization, prospects for achieving success under either scorecard were discouraging:

Constraints to accomplishment of the goals identified by the report included "the adverse development context, the restricted institutional framework, severe resource limitations and many other socio-economic, cultural, political and other constraints." (Government of Papua New Guinea & UNDP, 2004, p. 42) A "national strategy for responsible sustainable development in Papua New Guinea issued in 2014 made

Table 10.3 Composite index for Papua New Guinea

MDG	Achievement goal	State of supporting environment		
	MDG	MTDS	Policy/ Legislation	Implementation
1. Extreme poverty	Very unlikely	Potentially	Fair	Fair
2. Primary education	Very unlikely	Potentially	Fair	Fair
3. Gender equity	Very unlikely	Potentially	Fair	Weak
4. Child mortality	Very unlikely	Potentially	Fair	Fair
5. Maternal mortality	Very unlikely	Potentially	Fair	Weak
6. HIV/AIDS	Very unlikely	Very unlikely	Fair	Weak
7.Environmental sustainability	Very unlikely	Very unlikely	Strong	Very weak

only passing reference to the MDGs, and gave no hint of the forthcoming SDGs. (Government of Papua New Guinea. Department of National Planning and Monitoring, 2014).

At regional level, a process of "peer review", the Cairns Compact on Strengthening Development Coordination in the Pacific (Forum Compact) was established in 2009 to address concerns about slow progress toward achieving the MDGs. The Forum Leaders explained in an annex to their 2009 communique:

> Leaders expressed their deep concern that, despite continued high levels of development assistance over many years, the Pacific region remains off-track to achieve the Millennium Development Goals (MDGs) by 2015. Given the need to improve development outcomes, address the impact of the global economic downturn, and strengthen long-term economic resilience, Leaders agreed there was an urgent need to establish a new development compact for the Pacific. The key objective of this compact will be to drive more effective coordination of available development resources from both Forum Island Countries and all development partners, centred on the aim of achieving real progress against the MDGs. They also called on the International Financial Institutions to assist the Pacific Islands countries in responding to the global economic crisis, including through supporting better coordination mechanisms. (PIF, 2009, p. 1)

The Compact reviews examined each PIC's principles and processes for coordinated delivery, national planning, budgeting, public financial and aid management systems, and donor coordination. They generally comprised an expert group of senior Government representatives from two Pacific Islands plus a Monitoring and Evaluation specialist from the United Nations system. Tuvalu's 2011 review was completed by officials from Tonga, Vanuatu, and UNESCAP, with support from the Forum Secretariat. The Federated States of Micronesia's review in 2012 included officials from the Marshall Islands, Samoa, Vanuatu, and Mr. Asif Chida, the UNDP Pacific Centre in Fiji. In like manner all 14 independent states underwent peer reviews between 2010 and 2014: 2010 (Nauru and Kiribati), 2011 (Vanuatu, Tuvalu, Niue, and Kiribati), 2012 (Federated States of Micronesia, Marshall Islands, Tonga, and Papua New Guinea), 2013 (Palau, Solomon Islands, and Cook Islands) and 2014

(Samoa).[5] Although hailed as a successful initiative, the Cairns review process was not carried into peer assessment of SDG progress.

Progress toward the MDGs was assessed at global conferences in 2005, 2010, and 2015. In 2010, to take an instance, several summits were convened in New York adjacent to the annual meeting of the General Assembly and in May the Forum Leaders issued the *Port Vila Declaration on Accelerating progress on the achievement of the MDGs*, prior to the Millennium Development Goals United Nations Summit on 20–22 September and the five-year high-level review of the Mauritius Strategy for Implementation (MSI + 5) on 24–25 September. The declaration affirmed the leaders' commitment to achieving the MDGs but also listed concerns and requests. The concerns included "...impediments to stronger and broader based development posed by the poor coordination of development resources," and "the failure to better harmonise efforts to achieve outcomes and the necessity of mutual accountability and respect among partners." (PIFS, 2010) Requests made of development partners included: help to identify and develop major new activities and programs to expedite achievement of the MDGs; deliver on pledges to scale up aid and to channel increased resources to PICs; honour existing commitments under the *Paris Declaration on Aid Effectiveness*, the *Accra Agenda for Action*, and the *Pacific Principles on Aid Effectiveness*; and actively engage with regional efforts such as the *Cairns Compact on Strengthening Development Coordination in the Pacific*. The declaration also asked donors to mainstream the MSI and BPoA into programmatic work; develop vulnerability and resilience indices; help strengthen PIC data collection capability, including disaggregated data, analysis and dissemination.

Given its location near the centre of the South Pacific, and being a major transport hub, Fiji has been the location for innumerable preparatory committee meetings and regional meetings. In June 2012, on the initiative of PIF Secretary General Tuiloma Neroni Slade, the heads of CROP agencies and United Nations missions in the Pacific met simultaneously for the first time, and agreed to establish a working group to facilitate their collaboration on meeting the objectives of the Forum's Pacific Plan and the UN's Millennium Development Goals (MDGs).

By 2015 only Cook Islands and Niue had achieved all their MDG-related objectives; three countries achieved at least half (Fiji, Palau and Tonga), and five countries achieved less than half of their goals (Nauru, Federated States of Micronesia, Marshall Islands, Tuvalu and Vanuatu), while Kiribati, Papua New Guinea and Solomon Islands failed to achieve any of their MDG objectives. (PIF, 2015, p. 8) These results mean that whereas most PICs reduced child mortality and achieve universal primary education during the period 2000–2015, few were successful in reducing poverty or improving gender equality and the empowerment of women. Nor was there significant improvement in maternal health, or in combating HIV/AIDS, malaria and tuberculosis.

Despite this failure to meet all targets in the PICs, the MDGs were beneficial in a number of ways. They raised international awareness of the immensity of the

[5] https://www.forumsec.org/regional-reports-and-frameworks-of-the-pacific-islands-forum/

challenges in child mortality, maternal health, and education. International aid became more coordinated, and measures for tracking progress more sophisticated. Individual countries improved their data-collecting infrastructure capacities. In 2015 each of these capabilities were tested further through the launch of a subsequent global development plan, which more than doubled the size of the assignment: the Sustainable Development Goals.

10.3 Sustainable Development Goals

The commencement of a new global development plan for 2015–2030, the *2030 Agenda for Sustainable Development* (United Nations General Assembly, 2015), provided the United Nations as well as PICS and Pacific regional institutions with the opportunity to learn from the MDG experience and do things differently where appropriate. Although both the MDGs and the SDGs can be considered global development plans, there are significant differences in how they were conceived, the challenges they address, and their scope. Whereas the MDGs were determined through a "top-down" approach which resulted in the eight goals being put to the General Assembly for approval without consultation, the SDGs were formulated through broad and inclusive consultation processes that commenced in 2012. Both state and non-state stakeholders were involved, from local to global levels.

Much greater use was made of ICTs to obtain the general public's preferences, such as through the website worldwewant2015.org. One consequence of this more deliberative and comprehensive approach was expansion of the number of goals to 17 (see Table 10.4), which are being pursued by all member-states rather than just the least developed and developing states. The Forum Leaders expressed concern about the number of goals and targets—a concern shared by many who were then in the final stages of tracking the 8 MDGs. (PIF, 2015, p. 103) The 2030 Agenda globalised the notion of development, and poverty reduction became one goal amongst others rather than the main objective, as it was in the MDGs. The lack of significant accomplishment of the MDGs between 2000–2015 provoked much reflection on how collaboration and cooperation could be strengthened to do better with the Goals set for the period 2015–2030.

Three contrasting features of the two periods stand out. First, are the early moves by PICs and regional organizations to engage with the SDGs' preparatory phase. The PICs lobbied successfully, for instance, for Oceans to have a stand-alone goal (Goal 14). (PIFS, 2015, p. 102; Quirk & Hanich, 2016) The Forum Leaders' communiques of 2012, 2013 and 2014 tasked the Forum Secretariat to work in collaboration with CROP and UN agencies to develop a Pacific position on the pending SDGs, and to contribute effectively to relevant global processes, such as the 68th UNGA of September 2013. They also called for Pacific representation on the SDG working group and finance working group. These steps were set out in a policy paper prepared by PIFS for a meeting with its Development Partners in Suva in July 2012:

Table 10.4 The Sustainable Development Goals

The seventeen SDG goals include the eight MDG goals, plus more:	
01	End poverty in all its forms everywhere.
02	End hunger, achieve food security and improved nutrition and promote sustainable agriculture.
03	Ensure healthy lives and promote well-being for all at all ages.
04	Ensure inclusive and equitable quality education and promote lifelong learning opportunities for all.
05	Achieve gender equality and empower all women and girls.
06	Ensure availability and sustainable management of water and sanitation for all.
07	Ensure access to affordable, reliable, sustainable and modern energy for all.
08	Promote sustained, inclusive and sustainable economic growth, full and productive employment and decent work for all.
09	Build resilient infrastructure, promote inclusive and sustainable industrialization and foster innovation.
10	Reduce inequality within and among countries.
11	Make cities and human settlements inclusive, safe, resilient and sustainable.
12	Ensure sustainable consumption and production patterns.
13	Take urgent action to combat climate change and its impacts.
14	Conserve and sustainably use the oceans, seas and marine resources for sustainable development.
15	Protect, restore and promote sustainable use of terrestrial ecosystems, sustainably manage forests, combat desertification, and halt and reverse land degradation and halt biodiversity loss.
16	Promote peaceful and inclusive societies for sustainable development, provide access to justice for all and build effective, accountable and inclusive institutions at all levels.
17	Strengthen the means of implementation and revitalize the global partnership for sustainable development.

> These post-2015 and SDGs processes are an important opportunity for Forum Island countries (FICs) to influence the shape of the post-2015 framework and the SDGs. To do this, the Pacific must articulate its position on the post-2015 agenda and the SDGs development process.
>
> There are two crucial issues that the Pacific needs to tackle in order to form the regional position: (i) critical assessment of the MDG framework (what has worked, what needs improvement and what are the gaps, and (ii) identification of the key development challenges in the Pacific (reflecting the Pacific Plan, national priorities and emerging issues). The Pacific position must also equally incorporate the three pillars of sustainable development—*economic* development, *social* progress and *environmental* responsibility. (PIFS, 2012)

Immediately consequent to the agreement to develop the SDGs at the 2012 UN Conference on Sustainable Development (the "Rio + 20" Conference), Forum Leaders recognised the importance of actively participating in the global discussions, subsequently providing high-level direction each year, including for the post-2015 intergovernmental negotiations. (PIFS, 2015, p. 102) Preparation for the SIDS III conference in Samoa in September 2014 was also viewed as a preparatory exercise for the SDGs that was to begin the following year.

Fig. 10.2 The Sustainable Development Goals

In 2017 the Forum Leaders endorsed the *Pacific Roadmap for Sustainable Development*, (Pacific Sustainable Development Goals Taskforce, 2017) a framework that sets out 132 Pacific Sustainable Development Indicators (PSDI), that cover not only SDG goals but the *SIDS Accelerated Modalities for Action (SAMOA) Pathway*, the *Framework for Pacific Regionalism* (FPR) and the *Pacific Leaders Gender Equality Declaration* (PLGED) (Fig. 10.2).

A second contrast with the MDG experience was an increased desire by PICs to shape the policy framework and processes. In 2018, to provide one instance, FSM requested UN assistance in establishing locally relevant and contextualized targets and indicators by that emerged from a localization workshop in Phonpei: "As part of the key goals of the Global 2030 Agenda to select relevant and contextualized indicators to measure SDG progress," FSM's Secretary for Foreign Affairs wrote to the UNDP's Resident Representative in Suva:

> …it was unanimously decided that, in order to advance efforts towards SDG 1 (NO POVERTY), the country requires adequate and alternate measure which step away from traditional ones which focus on understanding poverty or hardship based on monetary basis. (Robert, 2018)

A third contrast with the MDGs is the early filing of voluntary national and regional progress reports to the High-Level Political Forum on Sustainable Development (HLPF). PIF coordinated the first Quadrennial Pacific Sustainable Development Report in 2018. (PIFS, 2018). The report emphasised Forum Leaders' commitment to a coherent, streamlined, and integrated approach to implementing and accounting for sustainable development initiatives. Although it emphasised increasing vulnerabilities across the region (deepening inequalities and limited access to infrastructure and basic services), the Report also identified opportunities for accelerating

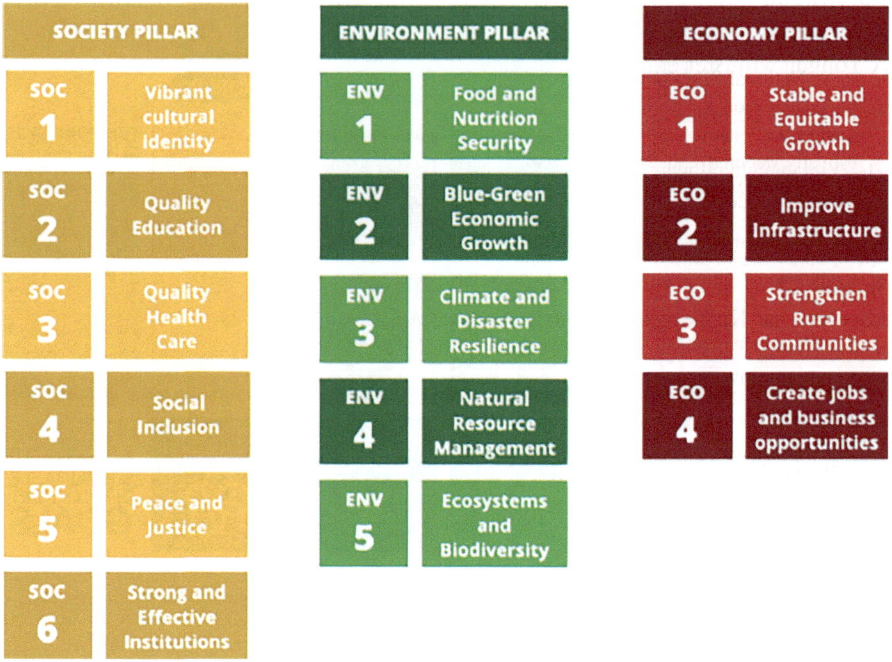

Fig. 10.3 Vanuatu's National Sustainable Development Goals

development through Pacific social adaptability, significant biodiversity and oceanic resources, desire to work as a "Blue Pacific collective", and increasing access to and use of ICTs and connectivity.

In addition to such findings, the report emphasised the scope of research on which it was based: two years of collaboration with member countries and territories, CROP agencies, and United Nations agencies in the Pacific; datasets for the 132 Pacific SDG Indicators generated by SPC, SPREP, PIFS, FFA, UN agencies, World Bank, ADB, and PICs; updates on Pacific SIDS/S.A.M.O.A Pathway partnerships; and regional sustainability briefs prepared by CROP and UN agencies in the Pacific. In sum: the report resulted from close collaboration between a comprehensive group of CROP and UN technical specialists covering society, environment, economic development, and means of implementation. This 2018 quadrennial report has since been updated by the first biennial report, (PIFS, 2020) and the 2022 second quadrennial report. (PIFS, 2022).

A fourth contrast with the MDGs experience was more effective integration of SDGs in PIC strategic plans. Vanuatu, for instance, re-worked the global development goals into fifteen National Sustainable Development Goals, to align more closely with national priorities and preferences (see Fig. 10.3). (Government of Vanuatu, 2016).

A comparison of Vanuatu's fifteen Sustainable Development Goals and the UN's 17 SDGs, as set out in Table 10.5, suggests that many of the latter have been

Table 10.5 Comparison between Vanuatu's National Sustainable Development Goals and the SDGs

Vanuatu National Sustainable Development Goals	Sustainable Development Goals
Social pillar	
01—Vibrant cultural identity	
02—Quality education	04—Ensure inclusive and equitable quality education and promote lifelong learning opportunities for all.
03—Quality health care	03—Ensure healthy lives and promote Well-being for all at all ages.
04—Social inclusion	05—Achieve gender equality and empower all women and girls.
05—Peace and justice	16—Promote peaceful and inclusive societies for sustainable development, provide access to justice for all and build effective, accountable and inclusive institutions at all levels.
06—Strong and effective institutions	16—Promote peaceful and inclusive societies for sustainable development, provide access to justice for all and build effective, accountable and inclusive institutions at all levels.
Environment pillar	
01—Food and nutrition security	01—End poverty in all its forms everywhere 02—End hunger, achieve food security and improved nutrition and promote sustainable agriculture.
	08—Promote sustained, inclusive and sustainable economic growth, full and productive employment and decent work for all.
02—Blue-green economic growth	
03—Climate and disaster resilience	13—Take urgent action to combat climate change and its impacts.
	10—Reduce inequality within and among countries.
04—Natural resource management	15—Protect, restore and promote sustainable use of terrestrial ecosystems, sustainably manage forests, combat desertification, and halt and reverse land degradation and halt biodiversity loss.
05—Ecosystems and biodiversity	14—Conserve and sustainably use the oceans, seas and marine resources for sustainable development.
Economy pillar	
01—Stable and equitable growth	12—Ensure sustainable consumption and production patterns.
02—Improve infrastructure	09—Build resilient infrastructure, promote inclusive and sustainable industrialization and foster innovation. 07—Ensure access to affordable, reliable, sustainable and modern energy for all. 06—Ensure availability and sustainable management of water and sanitation for all.
03—Strengthen rural communities	11—Make cities and human settlements inclusive, safe, resilient and sustainable.
04—Create jobs and business opportunities	
	17—Strengthen the means of implementation and revitalize the global partnership for sustainable development.

re-prioritized and re-phrased. Articulation of a "social pillar" and then an "environment pillar" before the "economy pillar" demonstrates the country's prioritization of culture and society, and the environment in which it operates, over material production and economic growth.

10.4 Millennium Development Goals and Sustainable Development Goals Governance Challenges

The frenetic pace in the development field during the years of the MDGs and now during the SDGs has raised numerous governance and policy challenges. One of these is determining appropriate levels of consultation within agencies, and amongst agencies and PICs. While there is always the danger of inadequate consultation with stakeholders, superfluous consultation can also occur. When a new global campaign such as the SDGs commences, the World Bank, the ADB, UNDP, WHO, UN-Habitat and very many other large development institutions and programmes convene their own regional consultations to ensure alignment of all parties with their plan of action. Whereas each such regional event is as unique from the perspective of the convening agency, it can be experienced as repetitive by PICs whose development officials are required to attend each of them. One sign of "meeting fatigue" is the delegation of attendance to lesser-ranked officials: the lower the rank the lower the priority that their department has attached to the event.

Viewed from a different perspective, that of the citizenry, questions remain concerning the democratic element in global development campaigns. If a concerned citizen wished to participate in some way, how would this occur? And if a citizen sought information about its government's progress with an SDG goal, how would she find out? The broad objectives are easily found on the internet, but progress reports and other sources of evaluation, are not. Undertaking development *with* the people rather than *for* the people continues to be one of the hardest aspects of the development process. This book is being published at the mid-point of the SDGs, and it is far to early to predict the extent to which they will be "owned" by Pacific societies.

References

Annan, K. (2000). *'We the Peoples': The role of the United Nations, 21st century*. UN Department of Public Information.

Farrell, M. (2017). Group politics in global development policy: From the millennium development goals to the Post-2015 development agenda. *The Hague Journal of Diplomacy, 12*(2–3), 221–248. https://doi.org/10.1163/1871191X-12341367

Government of the Federated States of Micronesia. (2005). *Address by H.E. Joseph J. Urusemal, President of the Federated States of Micronesia before the High-Level Plenary Meeting of the 60th Session of the United Nations General Assembly, Friday 16 September.*

Government of Papua New Guinea, & United Nations Development Programme. (2004). *Millennium development goals. Progress report for Papua New Guinea 2004*. Government of Papua New Guinea and United Nations in Papua New Guinea.
Government of Papua New Guinea. Department of National Planning and Monitoring. (2014). *National Strategy for responsible Sustainble development for Papua New Guinea StaRS* (2nd ed.).
Government of Vanuatu. (2016). Vanuatu 2030 the People's plan: National Sustainable Development Plan 2016–2030.
Micronesian Chief Executives Summit. (2003). Joint communique of the third Micronesian President's Summit.
Pacific Islands Forum. (2003). *Auckland Communiqué*. Pacific Islands Forum.
Pacific Islands Forum. (2009, August). *Forum communique Fortieth Pacific Islands forum. Annex B: Cairns compact on strengthening development coordination in the Pacific.*
Pacific Islands Forum. (2015). *Pacific regional MDGs tracking report*. Pacific Islands Forum Secretariat.
Pacific Islands Forum Secretariat. (1999). *Eighth meeting of smaller Island states leaders, Koror, Republic of Palau. Summary record*. (99) FSCOL. 17, 8 p.
Pacific Islands Forum Secretariat. (2010). *Forty-first Pacific Islands forum Port Vila, Vanuatu. Agreed record. PIFS(10)REP*.
Pacific Islands Forum Secretariat. (2012, July 19-20). *Pacific Island countries/development partners meeting. Annotated agenda. Session 3: Implications of the Post-2015 development agenda/ sustainable development goals on the Pacific*, PIFS(12) PICS.04(B).
Pacific Islands Forum Secretariat. (2015). Pacific regional MDGs tracking report.
Pacific Islands Forum Secretariat. (2018). *First quadrennial Pacific sustainable development: 2018 report*. Pacific Islands Forum Secretariat.
Pacific Islands Forum Secretariat. (2020). *The 2020 biennial Pacific sustainable development report*. Pacific Islands Forum Secretariat.
Pacific Islands Forum Secretariat. (2022). *Second quadrennial Pacific sustainable development: Report 2022*. Pacific Islands Forum Secretariat.
Pacific Sustainable Development Goals Taskforce. (2017). *Pacific roadmap for sustainable Develoment*. Pacific Islands Forum Secretariat.
Quirk, G., & Hanich, Q. A. (2016). Ocean diplomacy: The Pacific Island Countries' campaign to the UN for an ocean sustainable development goal. *Asia Pacific Journal of Ocean Law and Policy, 1*(1), 68–95.
Republic of Kiribati, & Secretariat of the Pacific Community. (2008). *Republic of Kiribati and Secretariat of the Pacific Community Joint Country Strategy 2008–2011*.
Robert, L. (2018). *Letter to Osnat Lubrani*. SPC Records.
South Pacific Forum. (1999). *Palau Communiqué*.
UNDP, & Pacific Island Forum Secretariat. (1994). *Putting people first: Regional meeting of ministers of planning and senior planners on human development strategies in the Pacific*. UNDP and Pacific Islands Forum Secretariat.
UNDP Contributes more resources to Pacific. (2008, April 08). *PACNEWS*.
United nations development programme (UNDP). (1994). *Pacific human development report: 'Putting people first'*. United nations development programme.
United Nations General Assembly. (2000). *United Nations millennium declaration: Resolution*. UN.
United Nations General Assembly. (2015). *Transforming our world: The 2030 agenda for sustainable development: Resolution*. United Nations.

Chapter 11
The Emergence of Small Island Developing States (SIDS) Development Discourse

Engagement with UN dialogue over several decades helped the PICs identify and articulate their national and regional characteristics. When writing to UN secretary General Javier Perez de Cuellar (perhaps for the first time) in 1985 about the decolonization of New Caledonia, the SPEC Director also took the opportunity to suggest that the UN's economic and technical agencies place special emphasis on the needs of the Smaller Island Countries: Particular emphasis could be placed on regional programmes already in operation in the Pacific such as regional telecommunications under ITU, and bilateral or regional programmes implemented by UNDP as well as regional programmes run by UNESCO, WHO, ILO, UNFPA and others. (PIFS, 1985, p. 7).

The development needs of "small island developing states" became clearer through participation at the 1992 United Nations Conference on the Environment and Development (UNCED), the "earth summit", in Rio de Janeiro. UNCED noted that "small island developing states" share physical, social, economic, and infrastructural characteristics not only amongst themselves, "...but with small states in other regions of the globe: small populations, with small economics, low to medium rates of human development, in need of mature political and public sector leadership, in the face of rapid social change." This dialogue resulted in the establishment of a specific support unit within the United Nations.[1] The SIDS have subsequently held three global conferences: Barbados in 1994, Mauritius in 2005, and Apia in 2014—these have become known as SIDS conferences I, II, and III.

The outcome of SIDS I was the 14-point "Barbados Programme of Action for Sustainable Development" (see Table 11.1). In later years SIDS were supported by the United Nations Office of the High Representative for the Least Developed Countries, Landlocked Developing Counties and Small Island Developing States

[1] http://unohrlls.org/

G. Hassall, *The United Nations and the Pacific Islands*, United Nations University Series on Regionalism 24, https://doi.org/10.1007/978-3-031-34155-7_11

Table 11.1 Barbados Programme of Action for the sustainable development of small island states

Barbados Programme of action for the sustainable development of Small Island states
1. Climate change and sea- level rise
2. Natural and environmental disasters
3. Management of wastes
4. Coastal and marine resources
5. Freshwater resources
6. Resources
7. Energy resources
8. Tourism resources
9. Biodiversity resources
10. National institutions and administrative capacity
11. Regional institutions and technical cooperation
12. Transport and communication
13. Science and technology
14. Human resource development

(UN-OHRLLS—http://unohrlls.org/), an entity established in 2001. UN-OHRLLS now assists 46 Least Developed Countries (LDCs), 32 Landlocked Developing Countries (LLDCs) and 38 Small Island Developing States (SIDS). Some twenty of these SIDS are PICs including the dependant territories American Samoa, Cook Islands, New Caledonia, Northern Mariana Islands, and Niue.

A second SIDs conference, convened in Mauritius in 2005, resulted in the Mauritius Strategy of Implementation (MSI). As noted elsewhere in this book, implementation of the BPOA was not judged to have been adequate. SIDS' unique vulnerabilities were again highlighted in *The Future We Want*, a report adopted at the United Nations Conference on Sustainable Development (Rio + 20) held in June 2012, once again in Rio de Janeiro.

11.1 SIDS III

The Third International Conference on SIDS held in Apia, Samoa, in September 2014, was themed "The sustainable development of small island developing States through genuine and durable partnerships."[2] The Pacific was keen to host this key UN meeting to further develop a unique, and substantively significant UN programme. It was also a way for Samoa to demonstrate its progress and emergence

[2]https://sustainabledevelopment.un.org/sids2014

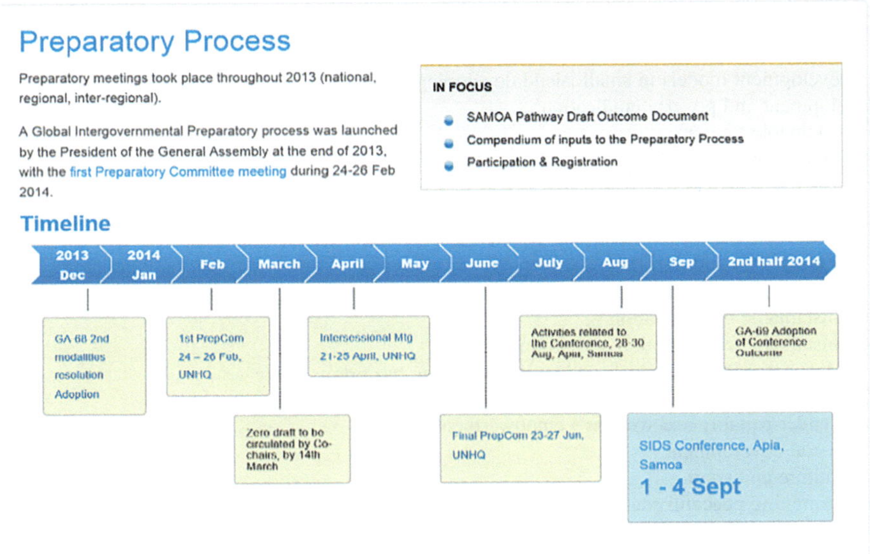

Fig. 11.1 Preparatory process for SIDS III

out of LDC status. Preparation followed the well-established format of preparatory meetings, as depicted in Fig. 11.1, prior to the Apia Conference, a clear statement of outcomes, together with a plan to monitor follow-up action. The SIDS Accelerated Modalities of Action (SAMOA) Pathway[3] reaffirmed the commitment of the Governments of Small Island Developing States to address sustainable development issues, reaffirmed commitments made at previous United Nations summits on sustainable development and registered over three hundred partnerships established on an online Partnership Platform. The UN's Department of Economic and Social Affairs (UNDESA) was appointed to lead inter-agency coordination both within and beyond the United Nations system [4] and to monitor the implementation of BPOA, MSI, and the Samoa Pathway, and monitor progress made by SIDS partnerships.

11.1.1 Samoa Pathway

The SIDS SAMOA Pathway recognized the need for partnerships comprising both state and non-state actors, and processes that focus not merely on problem analysis

[3]https://sustainabledevelopment.un.org/sids2014/samoapathway

[4]Within the UN through the Executive Committee of Economic and Social Affairs Plus (ECESA Plus), and among UN and non-UN entities active on SIDS issues through the Inter-Agency Consultative Group (IACG) on SIDS.

Table 11.2 Critical challenges addressed in the SAMOA Pathway

1. Sustained and sustainable, inclusive and equitable economic growth with decent work for all
2. Development models in small island developing states for the implementation of sustainable development and poverty eradication
3. Sustainable tourism
4. Climate change
5. Sustainable energy
6. Disaster risk reduction
7. Oceans and seas
8. Food security and nutrition
9. Water and sanitation
10. Sustainable transportation
11. Sustainable consumption and production
12. Management of chemicals and waste, including hazardous waste
13. Health and non-communicable diseases
14. Gender equality and women's empowerment
15. Social development
16. Culture and sport
17. Promoting peaceful societies and safe communities
18. Education
19. Biodiversity
20. Desertification, land degradation and drought
21. Forests
22. Invasive alien species

but on effective implementation, and effective monitoring and accountability. The statement consists of 124 clauses and refers to more than 20 critical policy challenges facing the Pacific and other Small Island Developing Countries (see Table 11.2). As with all statements of this genre, the early paragraphs recall the pathway leading to the current moment.

To take the third Paragraph as an instance:

> 3. We Recall As Well our Commitments in the Outcomes of all the Major United Nations Conferences and Summits in the Economic, Social and Environmental Fields, Including the United Nations Millennium Declaration, the 2005 World Summit Outcome, the Monterrey Consensus of the International Conference on Financing for Development, the Doha Declaration on Financing for Development: Outcome Document of the Follow-Up International Conference on Financing for Development to Review the Implementation of the Monterrey Consensus, the Outcome Document of the High-Level Plenary Meeting of the General Assembly on the Millennium Development Goals, the Programme of Action of the International Conference on Population and Development, the Key Actions for the Further Implementation of the Programme of Action of the International Conference on Population and Development[and the Beijing Declaration and Platform for Action... (United Nations. General Assembly, 2014, p. 2)

Paragraph 14 acknowledged that,

> ...in spite of the considerable efforts of small island developing States and the mobilization of their limited resources, their progress in the attainment of the internationally agreed development goals, including the Millennium Development Goals, and in implementing

> the Barbados Programme of Action and the Mauritius Strategy has been uneven, and some have regressed economically. A number of significant challenges remain. . . (United Nations. General Assembly, 2014, p. 4).

11.1.2 SIDS III: Partnerships

Given this recognition that more progress was required, SIDS III generated more than 500 partnerships (300 at the event and a further 200 by 2019, plus an additional 1400 partnerships registered at the 2017 Ocean Conference). More than half of all partnerships were initiated by UN agencies, while other actors included national governments, regional organizations, foundations, and other philanthropic organizations, NGOs, and academies. 147 of these partnerships were global, with others focused on one or another of the regions: the Pacific, the Caribbean, the Atlantic Ocean, Indian Ocean, and South China Sea.

In the Pacific context, UN agencies were participating in 77 partnerships, 33 as the lead agency. The partnerships addressed key priority areas: Sustainable Economic Development; Climate Change; Disaster Risk management; Social Development; Health and NCDs; Youth and Women; Sustainable Energy, Oceans, Seas; Biodiversity; Water and Sanitation; Food Security; and Waste Management. Commencing in 2016 UNDESA has sponsored SIDS Multi-stakeholder partnership dialogues in the margins of the General Assembly or the High-Level Political Forum on Sustainable Development.

For the Pacific region, a total of 287 partnerships were registered, of which 64 had completed their projects by 2018. Although 223 Pacific projects remained “active” by 2018, just 13 participated in an inter-regional dialogue in Apia. (Goransson et al., 2019, p. 27) A 2019 review found direct impact on beneficiaries and on the environment across a majority of the SAMOA pathway’s key issues. More than one million customers, for instance, had been enrolled in appropriate financial services (meeting goals of the global financial inclusion campaign toward sustainable, inclusive and equitable economic growth and decent work for all); 22 renewable energy projects had been completed; and more than 600 villages across 15 PICs had established over 420 locally managed marine areas (LMMAs). (Goransson et al., 2019, pp. 38–42).

The 2019 review found, on the other hand, gaps in SIDs partnerships in five dimensions: (1) poverty, social protection, equality, sustainable consumption and production, water and sanitation, sustainable transportation; (2) Technology transfer for surveillance and monitoring of EEZs, including as it relates to illegal fishing and piracy; (3) Technology as a driver of sustainable development; (4) Participation of women in parliament; and (5) Youth, marginalized groups, including people with disabilities. (Goransson et al., 2019, p. 51).

Clearly, the SIDS framework is beneficial to the small states—but there are nonetheless some concerns. There is, for instance, the fact that the development deficits in SIDS are for the most part due to lack of implementation rather than lack of issue awareness, policy, or funding. The issues had been identified and assessed by 1992 or even earlier, and since that time the machinery of development assistance has become highly evolved but with minimal impact. A second concern is potential overlap between SIDS dialogue and SGDs dialogue: why is it necessary to maintain two dialogue streams given that each of the issues elaborated in the SAMOA pathway is also included in one or other of the SDGs? This comparison between the two dialogue processes is set out in Table 11.3.

The prospect of duplication of focus is a valid concern. But retaining a separate SIDS stream still has merit, since it establishes a sphere within global policy discourse for consideration of small states' specific needs. These topics are discussed within other global spaces, but in those larger spaces their voices are overpowered by those of more numerous and much larger states. In addition, the SIDS platform facilitates the establishment of "south-south" connections with other states facing similar issues. Finally, it brings coherence to small states' requests to the major UN bodies—requests that might otherwise be overlooked within the considerable number of presentations made across the multiple UN platforms.

One of the most distinctive features of SIDS discourse concerns consolidation of an "Oceans narrative." Preservation of the Ocean has evolved as a central theme of policy at regional and global levels. Within the Pacific, it has been spurred by a flurry of technical reports and policy position statements by SPC, SPREP, and Forum Leaders (Pratt & Govan, 2010). The Pacific regional institutions have honed the "Blue Pacific Continent" narrative (see Fig. 11.2), which integrates history, identity, and culture into environmental, commercial, legal, and governance aspects of oceans management. These are reflected in a time sequence embedded in the Forum's 2019 *State of Pacific Regionalism* report (Pacific Islands Forum Secretariat, 2019, p. 5), which includes key dates for regional institutions, treaties and policy frameworks, and contributions from key Pacific thinkers and writers.

At global level, an Ocean Conference held at UN headquarters in New York in June 2017 identified some 1400 voluntary commitments for concrete action by governments, United Nations departments, civil society organizations, academia, the scientific community, and the private sector, to advance implementation of SDG 14. The conference outcome statement *Our Ocean, Our Future: Call for Action*, laid out the steps required for sustainable management and conservation of oceans, seas and marine resources. A follow-up conference in Oslo in 2019 attracted 600 participants. No PICs appear to have attended the conference, but it did elicit additional commitments to action from the Pacific Community (Enhancing climate finance of Pacific Islands Countries, Implementing social and environmental responsibility, and coordinating ocean science, knowledge and innovation). (Government of Norway. Ministry of Foreign Affairs, 2019).

Table 11.3 Issues common to the SAMOA pathway and the SDGs

Issues common to the SAMOA pathway and the SDGs	
Sustained and sustainable, inclusive and equitable economic growth with decent work for all	08 Promote sustained, inclusive and sustainable economic growth, full and productive employment and decent work for all.
Development models in small island developing states for the implementation of sustainable development and poverty eradication	01 end poverty in all its forms everywhere.
Climate change	13 take urgent action to combat climate change and its impacts
Sustainable energy	07 ensure access to affordable, reliable, sustainable and modern energy for all.
Oceans and seas	14 conserve and sustainably use the oceans, seas and marine resources for sustainable development.
Food security and nutrition	02 end hunger, achieve food security and improved nutrition and promote sustainable agriculture.
Water and sanitation	06 ensure availability and sustainable management of water and sanitation for all.
Health and non-communicable diseases	03 ensure healthy lives and promote Well-being for all at all ages.
Education	04 ensure inclusive and equitable quality education and promote lifelong learning opportunities for all.
Gender equality and women's empowerment	05 achieve gender equality and empower all women and girls.
Social development	09 build resilient infrastructure, promote inclusive and sustainable industrialization and foster innovation. 10 reduce inequality within and among countries. 11 make cities and human settlements inclusive, safe, resilient and sustainable. 12 ensure sustainable consumption and production patterns.
Management of chemicals and waste, including hazardous waste Sustainable transportation Sustainable consumption and production Biodiversity Desertification, land degradation and drought Forests Invasive alien species Promoting peaceful societies and safe communities	15 protect, restore and promote sustainable use of terrestrial ecosystems, sustainably manage forests, combat desertification, and halt and reverse land degradation and halt biodiversity loss. 16 promote peaceful and inclusive societies for sustainable development, provide access to justice for all and build effective, accountable and inclusive institutions at all levels. 17 strengthen the means of implementation and revitalize the global partnership for sustainable development.
Cross-cutting themes	
Culture and sport Sustainable tourism Disaster risk reduction	These goals are embedded throughout the 17 SDGs

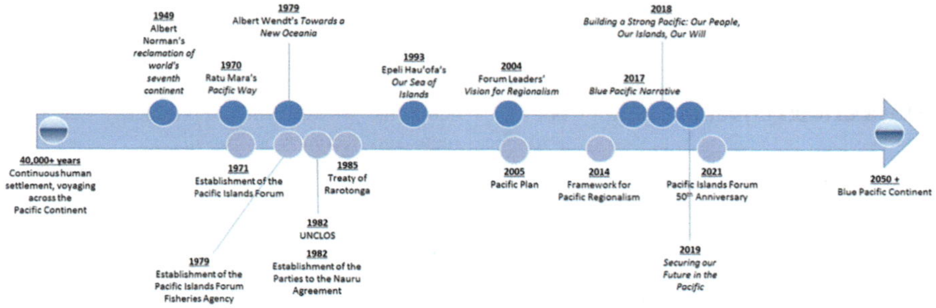

Fig. 11.2 Towards a Blue Pacific Continent

11.2 Discussion

The small island developing states face three broad challenges: (1) economic volatility, vulnerability and natural disasters; (2) building transitions to the changing global trade regime; and (3) strengthening their capacities and capability. They face these challenges as small, independent nations, as well as through regional groupings, to navigate a labyrinth of complex international systems for development finance, security, law, and governance.

References

Goransson, O., Vierros, M., & Borrevik, C. (2019). *Partnerships for Small Island Develoing states*. United Nations. Department of Economic and Social Affairs. Division for Sustainable Development Goals.

Government of Norway. Ministry of Foreign Affairs. (2019). *Our ocean 2019: Learning, sharing, acting*. Government of Norway.

Pacific Islands Forum Secretariat. (1985). *Director's statement and progress report*. SPEC Committee Budget Session, Suva 26–27 November 1985. Meeting Papers. Folder 1. *(85) MP. 9*.

Pacific Islands Forum Secretariat. (2019). *State of Pacific regionalism report 2019*. Pacific Islands Forum Secretariat.

Pratt, C., & Govan, H. (2010). *Our sea of islands–our livelihoods–our Oceania: Framework for a Pacific Oceanscape–a catalyst for implementation of ocean policy*. SPREP.

United Nations. General Assembly. (2014). *Resolution adopted by the General Assembly on 14 November 2014*. SIDS Accelerated Modalities of Action (SAMOA) Pathway. *A/RES/68/15*.

Chapter 12
Small States as Good States, Smart States, in the Global Era

This book set out to explore the impact of the UN on the development of PICs, the impact of PICs within the UN, and the extent to which UN facilitates small-state engagement in global governance in the twenty-first Century. UN agencies assisted the PICs through the processes of self-determination and decolonization, then with national development, and then with support for engagement in the international arena. Whereas the push for nationalism and independence in the Pacific cannot be compared to the revolutionary struggles undertaken by a considerable number of colonies in Africa and Asia, independence held no less significance for Pacific peoples as one of identity formation. The metropolitan powers had to be convinced to leave, and UN agencies and processes played substantial roles in achieving this.

Having achieved national independence, the new Pacific states engaged energetically in nation-building and searched for forms of public administration and political life that suitably expressed their cultural traditions and exercised full autonomy. In the initial post-independence years "development" referred primarily to economic expansion and the establishment of basic layers of physical infrastructure, such as inter-island communications and transportation, and health and educational facilities. UN agencies entered the Pacific region in this period and gradually consolidated their positions as essential partners in development.

Given that the political ethos and mentality into which Pacific states emerged from colonial rule was premised on "national sovereignty", the pursuit of capacity to act as sovereign and independent nations was understandable. Apart from the French overseas territories (French Polynesia and New Caledonia) and the American incorporated and unincorporated territories (American Samoa, Guam, and some smaller pieces of Pacific real estate), all other territories under British, Australian, New Zealand, and North American authority came to forms of independence recognized by the United Nations and were thus admitted to the global comity of nation-states. However, the process left several PICs devoid of status.

During this same period the UN evolved from a 56-member intergovernmental organization built as a security mechanism in response to the ravages of WWII, into a 193-member organization aspiring to respond to global challenges unimagined at

G. Hassall, *The United Nations and the Pacific Islands*, United Nations University Series on Regionalism 24, https://doi.org/10.1007/978-3-031-34155-7_12

its foundation. The United Nations Organization and the various Pacific Island Countries thus originated in the same post-WWII environment, and in pursuit of their diverse objectives and missions have made significant contributions to each other's growth and maturation. However, the UN was not established with the interests of "small states" in mind, nor was it designed to accommodate governance arrangements at regional level, nor accommodate the multiple voices of civil society, nor meet the demands of global governance (as distinct from international coordination).

Notions of "security" in the Pacific have moved toward "human security" and due to the special significance of the natural environment for the islands of the Pacific, toward "comprehensive security". A growing awareness of the responsibilities nation-states have for fostering human development has encouraged public institutions to focus on health, education, and other social services. On the other hand, the real challenges of governing small states have become apparent, and what could be termed "micro-state realism" has emerged: economies have not expanded as rapidly as had been hoped; public sector expenditure has expanded beyond state's ability to pay; and public expectations of improved living conditions have made reform programs politically difficult to initiate. Increasingly aware of the limits to micro-state capacities, the Pacific states embraced multi-level governance and collaborative regionalism. The PICS evolved from separate, dependent, island territories (which implies capability constraints) into small island—large ocean—developing states, which are still vulnerable, but which gain strength from collaborating across multiple, flexible, regional networks, in a globalized and interdependent world. Global policy networks overlay the nation-state and traditional modes of life (Hassall, 2021).

12.1 The United Nations in the Pacific Islands

The influence of the United Nations Organization on the Pacific Islands can best be observed if one steps back from examination of specific projects and campaigns, and the tenure of specific administrators and political leaders. Observation of this influence across several decades highlights the progress of virtually all Pacific countries and territories from colonies to independent nations. The human rights of individuals have been enhanced, and specific sections within society have had their rights more specifically prescribed: women, children, and those with disabilities. The "human development" agenda has influenced PIC policy making and assessment of government performance. If the UN had not taken on the role of advocacy for self-determination, for human rights, for good governance and human development, and for social policy improvements such as more inclusive societies, which other institution or agency would have done so?

In addition, the UN has played a key role in "harmonisation" of Pacific development activities. Systems of governance have been improved, as have the quality of legislation and parliamentary functioning, and standards of transparency and accountability. The UN has helped improve public sector functioning—notably in health, education, and agriculture. It has helped raise labour standards, and improve data collection, analysis, and accessibility. It has encouraged fiscal planning and scientific and technological innovation—especially in communications. It has advocated environmental protection and contributed in some measure to conflict prevention and peacebuilding.

Post-independence, PICs as "developing countries" share the need to navigate the arena of international relations and international intergovernmental organizations. They also share with all other countries—from the "least developed" to the "advanced industrial"—the need to establish effective methods for global governance and to generate global public goods. But the UN was established as an organization of sovereign nation-states. It has not been a source of encouragement to the small Pacific states to consider establishing a larger union, akin to the European Union established by the European member countries. It did not, it seems, have anything to contribute to the decades-long dilemma concerning the shape and degree of integration of Pacific regionalism. Inability to establish a single regional organization (SRO) to coordinate policymaking amongst the Pacific's myriad intergovernmental organizations hindered the region's functioning for more than a half-century.

12.2 The Pacific Islands in the United Nations

Although opening speeches at international meetings may at times be perfunctory or performative, the vote of thanks made by Faysal Abdel-Gadir, UNFPA Director for the South Pacific, at the closing of an NGO regional conference in 1994, indicated the emerging presence of the region in policy discourse:

> This Conference of Pacific NGOs is but another strong commitment by this region to set a stage and further its contribution towards the global issues of Population and Sustainable Development. Following the United Nations Conference on Environment and Development this region has broadened its steps to shape the survival issues of the islanders. That was no coincidence. For one thing, the islands represent a parameter, a bell weather, for our physical world. they reflect temperature changes, pollution and other changes in Sustainable Development and a warning that what happens to them first may happen to the rest of the world later.
>
> From UNCED to Cairo the march is going ahead to address the inter-linked survival issues. The region's plight has been subject of series of dialogue, setting the stage. Right in mind are the recent meetings of Officials and Ministers in Vanuatu on Population and Sustainable Development which concluded not only with statement of positions but also making the message clear and visible. The Port Vila Declaration as you all know was a distinct document not only as a contribution to the global concerns on Population and Sustainable Development but also reflected the conviction of this region by its countries and territories in obtaining a meaningful dialogue and the strong potential of building a broad base of support for sound population policies and programmes within the context and realities of the Pacific region. Unique as it is, it gained the desired recognition and is now

> widely circulated as a UN document. (Pacific Islands Association of Non-Governmental Organizations, 1994, p. 17 Faysal Abdel-Gadir, vote of thanks)

Such statements point to the contributions to global policy dialogue steadily but quietly made by the small states of the Pacific. They may not possess a loud voice, but it has been a reliable one, reporting how island peoples are coping with the circumstances they encounter in their post-colonial, modern, social and economic circumstances. The contributions being made by PICs to the functioning of UN organs, agencies, and programmes are becoming evident. They have endorsed most UN values and standards (but not all) and now contribute broadly to UN discourse. PIC UN member states, officials, and civil society have combined in various networks to advance one or other of the UN's fundamental purposes in the realms of peace and security, human rights, social and economic development, and multilateral coordination—notably concerning oceans and climate change.

Establishing appropriate climate governance regimes is particularly demanding. The climate is in the realm of the "global commons" and cannot be handled by any nation or group of nations on their own. The human-induced aspects of climate change are disproportionally reliant on changes in the behaviour of large and more powerful states and economies than those of the SIDs, including the PICs. Responding appropriately to the challenges presented by climate change nonetheless has the potential to improve the governance of PICs. First, responding to the imperatives of climate advocacy, they have already become conversant with global public policy networks and processes, and the mechanisms of global governance; second, development of mitigation strategies is obliging them to strengthen "whole of government" processes domestically; third, the requirements for monitoring and evaluation are stimulating more rapid adoption of ICTs/digital government; fourth has been the necessity of incorporating evidence-based, i.e. scientific knowledge, into policy-making; fifth, climate advocacy has built diplomatic skills which are transferable to other domains of international advocacy; and sixth, domestic efforts to reach climate targets have built coalitions with other stake-holders, especially the private sector. As Marshall Islands President Kessai H. Note stated at a High-Level Meeting of GA on 15 September 2005, the power of Pacific Island states "lies not in our economic strength, but in our ability to cooperate."

On the basis of this consistent engagement with global policy issues within the UN framework, the PICs have formed viewpoints on the adequacy or otherwise of its institutions and procedures, and have contributed their views on necessary reforms—speaking, as the saying goes—"truth to power". The Pacific States have experienced the effects of nuclear weapons testing and have rallied in opposition to them for close to a century. They initiated an international treaty that established the South Pacific Nuclear Free Zone, and they supported Marshall Islands' petitions to the International Court of Justice seeking relief from the long-term effects of such testing. In April 2013 Palau presided over the final vote when 154 member states agreed to the regulation of international trade in arms and conventional weapons. This event was a striking example of the engagement of the smallest of states in the governance of a global public policy matter.

The challenge for the Pacific states now is to establish a unified position in relation to the legality of nuclear weapons, by encouraging five remaining states to sign and ratify the *Treaty on the Prohibition of Nuclear Weapons*. The Pacific States could enhance the mission of the United Nations by completing their recognition of the Rome Statute and recognize the International Criminal Court: of the 123 Parties to the primary Rome Statute, eight are PICs: Cook Islands, Fiji, Kiribati, Marshall Islands, Nauru, Samoa, Solomon Islands and Vanuatu: is it not possible for Niue, Palau, Tuvalu, Federated States of Micronesia, Papua New Guinea, and Tonga to also sign and ratify? At present Samoa is the lone PIC amongst 42 parties to the 2010 Kampala Amendment on Aggression (2010): additional PIC signatures would surely consolidate the region's reputation for valuing world peace.

The Pacific States have expressed their views on UN reform. Papua New Guinea Ambassador Vaqi addressed UNGA in 1986, the International Year of Peace:

> Under the United Nations Charter, all United Nations members subscribe to peace. All of us say that we favour peace. But what have we really done to give effect to our common commitment? The United Nations, formed in the aftermath of the Second World War, was intended by its founders to be a vehicle for securing peace—genuine and lasting peace. Its founders recognized, as the Charter shows, that such peace involves more than the absence of war. They saw that genuine and lasting international peace can come about only if we attend to the underlying causes of international conflict.
>
> Iniquity, the denial of basic human rights and the arms race are just as surely threats to peace as is naked military aggression . . . Yet, year after year, leaders from all parts of the world stand in this Hall and speak in favour of peace, but little or nothing happens. In fact, international tensions are increasing. The United Nations is now more than a generation old can we, the next generation, do no more? Surely, we must—for ourselves, for our children and, in humanity is to survive, for their successors too. . . . Papua New Guinea is a full, committed and—within the scope of available resources—active member of the United Nations. For countries like mine, the United Nations is a vital—indeed, irreplaceable—means of securing national, regional and wider interests. (United Nations General Assembly, 1985, p. 17).

The Pacific states have shared their views on Security Council reform individually as well as collectively. In 1993 Fiji supported removal of the P5 veto power and in 1996 proposed an increase in the number of non-permanent seats from 10 to 17, together with an equitable system for rotation of seats amongst regional groups (Fassbender, 2020, p. 1250). In 2005 Kiribati expressed support for increasing the numbers of both permanent and non-permanent members, but also sought a deadline for finalization of UNSC reforms similar to deadlines in other areas, such as establishment of the Human Rights Council (Republic of Kiribati, 2005). Tonga indicated support for Japan gaining a seat on an enhanced Council (Kingdom of Tonga, 2005). The Marshall Islands has called for review toward more "fair democratic representation" (United Nations General Assembly, 1995, p. 2), and supports establishing seats for Japan and for a developing country (Republic of the Marshall Islands, 2004). Samoa has advocated including Japan and Germany (Fassbender, 2020, p. 666). In 2009 the PICs expressed a wish for improved working methods, more representative distribution of seats, and an enhanced number of non-permanent seats so that small states could have an opportunity to serve (Fassbender, 2020, pp. 1030–1031).

In September 2005, 150 Heads of State and Government attended the 60th GA Plenary meeting to consider the Secretary-General's proposal for UN reform, *In Larger Freedom* (United Nations. Secretary General, 2005). The Summit adopted an Outcomes document focused on development, peace and collective security, human rights and the rule of law, and strengthening of the United Nations. The Pacific region's perspective on reforms required with the UN system reflect criticisms by the global South: marginalization of the GA and ECOSOC, the financial role of UN, and the issue of representation in the Security Council. Setting aside concerns about last minute amendments to the document, the Pacific States had a hand in ensuring that it included reference to SIDS and Least developed Countries and reaffirmed the need and resolve to help implement commitments in the Brussels Program of Action for the Least Developing Countries, the Mauritius Strategy, the Barbados Programme of Action and the outcome of the twenty-second special session of the General Assembly. The *Outcome* also noted the special adaptation needs of SIDS to effects of climate change and, that "cessation of the transport of radioactive materials through the regions of small island developing States is an ultimate desired goal of small island developing states and some other countries civil society. In 2013 the PICa urged faster resolution on reform issues. "Above all else," Papua New Guinea representative Aisi stated, "…the principal reason for the Pacific SIDS to remain in this debate is that, like all the United Nations reform issues we have all committed to, it is a fundamental issue of governance." (Fassbender, 2020, p. 1095).

12.3 Small States as "Good States" in the Global Era

The PICs, although "small states", can increase their influence in world affairs through consolidating their reputations as "good" states. They need to avoid the charge of hypocrisy that would stem from, for instance, ardently preaching environmental protection from international platforms while failing to deliver environmental protection domestically. The first factor, thus, in establishing PICs as "good states" is their ability and willingness to align international statements and commitments with law and policy implementation domestically. Good states advocate cooperation rather than conflict, prefer collaboration and partnership rather than competition, cultivate self-confidence as much as self-reliance, operate open government rather than secret government, and pursue "track II and "soft" diplomacy as well as track I (official) diplomacy. Being "smart" implies applying information and communication technologies to the production of information societies, knowledge societies. Good states aim to be "intelligent" states in the sense that they are agile, and capable of learning from past errors or ineffective policy settings.

In agile states, governments can change rapidly in response to new ideas and approaches. They can use multi-stakeholder groups, partnerships, and other platforms to advance policy dialogue. Such a transformation requires much more openness in government decision-making, and in discussion of results and outcomes. This is certainly a condition that the UN has encouraged, by urging PICs to adopt UN standards for human rights, rule of law, transparency, accountability, democratic governance, and human development, into domestic law and practice and abide by international norms and agreements in their international relations. This may include a shift to more inclusive consultative processes, and openness to more systemic constitutional review. Of such states Creutz writes:

> Although the overall influence of small states may be limited, it is within specific issues that they can take initiatives and show leadership. For example, Trinidad and Tobago was one of the key advocates for the establishment of a permanent International Criminal Court, and Finland has demonstrated agency in the governance of the Arctic. Small states can thus show leverage within their own specialist field, and should not shy away from such a pursuit, even in informal institutions where they remain an outsider. (Creutz, 2017, p. 7)

Scenarios for future engagement between the United Nations and the Pacific Islands region range from status quo, to deterioration, to transformation. The first two possibilities are not desirable options but will nonetheless be sketched in brief. In the "business as usual", or "status quo", scenario, governance neither significantly declines in coming years, nor does it significantly improve. This outcome will result if these Pacific nations continue with the political culture as currently practiced. Elections will continue to be held as scheduled, but due to the instability of political coalitions, parliaments will continue to sit less often than is required to undertake adequate oversight of the executive. Legislative programmes will respond principally to international requests for compliance (such as reforms to security, finance, biosecurity requirements) more than to domestic imperatives. Health and Education budgets will remain static, and enhanced services will only be possible with continued donor support. PICs will continue to rely on development assistance, and at the regional level will, engage in low-level cooperation only. Ongoing lack of social services on outer islands will continue to fuel urban drift, and urban authorities will continue to struggle with the provision of urban infrastructure. Those living in villages and informal settlements will continue to be neglected. National governments will continue to call on CROP agencies for policy advice and look to foreign partners for material support. The CROP agencies, for their part, will continue to focus on programs that promote collaboration and capacity building—but also continue to struggle against duplication of effort. Some attention will be given to compliance with the SDGs, and Samoa Pathway, as well as initiatives borne out of the *Framework for Pacific Regionalism*—so long as external resources are provided. Climate change will more adversely affect coastal communities and internal migration issues will create tensions. Human Development will increase, but more slowly than on the continents of Asia and Africa and leave such Pacific countries with "Least developed country" status. The Pacific Countries will continue to send representatives to global conferences with an emphasis on identifying financial and other benefits for "less developed countries."

A second possible outcome in the coming years, one which few would want to consider, is deterioration in the quality of political governance that eventually leads to "state/regime failure. Solomon Islands came close to such a condition in the early

2000s, as did Bougainville during its separatist struggle with mainland Papua New Guinea. Failure will always be an option if a people and their governments do not remain vigilant. In such a scenario, the antagonisms that are embedded in opposing political parties increase, and those who hold executive power begin to stifle democratic processes to retain their authority. Elite interests capture executive power and focus on exploitation of rents. Unsustainable natural resource exploitation depletes future revenue potential and speeds up environmental degradation. The role of parliament is minimized, and public sector leadership is politicized. The institutions of accountability, such as Ombudsman, Public Accounts Committee, and Auditor General, are suppressed. Poor governance leads to a fall in government revenues, and this leads to a loss of investor confidence. Media and civil society are intimidated. Material and human capital takes flight, and development agencies ease up their commitment to longer-term projects. Neglect of social services leads to decline in human development outcomes, rise in social unrest and calls for separatism. The younger generations, who have emigrated, send ever-decreasing remittances. Adverse weather events threaten food security and degrade physical infrastructure, including transport and communications networks. The climate of uncertainty keeps tourists away. Leaders continue to participate in global meetings, but mostly to justify their circumstances and to seek funds and resources rather than to engage in global policy dialogue. The state or states experiencing this upheaval begin to consume the time and resources of the regional and UN agencies, which in turn draws resources from positive region-building activities. On balance, the small scale of intra-state and inter-state conflict within and between PICs draws little attention from the UN's Peace and Security mechanisms.

The third potential outcome, which serves as a higher aspiration for the Pacific Islands region and is more in keeping with the UN's fundamental tenets, is transformation. In general terms, this implies grounding all policies on values of human rights, social justice and sustainability, and ensuring that the apparatus of government serves all citizens equally, and to high standards. Public values of transparency, accountability, social inclusion, and sustainability are not only articulated but pursued. All major human rights covenants are signed, ratified, and translated into domestic legislation. Transformative countries, individually and as a region, respond to opportunities provided by the SDGs and SIDS agendas, and "own" development activities and results (both successes and failures, from which they learn and adapt). Acknowledgement of the benefits of more open information sharing and enhanced public consultation drive the expansion of E-Government services, and online transactions benefit citizens and business, as well as the many branches of government. Representatives attend global meetings to learn of best practices and to share the Pacific's experience. Pacific regionalism builds on lessons of regional organization from other parts of the world: far from being "one-size fits all", it allows countries to opt in to shared services and decision-making in sectors in which joint services more adequately meet their citizen's needs and expectations. The UN consolidates its partnerships with the Pacific's peak regional organization. The Pacific region is identified less with qualities of vulnerability and isolation than with those of high human development and global engagement in peacebuilding, collaboration, and agility. Civil society and the media appreciate such development aspirations as are achieved.

Realization of such transformation would honour the outcomes anticipated by the founders of the Pacific nations and regional institutions, and when accomplished in equal partnership with United Nations agencies would show that the UN does indeed adequately serve the interests of its smallest members. On current projections the immediate future for the Pacific region is more likely to resemble some combination of the first two scenarios sketched above, whilst the third remains the only one worth striving for and aspiring to.

References

Creutz, K. (2017). *Alternative ways of global governance: Informal institutions and the role of small states* (Vol. 215). Finnish Institute of International Affairs.

Fassbender, B. (2020). *Key documents on the reform of the UN security council 1991–2019*: Brill | Nijhoff.

Hassall, G. (2021). *Government and public policy in the Pacific Islands*. Emerald Publishing Limited.

Kingdom of Tonga. (2005). *Statement by Honourable Siosiua T 'Utoikamanu, Minister of Finance of the Kingdom of Tonga at the High-Level Plenary Meeting of the 60th Session of the United Nations General Assembly, Friday 16 September.*

Pacific Islands Association of Non-Governmental Organizations. (1994). Pacific region non-government Organizsations conference on population and sustainable development, Nadi, Fiji, 2–4 March. Conference Papers. Suva.

Republic of Kiribati. (2005). *Statement by His Excellency Anote Tong Beretitenti (President) of the Republic of Kiribati at the High-Level Plenary Meeting of the 60th Session of the United Nations General Assembly, Friday 16 September.*

Republic of the Marshall Islands. (2004). *Statement by H.E. Mr. Alfred Capelle, Permanent Representative of the Republic of the Marshall Islands to the United Nations. "Question of eqitable representation on and increase in the membership of the Security Council and Related maters"*. New York.

United Nations General Assembly. (1985). *Provisional verbatim record of the 7th meeting: Held on Tuesday 24 September 1985, New York, general assembly, 40th session*. UN.

United Nations General Assembly. (1995). *General assembly official records, 50th session: 29th plenary meeting, Wednesday, 11 October 1995*. UN.

United Nations. Secretary General. (2005). In larger freedom: Toward development, security and human rights for all. In *Report of the secretary general*. United Nations.

Appendix A: 2006 Statement on Cooperation between the United Nations and the Pacific Islands Forum

United Nations General Assembly

A/61/L.20/Rev.1

28 November 2006

Sixty-first session

Agenda item 108 (r)

Cooperation between the United Nations and regional and other organizations: cooperation between the United Nations and the Pacific Islands Forum

Armenia, Australia, Bangladesh, Cambodia, Cape Verde, Croatia, Czech Republic, Democratic Republic of the Congo, Fiji, Grenada, Guatemala, Haiti, India, Indonesia, Israel, Marshall Islands, Micronesia (Federated States of), Nauru, New Zealand, Palau, Papua New Guinea, Philippines, Romania, Saint Lucia, Samoa, Singapore, Solomon Islands, Sri Lanka, Thailand, Timor-Leste, Tonga, Tuvalu, Ukraine, United States of America, Vanuatu and Venezuela (Bolivarian Republic of): draft resolution

Cooperation between the United Nations and the Pacific Islands Forum

The General Assembly,

Recalling its resolutions 49/1 of 17 October 1994 and 59/20 of 8 November 2004,

Welcoming the ongoing efforts towards closer cooperation between the United Nations and the Pacific Islands Forum and its associated institutions,

Bearing in mind that the Pacific Islands Forum, established in 1971, promotes regional cooperation and integration among its members through trade, investment, economic development and political and international affairs, to achieve their shared goals of economic growth, sustainable development, good governance and security,

Recalling the importance of the internationally agreed development goals set out in the United Nations Millennium Declaration,[1] the Monterrey Consensus of the

[1] See resolution 55/2.

G. Hassall, *The United Nations and the Pacific Islands*, United Nations University Series on Regionalism 24, https://doi.org/10.1007/978-3-031-34155-7

International Conference on Financing for Development,[2] the Plan of Implementation of the World Summit on Sustainable Development ("Johannesburg Plan of Implementation")[3] and the Mauritius Strategy for the Further Implementation of the Programme of Action for the Sustainable Development of Small Island Developing States,[4]

Taking note of the special circumstances pertaining to the continued presence of radioactive contaminants in certain Pacific Islands Forum nations,[5]

Recalling that one of the purposes of the United Nations is to achieve international cooperation in addressing international problems of an economic, social, cultural or humanitarian character,

Reaffirming the commitment of leaders at the 2005 World Summit, and bearing in mind the call thereat for a stronger relationship between the United Nations and regional and subregional organizations,

Affirming the need to strengthen the cooperation that already exists between entities of the United Nations system and the Pacific Islands Forum in the areas of peace and security, sustainable development, environmental protection and good governance,

Noting that many Pacific Islands Forum nations depend upon the continued existence of sustainable marine ecosystems,

Welcoming the support and assistance given by the United Nations towards the maintenance of peace and security in the Pacific Islands Forum region,

Taking note of the Communiqué of the Thirty-seventh Pacific Islands Forum, held in Nadi, Fiji, on 24 and 25 October 2006,

Mindful of the need for coordinated and effective utilization of available resources in pursuing the common objectives of the two organizations,

Having considered the report of the Secretary-General on cooperation between the United Nations and regional and other organizations,[6]

1. *Takes note* of the report of the Secretary-General on cooperation between the United Nations and regional and other organizations,6 in particular part one, section XII, on cooperation between the United Nations and the Pacific Islands Forum, and encourages further such cooperation;

---

[2] *Report of the International Conference on Financing for Development, Monterrey, Mexico, 18-22 March 2002* (United Nations publication, Sales No. E.02.II.A.7), chap. I, resolution 1, annex.

[3] *Report of the World Summit on Sustainable Development, Johannesburg, South Africa, 26 August-4 September 2002* (United Nations publication, Sales No. E.03.II.A.1 and corrigendum), chap. I, resolution 2, annex.

[4] *Report of the International Meeting to Review the Implementation of the Programme of Action for the Sustainable Development of Small Island Developing States, Port Louis, Mauritius, 10-14 January 2005* (United Nations publication, Sales No. E.05.II.A.4 and corrigendum), chap. I, resolution 1, annex I.

[5] See Communiqué of the Thirty-seventh Pacific Islands Forum, held at Nadi, Fiji, on 24 and 25 October 2006.

[6] A/61/256.

2. *Notes with satisfaction* that regular consultations continue at all levels between the United Nations and the secretariat of the Pacific Islands Forum, including participation at the annual consultations between the Secretary-General and heads of regional organizations;
3. *Invites* the Secretary-General of the United Nations to take the necessary measures, in consultation with the Secretary-General of the Pacific Islands Forum, to promote and expand cooperation and coordination between the two secretariats in order to increase the capacity of the organizations to attain their common objectives;
4. *Welcomes* the ongoing work of various international organizations and United Nations agencies, funds and programmes in advancing knowledge in the key strategic areas related to governance, security, economic growth, trade and sustainable development, as well as in the implementation of the internationally agreed development goals, including those contained in the United Nations Millennium Declaration,[7] in the Pacific island countries;
5. *Recognizes* the challenges that Pacific island countries are facing in combating the HIV/AIDS pandemic, and, in this regard, urges the international community, including the United Nations system, to support Pacific island countries in their efforts to implement the Political Declaration on HIV/AIDS,[8] adopted by the General Assembly at its sixtieth session on 15 June 2006, including the commitment to set, in 2006, ambitious national targets that reflected the urgent need to scale up significantly towards the goal of universal access to comprehensive prevention programmes, treatment, care and support, by 2010;
6. *Urges* Governments and all relevant international and regional organizations, the Economic and Social Council, United Nations funds, programmes and regional economic commissions, specialized agencies, international financial institutions and the Global Environment Facility, as well as other intergovernmental organizations and major groups, to take timely actions to support Pacific island countries in their efforts to ensure the effective implementation of and follow-up to the Johannesburg Declaration on Sustainable Development,[9] the Johannesburg Plan of Implementation3 and the Mauritius Strategy for the Further Implementation of the Programme of Action for Sustainable Development of Small Islands Developing States;[10]

[7] See resolution 55/2.

[8] Resolution 60/262, annex.

[9] *Report of the World Summit on Sustainable Development, Johannesburg, South Africa 26 August-4 September 2004* (United Nations publication, Sales No. E.03.II.A.1 and corrigendum), chap. I, resolution 1, annex.

[10] *Report of the International Meeting to Review the Implementation of the Programme of Action for the Sustainable Development of Small Island Developing States, Port Louis, Mauritius, 10-14 January 2005* (United Nations publication, Sales No. E.05.II.A.4 and corrigendum), chap. I, resolution 1, annex I.

7. *Notes* the importance of the United Nations Global Counter-Terrorism Strategy adopted on 8 September 2006,[11] and, in this regard, calls for the support of the United Nations system and other international partners in assisting Pacific Island countries in their efforts to implement it;
8. *Welcomes* the ongoing efforts of the Pacific Islands Forum to promote, primarily through the Regional Security Committee, law enforcement cooperation, the rule of law and regional peace and security, including combating all types of terrorism, in implementing the core United Nations treaties on anti-terrorism, anti-money-laundering, transnational crime and the financing of terrorism;
9. *Requests*, in this regard, that the United Nations continue to assist the Pacific Islands Forum in facilitating among its members the timely implementation of relevant United Nations mandates, and invites States to contribute to the Biketawa Trust Fund, which is administered by the Pacific Islands Forum for confidence-building measures and conflict prevention;
10. *Welcomes* the significant efforts of the Pacific Islands Forum in enhancing peace and security in the region, including through the Regional Assistance Mission to Solomon Islands;
11. *Notes with appreciation* the role of the United Nations in the Bougainville peace process in Papua New Guinea and the steady progress being made by the parties;
12. *Welcomes* the establishment of the Peacebuilding Commission and the Peacebuilding Fund, and calls upon the Secretary-General and other relevant actors to give consideration to using these and other mechanisms to support post-conflict peacebuilding activities, reconstruction and institution-building efforts in Pacific island countries, particularly on the island of Bougainville in Papua New Guinea and in Solomon Islands;
13. *Requests* that the Department of Political Affairs of the Secretariat and the United Nations Development Programme, in cooperation with the Pacific Islands Forum, promote joint cooperative needs assessment missions in the region to determine additional support to enhance peacebuilding and reconciliation processes and to complement the activities of regional missions and mechanisms;
14. *Calls upon* the international community to provide, where appropriate, technical and financial support to Pacific island countries in combating the illicit traffic in small arms and light weapons through, inter alia, the implementation of the Programme of Action to Prevent, Combat and Eradicate the Illicit Trade in Small Arms and Light Weapons in All Its Aspects;[12]
15. *Notes* the importance of the United Nations field presence in Pacific island countries for enhancing the cooperation with the United Nations system and its development agencies needed to implement the internationally agreed

---

[11] Resolution 60/288.

[12] See *Report of the United Nations Conference on the Illicit Trade in Small Arms and Light Weapons in All Its Aspects, New York, 9-20 July 2001* (A/CONF.192/15), chap. IV, para. 24.

development goals, including the Millennium Development Goals and the Mauritius Strategy;

16. *Welcomes*, in this regard, the endorsement by the Secretary-General of the establishment of an expanded and joint in-country presence of the United Nations in Kiribati, the Marshall Islands, the Federated States of Micronesia, Nauru, Palau, Solomon Islands, Tuvalu and Vanuatu;
17. *Expresses*, in this regard, its appreciation for the support and cooperation of the Pacific Islands Forum members in fulfilling host-country obligations and other operational modalities;
18. *Notes* the efforts of the United Nations, in cooperation with the Pacific Islands Forum, in respect of considering ways to assist Nauru, and, in this regard, calls upon the United Nations system to support the implementation of the report of the United Nations Development Programme entitled "Republic of Nauru preparatory assistance project" as well as that country's national sustainable development strategy;
19. *Also notes* the participation of Pacific leaders at the special session of the Economic and Social Commission for Asia and the Pacific, held in Jakarta on 10 April 2006, and takes note of the project "Enhancing Pacific connectivity";[13]
20. *Welcomes* the adoption by Pacific Islands Forum Leaders of the Pacific Plan, at the Thirty-sixth Pacific Islands Forum in Madang, Papua New Guinea, during the Forum Leaders Retreat on 26 October 2005, which is aimed at enhancing regional integration and cooperation among its members and cooperation with the international community, including the United Nations system;
21. *Also welcomes* the leadership role of the Pacific Islands Forum in furthering the implementation of the 1995 Agreement for the Implementation of the Provisions of the United Nations Convention on the Law of the Sea of 10 December 1982 relating to the Conservation and Management of Straddling Fish Stocks and Highly Migratory Fish Stocks,[14] particularly in the convening of the negotiations for, and the adoption of, the Convention on the Conservation and Management of Highly Migratory Fish Stocks in the Western and Central Pacific Ocean;
22. *Further welcomes* the decision to convene the Pacific Regional Seminar on Decolonization in Nadi, Fiji, from 28 to 30 November 2006;
23. *Urges* the United Nations to support Pacific island countries in pursuing initiatives for enhancing South-South cooperation among themselves and also with other developing countries;
24. *Takes note* of the steps taken by the Pacific Islands Forum to solidify its partnership with non-State actors in the region in promoting governance and sustainable development issues;

[13] See A/61/256, para. 89.

[14] United Nations, *Treaty Series*, vol. 2167, No. 37924.

25. *Recognizes* the burden placed on small States by growing international reporting requirements, and encourages the investigation of innovative reporting modalities, including regional reporting, where appropriate;
26. *Calls upon* the Office of the United Nations High Commissioner for Human Rights to provide technical support to Pacific Islands Forum members to contribute to the regional efforts in promoting awareness and knowledge of all international human rights treaties and instruments;
27. *Requests* the Secretary-General to submit to the General Assembly at its sixty-third session a report on the implementation of the present resolution;
28. *Decides* to include in the provisional agenda of its sixty-third session the sub-item entitled "Cooperation between the United Nations and the Pacific Islands Forum."

Appendix B: 2012 Joint Statement by Pacific Leaders and the UN Secretary-General

Secretary-General
SG/2191

Department of Public Information • News and Media Division • New York

Joint Statement by Leaders of Pacific Islands Forum, UN Secretary-General

Following is a joint statement by leaders of the Pacific Islands Forum and the United Nations Secretary-General:

The leaders of Pacific Islands Forum countries and the Secretary-General of the United Nations met on 26 September 2012 on the margins of the sixty-seventh session of the United Nations General Assembly in New York.

They reaffirmed their commitment to the United Nations Charter and its principles.

At their meeting, Pacific Islands Forum leaders acknowledged the valuable contribution made by the United Nations system in the Pacific; and the Secretary-General of the United Nations acknowledged the key role played by the Pacific Islands Forum and its members in promoting sustainable development, good governance and peace and security through regional cooperation in the Pacific.

The Secretary-General of the United Nations welcomed the theme of the forty-third Pacific Islands Forum, "Large Ocean Island States: the Pacific Challenge", held in the Cook Islands on 27-31 August 2012, and welcomed the Communiqué and its Gender Equality Declaration issued by Pacific Islands Forum Leaders.

The Secretary-General of the United Nations and Pacific Islands Forum Leaders recalled their Joint Statement of 7 September 2011; reaffirmed the value of cooperation between the United Nations system and Pacific Islands Forum countries, the Pacific Islands Forum Secretariat, and its associated institutions; and reiterated their

G. Hassall, *The United Nations and the Pacific Islands*, United Nations University Series on Regionalism 24, https://doi.org/10.1007/978-3-031-34155-7

shared commitment to considering practical ways to further strengthen such cooperation. In this regard, the Secretary-General of the United Nations and Pacific Islands Forum Leaders:

Discussed key challenges facing the Pacific region and stressed the role of the United Nations system in assisting Pacific island countries address these challenges, and emphasised the need for greater accountability and reporting of the support provided by the United Nations system to Pacific small island developing States.

Reaffirmed their commitment to achieving the Millennium Development Goals in the Pacific by 2015, and discussed ways of enhancing the United Nations system's support for national and regional efforts in this regard, in line with national and regional priorities and relevant international frameworks, including the *Barbados Programme of Action*, the *Mauritius Strategy* and the *Istanbul Programme of Action*, noting in this regard progress made through ongoing programmes and projects.

Welcomed outcomes agreed at the United Nations Conference on Sustainable Development (Rio+20), noting in particular the importance to Pacific States of outcomes on climate change and ocean acidification, the conservation and sustainable management of oceans, coastal and fisheries resources, promotion of sustainable energy and transport, food and water security, and disaster risk reduction and emergency relief and recovery, and calling for their full, timely and effective implementation; stressed the importance of enhanced support for and investment in Pacific States' productive sectors; and agreed to consider financial, technological and other practical ways the United Nations system could support efforts to implement these outcomes in the Pacific, in line with national and regional priorities.

Stressed in this regard the critical importance of the sustainable development, management and conservation of the region's oceans, coastal and fisheries resources as a source of livelihoods and income for communities, industries and Governments, and called for the identification and mainstreaming of strategies by 2014 that further assist developing countries, in particular the least developed countries and small island developing States, in developing their national capacity to conserve, sustainably manage and realize the benefits of sustainable fisheries.

Recognized the unique and particular vulnerabilities and development needs of small island developing States and the continuing need to increase their resilience; welcomed the decision to hold a third International Conference on small island developing States in 2014, noted the support of Pacific leaders and members of the Alliance of Small Island States for it to be convened in the Pacific, and stressed the importance of ensuring all Pacific small island developing States are able to participate fully; and emphasized the importance of ensuring these vulnerabilities and needs are adequately reflected in discussions and decisions on the post-2015 United Nations development agenda, including agreement on the SDGs.

Stressed that climate change, ocean acidification and sea-level rise remained the greatest threat to the livelihoods, security and well-being of the peoples of the Pacific; reaffirmed the need for urgent international action to reduce emissions commensurate with the science and associated social, economic and security impacts, sufficient to enable the survival and viability of all Pacific small island developing States; stressed the need to address these impacts in all relevant international forums, including but not limited to the *United Nations Framework Convention on Climate Change*, the General Assembly and the Security Council; noted the proposal by the Ambassadors for Responsibility on Climate Change group to request through the General Assembly an advisory opinion from the International Court of Justice; and highlighted the role of the United Nations system in supporting national, regional and international efforts in this regard.

Reiterated the urgent need to address the serious and unprecedented threats to international peace and security caused by current and projected adverse impacts of climate change, including violation of territorial integrity, more frequent and severe climate-related disasters, including extreme and slow onset events, threats to water and food security, increased natural resource scarcity, and forced displacement; and emphasized the imperative for the international community to identify such threats and to assist vulnerable countries in assessing and addressing their own related security challenges resulting from climate change.

Reiterated the value of enhanced cooperation and experience-sharing on conflict prevention, peace-making, peacekeeping and peacebuilding, building on positive experiences in the Solomon Islands and Bougainville, Papua New Guinea, and looked forward to increasing dialogue in this regard and to considering collaborative activities in such areas as peacebuilding, democratic governance and transition, human rights and rule of law.

Noted a range of other areas in which cooperation between the United Nations and Pacific States would support the national and regional priorities identified by Pacific Islands Forum leaders and help strengthen the resilience of Pacific Island countries. These included, but were not limited to: increasing productive capacity of Pacific small island developing States, including to improve food, water and energy security; transportation; non-communicable diseases; human rights, including engagement with the universal periodic review process and other human rights mechanisms; and measures to address the economic and political empowerment of women and address gender-based violence; disaster risk reduction and emergency relief and recovery; non-proliferation and disarmament, including small arms and light weapons and the negotiation and regional implementation of an arms trade treaty and security sector reform; and addressing the ongoing impacts of nuclear testing in the Pacific.

Stressed the value of close cooperation and coordination between the programmes and activities of the United Nations system and Pacific Islands Forum countries, the Pacific Islands Forum Secretariat, and associated institutions; welcomed recent efforts by CROP Executives and United Nations Heads of Agencies in the Pacific to strengthen cooperation through joint activities, working groups and other means; and agreed to explore further practical ways of enhancing such cooperation and coordination, in line with national priorities and needs, including in implementation of the United Nations Development Assistance Framework and the Pacific Plan, and to report on progress to the next meeting of Council of Regional Organizations of the Pacific leaders.

Noted the continued commitment to regional cooperation and integration and the intention to review and update the Pacific Plan in 2013, as well as progress being made in implementing the Cairns Compact through tracking and reporting on development effectiveness and region-wide progress towards achieving the Millennium Development Goals; welcomed support from United Nations agencies and the Pacific Islands Forum Secretariat towards these efforts and towards effective Pacific engagement in relevant international meetings on sustainable development; and encouraged further such cooperation.

Noted the challenges faced by small island developing States in meeting international reporting obligations, and agreed to consider practical ways of facilitating enhanced reporting while reducing related compliance burdens and transaction costs.

Reaffirmed the importance of ensuring an enhanced and effective United Nations presence, particularly at the country level, in the Pacific region; and agreed to engage in dialogue between United Nations regional and country teams, Pacific States and other relevant stakeholders to consider practical ways in which progress could be made in this area.

Emphasized the importance of adequate collection of data and statistics for making accurate assessments and effective decision-making, and agreed to cooperate on practical ways of ensuring such data and statistics were available for all States, regardless of population size, as well as to build relevant national capacities in the Pacific.

Reaffirmed their commitment to regular dialogue between the United Nations system and Pacific Islands Forum members, including on the margins of the General Assembly, and agreed to meet again by 2014.

Appendix C: Key Pacific Regional Information Portals

Websites do not have the longevity of print publications. However, institutions are becoming better at establishing "permanent URLs" for online materials. The following is list of currently valid and useful "key portals." Websites that expire are sometimes still replicated at https://archive.org/web/

Pacific Islands Regional Portals

Pacific Data Hub https://pacificdata.org/
Pacific Islands Forum Secretariat https://www.forumsec.org/
Secretariat of the Pacific Regional Environment Programme https://www.sprep.org/

- Inform Regional Data Portal: https://pacific-data.sprep.org/
- Invasive Species Battler Resource Base (BRB): https://piln.sprep.org/
- Pacific Climate Change Portal (PCCP) / RTSM / APT / DST: https://www.pacificclimatechange.net/
- Pacific Islands Protected Area Portal (PIPAP): https://pipap.sprep.org/
- Pacific Meteorological Desk & Partnership (PMDP): https://www.pacificmet.net/
- Pacific Network for Environmental Assessment (PNEA): https://pnea.sprep.org/
- Regional Frameworks And Strategies Directory: https://www.sprep.org/node/987
- Lessons Learned And Best Practices In Environment Management: https://www.sprep.org/lessons-learned-and-best-practices-environment-management

G. Hassall, *The United Nations and the Pacific Islands*, United Nations University Series on Regionalism 24, https://doi.org/10.1007/978-3-031-34155-7

The United Nations Portals

The United Nations: https://www.un.org/en/
United Nations Human Development Report http://hdr.undp.org/
United Nations Statistics Division: https://unstats.un.org/home/
United Nations Evaluation Group: http://www.uneval.org/
United Nations Multi-country office for Cook Islands, Niue, Samoa and Tokelau https://samoa.un.org/
United Nations Multi-country office for Federated States of Micronesia, Fiji, Kiribati, Marshall Islands, Nauru, Palau, Solomon Islands, Tonga, Tuvalu and Vanuatu https://pacific.un.org/
United Nations System Chief Executives Board for Coordination https://unsceb.org/ United Nations
United Nations Country office for Papua New Guinea https://papuanewguinea.un.org/
UNDP Evaluation Resource Centre: https://erc.undp.org/
UNDP Multi-partner Trust Fund Office Gateway https://mptf.undp.org/

Index

G. Hassall, *The United Nations and the Pacific Islands*, United Nations University Series on Regionalism 24, https://doi.org/10.1007/978-3-031-34155-7

O

P